Recommender Systems

An Introduction

In this age of information overload, people use a variety of strategies to make choices about what to buy, how to spend their leisure time, and even whom to date. Recommender systems automate some of these strategies with the goal of providing affordable, personal, and high-quality recommendations. This book offers an overview of approaches to developing state-of-the-art recommender systems. The authors present current algorithmic approaches for generating personalized buying proposals, such as collaborative and content-based filtering, as well as more interactive and knowledge-based approaches. They also discuss how to measure the effectiveness of recommender systems and illustrate the methods with practical case studies. The authors also cover emerging topics such as recommender systems in the social web and consumer buying behavior theory. Suitable for computer science researchers and students interested in getting an overview of the field, this book will also be useful for professionals looking for the right technology to build real-world recommender systems.

DIETMAR JANNACH is a Chaired Professor of computer science at Technische Universität Dortmund, Germany. The author of more than one hundred scientific papers, he is a member of the editorial board of the *Applied Intelligence* journal and the review board of the *International Journal of Electronic Commerce*.

MARKUS ZANKER is an Assistant Professor in the Department for Applied Informatics and the director of the study program Information Management at Alpen-Adria Universität Klagenfurt, Austria. He is an associate editor of the *International Journal of Human-Computer Studies* and cofounder and director of ConfigWorks GmbH.

ALEXANDER FELFERNIG is University Professor at Technische Universität Graz, Austria. His research in recommender and configuration systems was honored in 2009 with the *Heinz Zemanek Award*. Felfernig has published more than 130 scientific papers, is a review board member of the *International Journal of Electronic Commerce*, and is a cofounder of ConfigWorks GmbH.

GERHARD FRIEDRICH is a Chaired Professor at Alpen-Adria Universität Klagenfurt, Austria, where he is head of the Institute of Applied Informatics and directs the Intelligent Systems and Business Informatics research group. He is an editor of *AI Communications* and an associate editor of the *International Journal of Mass Customisation*.

Recommender Systems
An Introduction

DIETMAR JANNACH
Technische Universität Dortmund

MARKUS ZANKER
Alpen-Adria Universität Klagenfurt

ALEXANDER FELFERNIG
Technische Universität Graz

GERHARD FRIEDRICH
Alpen-Adria Universität Klagenfurt

CAMBRIDGE
UNIVERSITY PRESS

CAMBRIDGE UNIVERSITY PRESS
Cambridge, New York, Melbourne, Madrid, Cape Town,
Singapore, São Paulo, Delhi, Mexico City

Cambridge University Press
32 Avenue of the Americas, New York, NY 10013-2473, USA

www.cambridge.org
Information on this title: www.cambridge.org/9780521493369

First published 2011
Reprinted 2012 (twice)

A catalog record for this publication is available from the British Library.

Library of Congress Cataloging in Publication Data

Recommender systems : an introduction / Dietmar Jannach . . . [et al.].
  p. cm.
Includes bibliographical references and index.
ISBN 978-0-521-49336-9 (hardback)
1. Personal communication service systems. Recommender systems
(Information filtering) I. Jannach, Dietmar, 1973– II. Title.
TK5103.485.R43 2010
006.3´3 – dc22   2010021870

ISBN 978-0-521-49336-9 Hardback

Additional resources for this publication at http:www.recommenderbook.net/

Contents

Foreword

It was a seductively simple idea that emerged in the early 1990s – to harness the opinions of millions of people online in an effort to help all of us find more useful and interesting content. And, indeed, in various domains and in various forms, this simple idea proved effective. The PARC Tapestry system (Goldberg et al. 1992) introduced the idea (and terminology) of collaborative filtering and showed how both explicit annotation data and implicit behavioral data could be collected into a queryable database and tapped by users to produce personal filters. Less than two years later, the GroupLens system (Resnick et al. 1994) showed that the collaborative filtering approach could be both distributed across a network and automated. Whereas GroupLens performed automated collaborative filtering to Usenet news messages, the Ringo system at Massachusetts Institute of Technology (MIT) (Shardanand and Maes 1995) did the same for music albums and artists and the Bellcore Video Recommender (Hill et al. 1995) did the same for movies. Each of these systems used similar automation techniques – algorithms that identified other users with similar tastes and combined their ratings together into a personalized, weighted average. This simple "k-nearest neighbor" algorithm proved so effective that it quickly became the gold standard against which all collaborative filtering algorithms were compared.

Systems-oriented exploration. With hindsight, it is now clear that these early collaborative filtering systems were important examples from the first of four overlapping phases of recommender systems advances. This systems-oriented exploration stage – through not only collaborative filtering but also knowledge-based systems such as the FindMe systems (Burke et al. 1996) – demonstrated the feasibility and efficacy of recommender systems and generated substantial excitement to move the field forward, in both research and commercial practice. (I do not mean to imply that these early research efforts did not also explore algorithms and design alternatives, but to a great extent we were so excited that "the dog sang" that we did not worry too much about whether it was perfectly in tune.)

A key event in this phase was the Collaborative Filtering Workshop at Berkeley in March 1996. This gathering helped coalesce the community, bringing together people working on personalized and nonpersonalized systems, on divergent algorithmic approaches (from statistical summaries to k-nearest neighbor to Bayesian clustering), and on different domains. By the end of the day, there was a consensus that these were all aspects of one larger problem – a problem that quickly became known as recommender systems, thanks in part to a special issue of *Communications of the ACM* that grew out of the workshop (Resnick and Varian 1997).

Rapid commercialization – the challenges of scale and value. Recommender systems emerged into a rapidly expanding Internet business climate, and commercialization was almost immediate. Pattie Maes's group at MIT founded Agents, Inc., in 1995 (later renamed Firefly Networks). Our GroupLens group at Minnesota founded Net Perceptions in 1996. Many other companies emerged as well. Quickly, we started to face real-world challenges largely unknown in the research lab. To succeed, companies had to move beyond demonstrating accurate predictions. We had to show that we could provide valuable recommendations – usually in the form of selecting a few particular products to recommend that would yield additional purchases – and that we could do so without slowing down existing web sites. These systems had to work at greater-than-research scales – handling millions of users and items and hundreds or thousands of transactions per second. It is perhaps no surprise that the first book on recommender systems, John Riedl's and my *Word of Mouse*, was targeted not at researchers but at marketing professionals.

Research at the time moved forward to address many of these technological challenges. New algorithms were developed to reduce online computation time, including item-based correlation algorithms and dimensionality-reduction approaches, both of which are still used today. Researchers became more interested in evaluating recommenders based on "top-n" recommendation list metrics. A wide set of research explored issues related to implicit ratings, startup issues for new users and new items, and issues related to user experience, including trust, explanation, and transparency.

Research explosion – recommenders go mainstream. Somewhere between 2000 and 2005, many of the recommender systems companies dried up, imploding with the Internet bubble or simply unable to compete with more mainstream companies that integrated recommendation into a broader set of business tools. As a technology, however, recommender systems were here to stay, with wide usage in e-commerce, broader retail, and a variety of knowledge management applications.

At the same time, research in recommender systems exploded with an infusion of people and approaches from many disciplines. From across the spectrum of artificial intelligence, information retrieval, data mining, security and privacy, and business and marketing research emerged new analyses and approaches to recommender systems. The algorithmic research was fueled by the availability of large datasets and then ignited by the 2006 announcement of the $1 million Netflix Prize for a 10 percent improvement in prediction accuracy.

Moving forward – recommenders in context. The excitement of the Netflix Prize brought many researchers together in heroic efforts to improve prediction accuracy. But even as these researchers closed in on success, a wave of researchers and practitioners were arguing for a step back toward the values of exploration and value. In 2006, MyStrands organized Recommenders06, a summer school on the present and future of recommender systems. In 2007, I organized the first ACM Recommender Systems Conference – a conference that has grown from 120 people to more than 300 in 2009. A look at the programs of these events shows increased interest in viewing recommendation in context, retooling research to ground it in an understanding of how people interact with organizations or businesses, and how recommendations can facilitate those interactions. Indeed, even though the field was nearly unanimously excited by the success of Netflix in bringing in new ideas, most of us also realized that an elaborate algorithm that improved predictions of just how much a user would dislike a set of bad movies did not help the user or Netflix. It is telling that the 2009 best-paper award went to a paper that demonstrated the flaws in the field's traditional "hold some data back" method of evaluating algorithms (Marlin and Zemel 2009), and that the most cited recent research paper on recommender systems is one that lays out how to match evaluation to user needs (Herlocker et al. 2004).

That brings us to this book. Behind the modest subtitle of "an introduction" lies the type of work the field needs to do to consolidate its learnings and move forward to address new challenges. Across the chapters that follow lies both a tour of what the field knows well – a diverse collection of algorithms and approaches to recommendation – and a snapshot of where the field is today, as new approaches derived from social computing and the semantic web find their place in the recommender systems toolbox. Let us all hope that this worthy effort spurs yet more creativity and innovation to help recommender systems move forward to new heights.

Joseph A. Konstan
Distinguished McKnight Professor
Department of Computer Science and Engineering
University of Minnesota

Preface

"Which digital camera should I buy? What is the best holiday for me and my family? Which is the best investment for supporting the education of my children? Which movie should I rent? Which web sites will I find interesting? Which book should I buy for my next vacation? Which degree and university are the best for my future?"

It is easy to expand this list with many examples in which people have to make decisions about how they want to spend their money or, on a broader level, about their future.

Traditionally, people have used a variety of strategies to solve such decision-making problems: conversations with friends, obtaining information from a trusted third party, hiring an expert team, consulting the Internet, using various methods from decision theory (if one tries to be rational), making a gut decision, or simply following the crowd.

However, almost everyone has experienced a situation in which the advice of a friendly sales rep was not really useful, in which the gut decision to follow the investments of our rich neighbor was not really in our interest, or in which spending endless hours on the Internet led to confusion rather than to quick and good decisions. To sum up, good advice is difficult to receive, is in most cases time-consuming or costly, and even then is often of questionable quality.

Wouldn't it be great to have an affordable personal advisor who helps us make good decisions efficiently?

The construction of systems that support users in their (online) decision making is the main goal of the field of recommender systems. In particular, the goal of recommender systems is to provide easily accessible, high-quality recommendations for a large user community.

This focus on volume and easy accessibility makes the technology very powerful. Although recommender systems aim at the individual decisions of users, these systems have a significant impact in a larger sense because of their mass application – as, for instance, Amazon.com's recommendation engines. Because of the far reach of the Internet market, this issue must not be underestimated, as the control of recommender systems allows markets themselves to be controlled in a broader sense. Consider, for example, a department store in which all the sales clerks follow orders to push only certain products.

One can argue that recommender systems are for the masses who cannot afford or are not willing to pay for high-quality advice provided by experts. This is partially true in some domains, such as financial services or medicine; however, the goal of making good decisions includes the aim of outperforming domain experts. Although this is clearly not possible and also not necessary in all domains, there are many cases in which the wisdom of the crowds can be exploited to improve decisions. Thus, given the huge information bases available on the Internet, can we develop systems that provide better recommendations than humans?

The challenge of providing affordable, personal, and high-quality recommendations is central to the field and generates many interesting follow-up goals on both a technical and a psychological level. Although, on the technical level, we are concerned with finding methods that exploit the available information and knowledge as effectively and efficiently as possible, psychological factors must be considered when designing the end-user interaction processes. The design of these communication processes greatly influences the trust in the subsequent recommendations and ultimately in the decisions themselves. Users rarely act as rational agents who know exactly what they want. Even the way a recommender agent asks for a customer's preferences or which decision options are presented will affect a customer's choice. Therefore, recommender systems cannot be reduced to simple decision theoretical concepts.

More than fifteen years have passed since the software systems that are now called "recommender systems" were first developed. Since then, researchers have continuously developed new approaches for implementing recommender systems, and today most of us are used to being supported by recommendation services such as the one found on Amazon.com. Historically, recommender systems have gained much attention by applying methods from artificial intelligence to information filtering – that is, to recommend web sites or to filter and rank news items. In fact, recommendation methods such as case-based or rule-based techniques have their roots in the expert systems of the 1980s. However, the application areas of recommender systems go far beyond pure information filtering methods, and nowadays recommendation technology is providing solutions in domains as diverse as financial products, real estate, electronic consumer products, movies, books, music, news, and web sites, just to name a few.

This book provides an introduction to the broad field of recommender systems technology, as well as an overview of recent improvements. It is aimed at both graduate students or new PhDs who are starting their own research in the field and practitioners and IT experts who are looking for a starting point for the design and implementation of real-world recommender applications. Additional advanced material can be found, for instance, in *Recommender Systems Handbook* (Ricci et al. 2010), which contains a comprehensive collection of contributions from leading researchers in the field.

This book is organized into two parts. In the first part, we start by summarizing the basic approaches to implementing recommender systems and discuss their individual advantages and shortcomings. In addition to describing how such systems are built, we focus on methods for evaluating the accuracy of recommenders and examining their effect on the behavior of online customers. The second part of the book focuses on recent developments and covers issues such as trust in recommender systems and emerging applications based on Web 2.0 and Semantic Web technologies. Teaching material to accompany the topics presented in this book is provided at the site http://www.recommenderbook.net/.

We would like to thank everyone who contributed to this book, in particular, Heather Bergman and Lauren Cowles from Cambridge University Press, who supported us throughout the editorial process. Particular thanks also go to Arthur Pitman, Kostyantyn Shchekotykhin, Carla Delgado-Battenfeld, and Fatih Gedikli for their great help in proofreading the manuscript, as well as to several scholar colleagues for their effort in reviewing and giving helpful feedback.

> *Dietmar Jannach*
> *Markus Zanker*
> *Alexander Felfernig*
> *Gerhard Friedrich*
> Dortmund, Klagenfurt, and Graz, 2010

1

Introduction

Most Internet users have come across a recommender system in one way or another. Imagine, for instance, that a friend recommended that you read a new book and that you subsequently visit your favorite online bookstore. After typing in the title of the book, it appears as just one of the results listed. In one area of the web page possibly called "Customers Who Bought This Item Also Bought," a list is shown of additional books that are supposedly of interest to you. If you are a regular user of the same online bookstore, such a personalized list of recommendations will appear automatically as soon as you enter the store. The software system that determines which books should be shown to a particular visitor is a *recommender system*.

The online bookstore scenario is useful for discussing several aspects of such software systems. First, notice that we are talking about *personalized* recommendations – in other words, every visitor sees a different list depending on his or her tastes. In contrast, many other online shops or news portals may simply inform you of their top-selling items or their most read articles. Theoretically, we could interpret this information as a sort of impersonal buying or reading recommendation as well and, in fact, very popular books will suit the interests and preferences of many users. Still, there will be also many people who do not like to read *Harry Potter* despite its strong sales in 2007 – in other words, for these people, recommending top-selling items is not very helpful. In this book, we will focus on systems that generate personalized recommendations.

The provision of personalized recommendations, however, requires that the system knows something about every user. Every recommender system must develop and maintain a *user model* or *user profile* that, for example, contains the user's preferences. In our bookstore example, the system could, for instance, remember which books a visitor has viewed or bought in the past to predict which other books might be of interest.

Although the existence of a user model is central to every recommender system, the way in which this information is acquired and exploited depends on the particular recommendation technique. User preferences can, for instance, be acquired *implicitly* by monitoring user behavior, but the recommender system might also *explicitly* ask the visitor about his or her preferences.

The other question in this context is what kind of *additional information* the system should exploit when it generates a list of personalized recommendations. The most prominent approach, which is actually used by many real online bookstores, is to take the behavior, opinions, and tastes of a large community of other users into account. These systems are often referred to as *community-based* or *collaborative* approaches.

This textbook is structured into two parts, reflecting the dynamic nature of the research field. Part I summarizes the well-developed aspects of recommendation systems research that have been widely accepted for several years. Therefore, Part I is structured in a canonical manner and introduces the basic paradigms of collaborative (Chapter 2), content-based (Chapter 3), and knowledge-based recommendation (Chapter 4), as well as hybridization methods (Chapter 5). Explaining the reasons for recommending an item (Chapter 6) as well as evaluating the quality of recommendation systems (Chapter 7) are also fundamental chapters. The first part concludes with an experimental evaluation (Chapter 8) that compares different recommendation algorithms in a mobile environment that can serve as a practical reference for further investigations. In contrast, Part II discusses very recent research topics within the field, such as how to cope with efforts to attack and manipulate a recommender system from outside (Chapter 9), supporting consumer decision making and potential persuasion strategies (Chapter 10), recommendation systems in the context of the social and semantic webs (Chapter 11), and the application of recommender systems to ubiquitous domains (Chapter 12). Consequently, chapters of the second part should be seen as a reference point for ongoing research.

1.1 Part I: Introduction to basic concepts

1.1.1 Collaborative recommendation

The basic idea of these systems is that if users shared the same interests in the past – if they viewed or bought the same books, for instance – they will also have similar tastes in the future. So, if, for example, user A and user B have a purchase history that overlaps strongly and user A has recently bought a book that *B* has not yet seen, the basic rationale is to propose this book also to *B*. Because this selection of hopefully interesting books involves filtering the most promising ones from a large set and because the users implicitly collaborate with one another, this technique is also called *collaborative filtering* (CF).

Today, systems of this kind are in wide use and have also been extensively studied over the past fifteen years. We cover the underlying techniques and open questions associated with collaborative filtering in detail in the next chapter of this book. Typical questions that arise in the context of collaborative approaches include the following:

- How do we find users with similar tastes to the user for whom we need a recommendation?
- How do we measure similarity?
- What should we do with new users, for whom a buying history is not yet available?
- How do we deal with new items that nobody has bought yet?
- What if we have only a few ratings that we can exploit?
- What other techniques besides looking for similar users can we use for making a prediction about whether a certain user will like an item?

Pure CF approaches do not exploit or require any knowledge about the items themselves. Continuing with the bookstore example, the recommender system, for instance, does not need to know what a book is about, its genre, or who wrote it. The obvious advantage of this strategy is that these data do not have to be entered into the system or maintained. On the other hand, using such characteristics to propose books that are actually similar to those the user liked in the past might be more effective.

1.1.2 Content-based recommendation

In general, recommender systems may serve two different purposes. On one hand, they can be used to stimulate users into doing something such as buying a specific book or watching a specific movie. On the other hand, recommender systems can also be seen as tools for dealing with *information overload*, as these systems aim to select the most interesting items from a larger set. Thus, recommender systems research is also strongly rooted in the fields of *information retrieval* and *information filtering*. In these areas, however, the focus lies mainly on the problem of discriminating between relevant and irrelevant *documents* (as opposed to the artifacts such as books or digital cameras recommended in traditional e-commerce domains). Many of the techniques developed in these areas exploit information derived from the documents' contents to rank them. These techniques will be discussed in the chapter on content-based recommendation[1].

At its core, content-based recommendation is based on the availability of (manually created or automatically extracted) item descriptions and a profile that assigns importance to these characteristics. If we think again of the bookstore example, the possible characteristics of books might include the genre, the specific topic, or the author. Similar to item descriptions, user profiles may also be automatically derived and "learned" either by analyzing user behavior and feedback or by asking explicitly about interests and preferences.

In the context of content-based recommendation, the following questions must be answered:

- How can systems automatically acquire and continuously improve user profiles?
- How do we determine which items match, or are at least similar to or compatible with, a user's interests?
- What techniques can be used to automatically extract or learn the item descriptions to reduce manual annotation?

When compared with the content-agnostic approaches described above, content-based recommendation has two advantages. First, it does not require large user groups to achieve reasonable recommendation accuracy. In addition, new items can be immediately recommended once item attributes are available. In some domains, such item descriptions can be automatically extracted (for instance, from text documents) or are already available in an electronic catalog. In many domains, however, the more subjective characteristics of an item – such as "ease of use" or "elegance of design" – would be useful in the recommendation process. These characteristics are hard to acquire automatically, however, meaning that such information must be manually entered into the system in a potentially expensive and error-prone process.

1.1.3 Knowledge-based recommendation

If we turn our attention to other application domains, such as consumer electronics, many involve large numbers of one-time buyers. This means that we cannot rely on the existence of a purchase history, a prerequisite for collaborative and content-based filtering approaches. However, more detailed and structured *content* may be available, including technical and quality features.

---

[1] Some authors use the term "content-based filtering" instead of content-based recommendation.

Take, for instance, a recommender system for digital cameras that should help the end user find a camera model that fits his or her particular requirements. Typical customers buy a new camera only once every few years, so the recommender system cannot construct a user profile or propose cameras that others liked, which – as a side note – would result in proposing only top-selling items.

Thus, a system is needed that exploits additional and means–end knowledge to generate recommendations. In such *knowledge-based approaches*, the recommender system typically makes use of additional, often manually provided, information about both the current user and the available items. *Constraint-based recommenders* are one example of such systems, which we will consider in our discussion of the different aspects of knowledge-based approaches. In the digital camera domain, a constraint-based system could use detailed knowledge about the features of the cameras, such as resolution, weight, or price. In addition, explicit constraints may be used to describe the context in which certain features are relevant for the customer, such as, for example, that a high-resolution camera is advantageous if the customer is interested in printing large pictures. Simply presenting products that fulfill a given set of requested features is not enough, as the aspect of personalization is missing, and every user (with the same set of requested features) will get the same set of recommendations. Thus, constraint-based recommender systems also need to maintain user profiles. In the digital camera scenario the system could, for instance, ask the user about the relative importance of features, such as whether resolution is more important than weight.

The other aspect covered in this chapter is "user interaction", as in many knowledge-based recommender systems, the user requirements must be elicited interactively. Considering the bookstore example and collaborative recommendation techniques once again, we see that users can interact with the system in only a limited number of ways. In fact, in many applications the only possibility for interaction is to rate the proposed items – for example, on a scale from 1 to 5 or in terms of a "like/dislike" statement. Think, however, about the digital camera recommender, which should also be able to serve first-time users. Therefore, more complex types of interaction are required to determine the user's needs and preferences, mostly because no purchase history is available that can be exploited. A simple approach would be to ask the user directly about his or her requirements, such as the maximum price, the minimum resolution, and so forth. Such an approach, however, not only requires detailed technical understanding of the item's features but also generates additional cognitive load in scenarios with a large number of item features. More elaborate approaches, therefore, try to implement more conversational interaction styles, in which the system tries to incrementally ascertain preferences within an interactive and personalized dialog.

Overall, the questions that are addressed in the chapter on knowledge-based recommender systems include the following:

- What kinds of domain knowledge can be represented in a knowledge base?
- What mechanisms can be used to select and rank the items based on the user's characteristics?
- How do we acquire the user profile in domains in which no purchase history is available, and how can we take the customer's explicit preferences into account?
- Which interaction patterns can be used in interactive recommender systems?
- Finally, in which dimensions can we personalize the dialog to maximize the precision of the preference elicitation process?

1.1.4 Hybrid approaches

We have already seen that the different approaches discussed so far have certain advantages and, of course, disadvantages depending on the problem setting. One obvious solution is to combine different techniques to generate better or more precise recommendations (we will discuss the question of what a "good" recommendation is later). If, for instance, community knowledge exists and detailed information about the individual items is available, a recommender system could be enhanced by hybridizing collaborative or social filtering with content-based techniques. In particular, such a design could be used to overcome the described ramp-up problems of pure collaborative approaches and rely on content analysis for new items or new users.

When combining different approaches within one recommender system, the following questions have to be answered and will be covered in the chapter on hybrid approaches:

- Which techniques can be combined, and what are the prerequisites for a given combination?
- Should proposals be calculated for two or more systems sequentially, or do other hybridization designs exist?
- How should the results of different techniques be weighted and can they be determined dynamically?

1.1.5 Explanations in recommender systems

Explanations aim to make a recommendation system's line of reasoning transparent to its users. This chapter outlines how the different recommendation strategies can be extended to provide reasons for the recommendations they propose to users. As knowledge-based recommendation systems have a long tradition of providing reasons to support their computed results, this chapter focuses on computing explanations for constraint-based and case-based recommender systems. In addition, efforts to explain collaborative filtering results are described to address the following topics:

- How can a recommender system explain its proposals while increasing the user's confidence in the system?
- How does the recommendation strategy affect the way recommendations can be explained?
- Can explanations be used to convince a user that the proposals made by the system are "fair" or unbiased?

1.1.6 Evaluating recommender systems

Research in recommender systems is strongly driven by the goal of improving the quality of the recommendations that are produced. The question that immediately arises is, of course, how can we actually measure the quality of the proposals made by a recommender system?

We start the chapter on evaluating recommender systems by reflecting on the general principles of empirical research and discuss the current state of practice in evaluating recommendation techniques. Based on the results of a small survey, we focus in particular on empirical evaluations on historical datasets and present different methodologies and metrics.

We also explore alternate evaluation approaches to address the necessity of, for instance, better capturing user experience or system goals. Evaluation approaches are classified into experimental, quasi-experimental, and nonexperimental research designs. Thus, the questions answered in the chapter include the following:

- Which research designs are applicable for evaluating recommender systems?
- How can recommender systems be evaluated using experiments on historical datasets?
- What metrics are applicable for different evaluation goals?

- What are the limitations of existing evaluation techniques, in particular when it comes to the conversational or business value aspects of recommender systems?

1.1.7 Case study

The final chapter of the book's first part is devoted to an experimental online evaluation that compares different personalized and impersonalized recommendation strategies on a mobile Internet portal. The purpose of this large-scale case study of a commercial recommender system is to address questions such as

- What is the business value of recommender systems?
- Do they help to increase sales or turn more visitors into buyers?
- Are there differences in the effectiveness of different recommendation algorithms? Which technique should be used in which situation?

1.2 Part II: Recent developments

Although many of the ideas and basic techniques that are used in today's recommender systems were developed more than a decade ago, the field is still an area of active research, in particular because the web itself has become an integral part of our everyday life and, at the same time, new technologies are constantly emerging.

In the second part of the book we will therefore focus – in the form of shorter chapters – on current research topics and recent advancements in the field. Among others, the following questions will be addressed:

- *Privacy and robustness.* How can we prevent malicious users from manipulating a recommender system – for instance, by inserting fake users or ratings into the system's database? How can we ensure the privacy of users?
- *Online consumer decision making.* Which consumer decision-making theories are the most relevant? Can the insights gained in traditional sales channels be transferred to the online channel, and in particular, how can this knowledge be encoded in a recommender system? Are there additional techniques or new models that can help us to improve the (business) value or acceptance of a recommendation service?
- *Recommender systems in the context of the social and the semantic web.* How can we exploit existing trust structures or social relationships between users to improve the recommender's accuracy? How do Semantic Web technologies

affect recommendation algorithms? What is the role of recommenders in Web 2.0?

- *Ubiquitous applications.* How do current technological advances, for instance in the area of mobile solutions, open new doors for building next-generation recommender systems? How do ubiquitous application domains affect recommendation algorithms – for instance, by placing more emphasis on contextual and situational parameters?

PART I

Introduction to basic concepts

2

Collaborative recommendation

The main idea of collaborative recommendation approaches is to exploit information about the past behavior or the opinions of an existing user community for predicting which items the current user of the system will most probably like or be interested in. These types of systems are in widespread industrial use today, in particular as a tool in online retail sites to customize the content to the needs of a particular customer and to thereby promote additional items and increase sales.

From a research perspective, these types of systems have been explored for many years, and their advantages, their performance, and their limitations are nowadays well understood. Over the years, various algorithms and techniques have been proposed and successfully evaluated on real-world and artificial test data.

Pure collaborative approaches take a matrix of given user–item ratings as the only input and typically produce the following types of output: (a) a (numerical) prediction indicating to what degree the current user will like or dislike a certain item and (b) a list of n recommended items. Such a *top-N* list should, of course, not contain items that the current user has already bought.

2.1 User-based nearest neighbor recommendation

The first approach we discuss here is also one of the earliest methods, called *user-based nearest neighbor recommendation*. The main idea is simply as follows: given a ratings database and the ID of the current (active) user as an input, identify other users (sometimes referred to as *peer users* or *nearest neighbors*) that had similar preferences to those of the active user in the past. Then, for every product p that the active user has not yet seen, a prediction is computed based on the ratings for p made by the peer users. The underlying

Table 2.1. *Ratings database for collaborative recommendation.*

	Item1	Item2	Item3	Item4	Item5
Alice	5	3	4	4	?
User1	3	1	2	3	3
User2	4	3	4	3	5
User3	3	3	1	5	4
User4	1	5	5	2	1

assumptions of such methods are that (a) if users had similar tastes in the past they will have similar tastes in the future and (b) user preferences remain stable and consistent over time.

2.1.1 First example

Let us examine a first example. Table 2.1 shows a database of ratings of the current user, Alice, and some other users. Alice has, for instance, rated "Item1" with a "5" on a 1-to-5 scale, which means that she strongly liked this item. The task of a recommender system in this simple example is to determine whether Alice will like or dislike "Item5", which Alice has not yet rated or seen. If we can predict that Alice will like this item very strongly, we should include it in Alice's recommendation list. To this purpose, we search for users whose taste is similar to Alice's and then take the ratings of this group for "Item5" to predict whether Alice will like this item.

Before we discuss the mathematical calculations required for these predictions in more detail, let us introduce the following conventions and symbols. We use $U = \{u_1, \ldots, u_n\}$ to denote the set of users, $P = \{p_1, \ldots, p_m\}$ for the set of products (items), and R as an $n \times m$ matrix of ratings $r_{i,j}$, with $i \in 1 \ldots n, j \in 1 \ldots m$. The possible rating values are defined on a numerical scale from 1 (strongly dislike) to 5 (strongly like). If a certain user i has not rated an item j, the corresponding matrix entry $r_{i,j}$ remains empty.

With respect to the determination of the set of similar users, one common measure used in recommender systems is Pearson's correlation coefficient. The similarity $sim(a, b)$ of users a and b, given the rating matrix R, is defined in Formula 2.1. The symbol $\overline{r_a}$ corresponds to the average rating of user a.

$$sim(a, b) = \frac{\sum_{p \in P}(r_{a,p} - \overline{r_a})(r_{b,p} - \overline{r_b})}{\sqrt{\sum_{p \in P}(r_{a,p} - \overline{r_a})^2}\sqrt{\sum_{p \in P}(r_{b,p} - \overline{r_b})^2}} \tag{2.1}$$

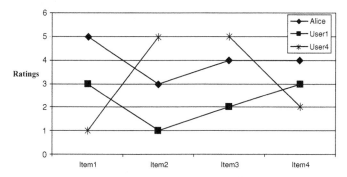

Figure 2.1. Comparing Alice with two other users.

The similarity of *Alice* to *User1* is thus as follows ($\overline{r_{Alice}} = \overline{r_a} = 4, \overline{r_{User1}} = \overline{r_b} = 2.4$):

$$\frac{(5 - \overline{r_a}) * (3 - \overline{r_b}) + (3 - \overline{r_a}) * (1 - \overline{r_b}) + \cdots + (4 - \overline{r_a}) * (3 - \overline{r_b}))}{\sqrt{(5 - \overline{r_a})^2 + (3 - \overline{r_a})^2 + \cdots} \sqrt{(3 - \overline{r_b})^2 + (1 - \overline{r_b})^2 + \cdots}} = 0.85$$

$$(2.2)$$

The Pearson correlation coefficient takes values from $+1$ (strong positive correlation) to -1 (strong negative correlation). The similarities to the other users, *User2* to *User4*, are 0.70, 0.00, and -0.79, respectively.

Based on these calculations, we observe that *User1* and *User2* were somehow similar to *Alice* in their rating behavior in the past. We also see that the Pearson measure considers the fact that users are different with respect to how they interpret the rating scale. Some users tend to give only high ratings, whereas others will never give a 5 to any item. The Pearson coefficient factors these averages out in the calculation to make users comparable – that is, although the absolute values of the ratings of *Alice* and *User1* are quite different, a rather clear linear correlation of the ratings and thus similarity of the users is detected.

This fact can also be seen in the visual representation in Figure 2.1, which both illustrates the similarity between *Alice* and *User1* and the differences in the ratings of *Alice* and *User4*.

To make a prediction for *Item5*, we now have to decide which of the neighbors' ratings we shall take into account and how strongly we shall value their opinions. In this example, an obvious choice would be to take *User1* and *User2* as peer users to predict Alice's rating. A possible formula for computing a prediction for the rating of user *a* for item *p* that also factors the relative *proximity* of the nearest neighbors N and a's average rating $\overline{r_a}$ is the following:

$$pred(a, p) = \overline{r_a} + \frac{\sum_{b \in N} sim(a, b) * (r_{b,p} - \overline{r_b})}{\sum_{b \in N} sim(a, b)} \qquad (2.3)$$

In the example, the prediction for Alice's rating for *Item5* based on the ratings of near neighbors *User1* and *User2* will be

$$4 + 1/(0.85 + 0.7) * (0.85 * (3 - 2.4) + 0.70 * (5 - 3.8)) = 4.87 \qquad (2.4)$$

Given these calculation schemes, we can now compute rating predictions for Alice for all items she has not yet seen and include the ones with the highest prediction values in the recommendation list. In the example, it will most probably be a good choice to include *Item5* in such a list.

The example rating database shown above is, of course, an idealization of the real world. In real-world applications, rating databases are much larger and can comprise thousands or even millions of users and items, which means that we must think about computational complexity. In addition, the rating matrix is typically very sparse, meaning that every user will rate only a very small subset of the available items. Finally, it is unclear what we can recommend to new users or how we deal with new items for which no ratings exist. We discuss these aspects in the following sections.

2.1.2 Better similarity and weighting metrics

In the example, we used Pearson's correlation coefficient to measure the similarity among users. In the literature, other metrics, such as *adjusted cosine similarity* (which will be discussed later in more detail), *Spearman's rank correlation coefficient*, or the *mean squared difference* measure have also been proposed to determine the proximity between users. However, empirical analyses show that for user-based recommender systems – and at least for the best studied recommendation domains – the Pearson coefficient outperforms other measures of comparing users (Herlocker et al. 1999). For the later-described item-based recommendation techniques, however, it has been reported that the cosine similarity measure consistently outperforms the Pearson correlation metric.

Still, using the "pure" Pearson measure alone for finding neighbors and for weighting the ratings of those neighbors may not be the best choice. Consider, for instance, the fact that in most domains there will exist some items that are liked by everyone. A similarity measure such as Pearson will not take into account that an agreement by two users on a more controversial item has more "value" than an agreement on a generally liked item. As a resolution to this,

Breese et al. (1998) proposed applying a transformation function to the item ratings, which reduces the relative importance of the agreement on universally liked items. In analogy to the original technique, which was developed in the information retrieval field, they called that factor the *inverse user frequency*. Herlocker et al. (1999) address the same problem through a *variance weighting factor* that increases the influence of items that have a high variance in the ratings – that is, items on which controversial opinions exist.

Our basic similarity measure used in the example also does not take into account whether two users have co-rated only a few items (on which they may agree by chance) or whether there are many items on which they agree. In fact, it has been shown that predictions based on the ratings of neighbors with which the active user has rated only a very few items in common are a bad choice and lead to poor predictions (Herlocker et al. 1999). Herlocker et al. (1999) therefore propose using another weighting factor, which they call *significance weighting*. Although the weighting scheme used in their experiments, reported by Herlocker et al. (1999, 2002), is a rather simple one, based on a linear reduction of the similarity weight when there are fewer than fifty co-rated items, the increases in prediction accuracy are significant. The question remains open, however, whether this weighting scheme and the heuristically determined thresholds are also helpful in real-world settings, in which the ratings database is smaller and we cannot expect to find many users who have co-rated fifty items.

Finally, another proposal for improving the accuracy of the recommendations by fine-tuning the prediction weights is termed *case amplification* (Breese et al. 1998). Case amplification refers to an adjustment of the weights of the neighbors in a way that values close to $+1$ and -1 are emphasized by multiplying the original weights by a constant factor ρ. Breese et al. (1998) used 2.5 for ρ in their experiments.

2.1.3 Neighborhood selection

In our example, we intuitively decided not to take all neighbors into account (*neighborhood selection*). For the calculation of the predictions, we included only those that had a positive correlation with the active user (and, of course, had rated the item for which we are looking for a prediction). If we included all users in the neighborhood, this would not only negatively influence the performance with respect to the required calculation time, but it would also have an effect on the accuracy of the recommendation, as the ratings of other users who are not really comparable would be taken into account.

The common techniques for reducing the size of the neighborhood are to define a specific minimum threshold of user similarity or to limit the size to a fixed number and to take only the k nearest neighbors into account. The potential problems of either technique are discussed by Anand and Mobasher (2005) and by Herlocker et al. (1999): if the similarity threshold is too high, the size of the neighborhood will be very small for many users, which in turn means that for many items no predictions can be made (*reduced coverage*). In contrast, when the threshold is too low, the neighborhood sizes are not significantly reduced.

The value chosen for k – the size of the neighborhood – does not influence coverage. However, the problem of finding a good value for k still exists: When the number of neighbors k taken into account is too high, too many neighbors with limited similarity bring additional "noise" into the predictions. When k is too small – for example, below 10 in the experiments from Herlocker et al. (1999) – the quality of the predictions may be negatively affected. An analysis of the MovieLens dataset indicates that "in most real-world situations, a neighborhood of 20 to 50 neighbors seems reasonable" (Herlocker et al. 2002).

A detailed analysis of the effects of using different weighting and similarity schemes, as well as different neighborhood sizes, can be found in Herlocker et al. (2002).

2.2 Item-based nearest neighbor recommendation

Although user-based CF approaches have been applied successfully in different domains, some serious challenges remain when it comes to large e-commerce sites, on which we must handle millions of users and millions of catalog items. In particular, the need to scan a vast number of potential neighbors makes it impossible to compute predictions in real time. Large-scale e-commerce sites, therefore, often implement a different technique, *item-based recommendation*, which is more apt for offline preprocessing[1] and thus allows for the computation of recommendations in real time even for a very large rating matrix (Sarwar et al. 2001).

The main idea of item-based algorithms is to compute predictions using the similarity between items and not the similarity between users. Let us examine our ratings database again and make a prediction for Alice for *Item5*. We first compare the rating vectors of the other items and look for items that have ratings similar to *Item5*. In the example, we see that the ratings for *Item5* (3, 5, 4, 1) are similar to the ratings of *Item1* (3, 4, 3, 1) and there is also a partial similarity

---

[1] Details about data preprocessing for item-based filtering are given in Section 2.2.2.

with *Item4* (3, 3, 5, 2). The idea of item-based recommendation is now to simply look at Alice's ratings for these similar items. Alice gave a "5" to *Item1* and a "4" to *Item4*. An item-based algorithm computes a weighted average of these other ratings and will predict a rating for *Item5* somewhere between 4 and 5.

2.2.1 The cosine similarity measure

To find similar items, a similarity measure must be defined. In item-based recommendation approaches, *cosine similarity* is established as the standard metric, as it has been shown that it produces the most accurate results. The metric measures the similarity between two n-dimensional vectors based on the angle between them. This measure is also commonly used in the fields of information retrieval and text mining to compare two text documents, in which documents are represented as vectors of terms.

The similarity between two items a and b – viewed as the corresponding rating vectors $\vec{a}$ and $\vec{b}$ – is formally defined as follows:

$$sim(\vec{a}, \vec{b}) = \frac{\vec{a} \cdot \vec{b}}{|\vec{a}| * |\vec{b}|} \qquad (2.5)$$

The $\cdot$ symbol is the dot product of vectors. $|\vec{a}|$ is the Euclidian length of the vector, which is defined as the square root of the dot product of the vector with itself.

The cosine similarity of *Item5* and *Item1* is therefore calculated as follows:

$$sim(I5, I1) = \frac{3 * 3 + 5 * 4 + 4 * 3 + 1 * 1}{\sqrt{3^2 + 5^2 + 4^2 + 1^2} * \sqrt{3^2 + 4^2 + 3^2 + 1^2}} = 0.99 \qquad (2.6)$$

The possible similarity values are between 0 and 1, where values near to 1 indicate a strong similarity. The basic cosine measure does not take the differences in the average rating behavior of the users into account. This problem is solved by using the *adjusted cosine* measure, which subtracts the user average from the ratings. The values for the adjusted cosine measure correspondingly range from -1 to $+1$, as in the Pearson measure.

Let U be the set of users that rated both items a and b. The adjusted cosine measure is then calculated as follows:

$$sim(a, b) = \frac{\sum_{u \in U}(r_{u,a} - \overline{r_u})(r_{u,b} - \overline{r_u})}{\sqrt{\sum_{u \in U}(r_{u,a} - \overline{r_u})^2}\sqrt{\sum_{u \in U}(r_{u,b} - \overline{r_u})^2}} \qquad (2.7)$$

We can therefore transform the original ratings database and replace the original rating values with their deviation from the average ratings as shown in Table 2.2.

Table 2.2. *Mean-adjusted ratings database.*

	Item1	Item2	Item3	Item4	Item5
Alice	1.00	−1.00	0.00	0.00	?
User1	0.60	−1.40	−0.40	0.60	0.60
User2	0.20	−0.80	0.20	−0.80	1.20
User3	−0.20	−0.20	−2.20	2.80	0.80
User4	−1.80	2.20	2.20	−0.80	−1.80

The adjusted cosine similarity value for *Item5* and *Item1* for the example is thus:

$$\frac{0.6 * 0.6 + 0.2 * 1.2 + (-0.2) * 0.80 + (-1.8) * (-1.8)}{\sqrt{(0.6^2 + 0.2^2 + (-0.2)^2 + (-1.8)^2} * \sqrt{0.6^2 + 1.2^2 + 0.8^2 + (-1.8)^2}} = 0.80$$

(2.8)

After the similarities between the items are determined we can predict a rating for Alice for *Item5* by calculating a weighted sum of Alice's ratings for the items that are similar to *Item5*. Formally, we can predict the rating for user u for a product p as follows:

$$pred(u, p) = \frac{\sum_{i \in ratedItems(u)} sim(i, p) * r_{u,i}}{\sum_{i \in ratedItems(a)} sim(i, p)}$$

(2.9)

As in the user-based approach, the size of the considered neighborhood is typically also limited to a specific size – that is, not all neighbors are taken into account for the prediction.

2.2.2 Preprocessing data for item-based filtering

Item-to-item collaborative filtering is the technique used by Amazon.com to recommend books or CDs to their customers. Linden et al. (2003) report on how this technique was implemented for Amazon's online shop, which, in 2003, had 29 million users and millions of catalog items. The main problem with traditional user-based CF is that the algorithm does not scale well for such large numbers of users and catalog items. Given M customers and N catalog items, in the worst case, all M records containing up to N items must be evaluated. For realistic scenarios, Linden et al. (2003) argue that the actual complexity is much lower because most of the customers have rated or bought only a very small number of items. Still, when the number of customers M is around several million, the calculation of predictions in real time is still infeasible, given the short response times that must be obeyed in the online environment.

For making item-based recommendation algorithms applicable also for large scale e-commerce sites without sacrificing recommendation accuracy, an approach based on offline precomputation of the data is typically chosen. The idea is to construct in advance the *item similarity matrix* that describes the pairwise similarity of all catalog items. At run time, a prediction for a product p and user u is made by determining the items that are most similar to i and by building the weighted sum of u's ratings for these items in the neighborhood. The number of neighbors to be taken into account is limited to the number of items that the active user has rated. As the number of such items is typically rather small, the computation of the prediction can be easily accomplished within the short time frame allowed in interactive online applications.

With respect to memory requirements, a full item similarity matrix for N items can theoretically have up to N^2 entries. In practice, however, the number of entries is significantly lower, and further techniques can be applied to reduce the complexity. The options are, for instance, to consider only items that have a minimum number of co-ratings or to memorize only a limited neighborhood for each item; this, however, increases the danger that no prediction can be made for a given item (Sarwar et al. 2001).

In principle, such an offline precomputation of neighborhoods is also possible for user-based approaches. Still, in practical scenarios the number of overlapping ratings for two users is relatively small, which means that a few additional ratings may quickly influence the similarity value between users. Compared with these user similarities, the item similarities are much more stable, such that precomputation does not affect the preciseness of the predictions too much (Sarwar et al. 2001).

Besides different preprocessing techniques used in so-called model-based approaches, it is an option to exploit only a certain fraction of the rating matrix to reduce the computational complexity. Basic techniques include *subsampling*, which can be accomplished by randomly choosing a subset of the data or by ignoring customer records that have only a very small set of ratings or that only contain very popular items. A more advanced and information-theoretic technique for filtering out the most "relevant" customers was also proposed by Yu et al. (2003). In general, although some computational speedup can be achieved with such techniques, the capability of the system to generate accurate predictions might deteriorate, as these recommendations are based on less information.

Further model-based and preprocessing-based approaches for complexity and dimensionality reduction will be discussed in Section 2.4.

2.3 About ratings

Before we discuss further techniques for reducing the computational complexity and present additional algorithms that operate solely on the basis of a user–item ratings matrix, we present a few general remarks on ratings in collaborative recommendation approaches.

2.3.1 Implicit and explicit ratings

Among the existing alternatives for gathering users' opinions, asking for explicit item ratings is probably the most precise one. In most cases, five-point or seven-point Likert response scales ranging from "Strongly dislike" to "Strongly like" are used; they are then internally transformed to numeric values so the previously mentioned similarity measures can be applied. Some aspects of the usage of different rating scales, such as how the users' rating behavior changes when different scales must be used and how the quality of recommendation changes when the granularity is increased, are discussed by Cosley et al. (2003). What has been observed is that in the movie domain, a five-point rating scale may be too narrow for users to express their opinions, and a ten-point scale was better accepted. An even more fine-grained scale was chosen in the joke recommender discussed by Goldberg et al. (2001), where a continuous scale (from -10 to $+10$) and a graphical input bar were used. The main arguments for this approach are that there is no precision loss from the discretization, user preferences can be captured at a finer granularity, and, finally, end users actually "like" the graphical interaction method, which also lets them express their rating more as a "gut reaction" on a visual level.

The question of how the recommendation accuracy is influenced and what is the "optimal" number of levels in the scaling system is, however, still open, as the results reported by Cosley et al. (2003) were developed on only a small user basis and for a single domain.

The main problems with explicit ratings are that such ratings require additional efforts from the users of the recommender system and users might not be willing to provide such ratings as long as the value cannot be easily seen. Thus, the number of available ratings could be too small, which in turn results in poor recommendation quality.

Still, Shafer et al. (2006) argue that the problem of gathering explicit ratings is not as hard as one would expect because only a small group of "early adopters" who provide ratings for many items is required in the beginning to get the system working.

Besides that, one can observe that in the last few years in particular, with the emergence of what is called Web 2.0, the role of online communities has changed and users are more willing to contribute actively to their community's knowledge. Still, in light of these recent developments, more research focusing on the development of techniques and measures that can be used to persuade the online user to provide more ratings is required.

Implicit ratings are typically collected by the web shop or application in which the recommender system is embedded. When a customer buys an item, for instance, many recommender systems interpret this behavior as a positive rating. The system could also monitor the user's browsing behavior. If the user retrieves a page with detailed item information and remains at this page for a longer period of time, for example, a recommender could interpret this behavior as a positive orientation toward the item.

Although implicit ratings can be collected constantly and do not require additional efforts from the side of the user, one cannot be sure whether the user behavior is correctly interpreted. A user might not like all the books he or she has bought; the user also might have bought a book for someone else. Still, if a sufficient number of ratings is available, these particular cases will be factored out by the high number of cases in which the interpretation of the behavior was right. In fact, Shafer et al. (2006) report that in some domains (such as personalized online radio stations) collecting the implicit feedback can even result in more accurate user models than can be done with explicit ratings.

A further discussion of costs and benefits of implicit ratings can be found in Nichols (1998).

2.3.2 Data sparsity and the cold-start problem

In the rating matrices used in the previous examples, ratings existed for all but one user–item combination. In real-world applications, of course, the rating matrices tend to be very sparse, as customers typically provide ratings for (or have bought) only a small fraction of the catalog items.

In general, the challenge in that context is thus to compute good predictions when there are relatively few ratings available. One straightforward option for dealing with this problem is to exploit additional information about the users, such as gender, age, education, interests, or other available information that can help to classify the user. The set of similar users (neighbors) is thus based not only on the analysis of the explicit and implicit ratings, but also on information external to the ratings matrix. These systems – such as the hybrid one mentioned by Pazzani (1999b), which exploits demographic information – are,

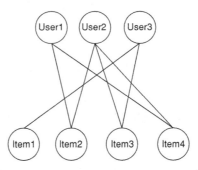

Figure 2.2. Graphical representation of user–item relationships.

however, no longer "purely" collaborative, and new questions of how to acquire the additional information and how to combine the different classifiers arise. Still, to reach the critical mass of users needed in a collaborative approach, such techniques might be helpful in the ramp-up phase of a newly installed recommendation service.

Over the years, several approaches to deal with the cold-start and data sparsity problems have been proposed. Here, we discuss one graph-based method proposed by Huang et al. (2004) as one example in more detail. The main idea of their approach is to exploit the supposed "transitivity" of customer tastes and thereby augment the matrix with additional information[2].

Consider the user-item relationship graph in Figure 2.2, which can be inferred from the binary ratings matrix in Table 2.3 (adapted from Huang et al. (2004)).

A 0 in this matrix should not be interpreted as an explicit (poor) rating, but rather as a missing rating. Assume that we are looking for a recommendation for *User1*. When using a standard CF approach, *User2* will be considered a peer for *User1* because they both bought *Item2* and *Item4*. Thus *Item3* will be recommended to *User1* because the nearest neighbor, *User2*, also bought or liked it. Huang et al. (2004) view the recommendation problem as a graph analysis problem, in which recommendations are determined by determining paths between users and items. In a standard user-based or item-based CF approach, paths of length 3 will be considered – that is, *Item3* is relevant for *User1* because there exists a three-step path (*User1–Item2–User2–Item3*) between them. Because the number of such paths of length 3 is small in sparse rating databases, the idea is to also consider longer paths (indirect associations) to compute recommendations. Using path length 5, for instance, would allow

---

[2] A similar idea of exploiting the neighborhood relationships in a recursive way was proposed by Zhang and Pu (2007).

Table 2.3. *Ratings database for spreading activation approach.*

	Item1	Item2	Item3	Item4
User1	0	1	0	1
User2	0	1	1	1
User3	1	0	1	0

for the recommendation also of *Item1*, as two five-step paths exist that connect *User1* and *Item1*.

Because the computation of these distant relationships is computationally expensive, Huang et al. (2004) propose transforming the rating matrix into a bipartite graph of users and items. Then, a specific graph-exploring approach called *spreading activation* is used to analyze the graph in an efficient manner. A comparison with the standard user-based and item-based algorithms shows that the quality of the recommendations can be significantly improved with the proposed technique based on indirect relationships, in particular when the ratings matrix is sparse. Also, for new users, the algorithm leads to measurable performance increases when compared with standard collaborative filtering techniques. When the rating matrix reaches a certain density, however, the quality of recommendations can also decrease when compared with standard algorithms. Still, the computation of distant relationships remains computationally expensive; it has not yet been shown how the approach can be applied to large ratings databases.

Default voting, as described by Breese et al. (1998), is another technique of dealing with sparse ratings databases. Remember that standard similarity measures take into account only items for which both the active user and the user to be compared will have submitted ratings. When this number is very small, coincidental rating commonalities and differences influence the similarity measure too much. The idea is therefore to assign default values to items that only one of the two users has rated (and possibly also to some additional items) to improve the prediction quality of sparse rating databases (Breese et al. 1998). These artificial default votes act as a sort of damping mechanism that reduces the effects of individual and coincidental rating similarities.

More recently, another approach to deal with the data sparsity problem was proposed by Wang et al. (2006). Based on the observation that most collaborative recommenders use only a certain part of the information – either user similarities or item similarities – in the ratings databases, they suggest combining the two different similarity types to improve the prediction accuracy.

In addition, a third type of information ("similar item ratings made by similar users"), which is not taken into account in previous approaches, is exploited in their prediction function. The "fusion" and smoothing of the different predictions from the different sources is accomplished in a probabilistic framework; first experiments show that the prediction accuracy increases, particularly when it comes to sparse rating databases.

The cold-start problem can be viewed as a special case of this sparsity problem (Huang et al. 2004). The questions here are (a) how to make recommendations to new users that have not rated any item yet and (b) how to deal with items that have not been rated or bought yet. Both problems can be addressed with the help of hybrid approaches – that is, with the help of additional, external information (Adomavicius and Tuzhilin 2005). For the new-users problem, other strategies are also possible. One option could be to ask the user for a minimum number of ratings before the service can be used. In such situations the system could intelligently ask for ratings for items that, from the viewpoint of information theory, carry the most information (Rashid et al. 2002). A similar strategy of asking the user for a *gauge set* of ratings is used for the Eigentaste algorithm presented by Goldberg et al. (2001).

2.4 Further model-based and preprocessing-based approaches

Collaborative recommendation techniques are often classified as being either *memory-based* or *model-based*. The traditional user-based technique is said to be memory-based because the original rating database is held in memory and used directly for generating the recommendations. In model-based approaches, on the other hand, the raw data are first processed offline, as described for item-based filtering or some dimensionality reduction techniques. At run time, only the precomputed or "learned" model is required to make predictions. Although memory-based approaches are theoretically more precise because full data are available for generating recommendations, such systems face problems of scalability if we think again of databases of tens of millions of users and millions of items.

In the next sections, we discuss some more model-based recommendation approaches before we conclude with a recent practical-oriented approach.

2.4.1 Matrix factorization/latent factor models

The Netflix Prize competition, which was completed in 2009, showed that advanced matrix factorization methods, which were employed by many participating teams, can be particularly helpful to improve the predictive accuracy of recommender systems[3].

Roughly speaking, matrix factorization methods can be used in recommender systems to derive a set of latent (hidden) factors from the rating patterns and characterize both users and items by such vectors of factors. In the movie domain, such automatically identified factors can correspond to obvious aspects of a movie such as the genre or the type (drama or action), but they can also be uninterpretable. A recommendation for an item i is made when the active user and the item i are similar with respect to these factors (Koren et al. 2009).

This general idea of exploiting latent "semantic" factors has been successfully applied in the context of information retrieval since the late 1980s. Specifically, Deerwester et al. (1990) proposed using *singular value decomposition* (SVD) as a method to discover the latent factors in documents; in information retrieval settings, this *latent semantic analysis* (LSA) technique is also referred to as *latent semantic indexing* (LSI).

In information retrieval scenarios, the problem usually consists of finding a set of documents, given a query by a user. As described in more detail in Chapter 3, both the existing documents and the user's query are encoded as a vector of terms. A basic retrieval method could simply measure the overlap of terms in the documents and the query. However, such a retrieval method does not work well when there are synonyms such as "car" and "automobile" and polysemous words such as "chip" or "model" in the documents or the query. With the help of SVD, the usually large matrix of document vectors can be collapsed into a smaller-rank approximation in which highly correlated and co-occurring terms are captured in a single factor. Thus, LSI-based retrieval makes it possible to retrieve relevant documents even if it does not contain (many) words of the user's query.

The idea of exploiting latent relationships in the data and using matrix factorization techniques such as SVD or principal component analysis was relatively soon transferred to the domain of recommender systems (Sarwar et al. 2000b; Goldberg et al. 2001; Canny 2002b). In the next section, we will show an example of how SVD can be used to generate recommendations; the example is adapted from the one given in the introduction to SVD-based recommendation by Grigorik (2007).

---

[3] The DVD rental company Netflix started this open competition in 2006. A $1 million prize was awarded for the development of a CF algorithm that is better than Netflix's own recommendation system by 10 percent; see http://www.netflixprize.com.

Table 2.4. *Ratings database for SVD-based recommendation.*

	User1	User2	User3	User4
Item1	3	4	3	1
Item2	1	3	2	6
Item3	2	4	1	5
Item4	3	3	5	2

Example for SVD-based recommendation. Consider again our rating matrix from Table 2.1, from which we remove Alice and that we transpose so we can show the different operations more clearly (see Table 2.4).

Informally, the SVD theorem (Golub and Kahan 1965) states that a given matrix M can be decomposed into a product of three matrices as follows, where U and V are called *left* and *right singular vectors* and the values of the diagonal of Σ are called the *singular values*.

$$M = U \Sigma V^T \tag{2.10}$$

Because the 4×4-matrix M in Table 2.4 is quadratic, U, Σ, and V are also quadratic 4×4 matrices. The main point of this decomposition is that we can approximate the full matrix by observing only the most important features – those with the largest singular values. In the example, we calculate U, V, and Σ (with the help of some linear algebra software) but retain only the two most important features by taking only the first two columns of U and V^T, see Table 2.5.

The projection of U and V^T in the two-dimensional space (U_2, V_2^T) is shown in Figure 2.3. Matrix V corresponds to the users and matrix U to the catalog items. Although in our particular example we cannot observe any clusters of users, we see that the items from U build two groups (above and below the x-axis). When looking at the original ratings, one can observe that *Item1* and

Table 2.5. *First two columns of decomposed matrix and singular values Σ.*

U_2		V_2		Σ_2	
−0.4312452	0.4931501	−0.3593326	0.36767659	12.2215	0
−0.5327375	−0.5305257	−0.5675075	0.08799758	0	4.9282
−0.5237456	−0.4052007	−0.4428526	0.56862492		
−0.5058743	0.5578152	−0.5938829	−0.73057242		

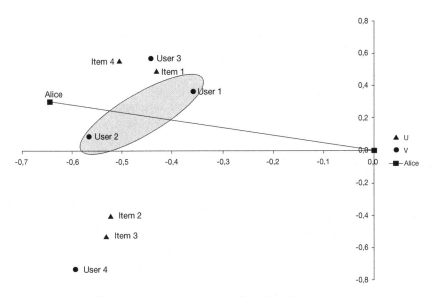

Figure 2.3. SVD-based projection in two-dimensional space.

Item4 received somewhat similar ratings. The same holds for *Item2* and *Item3*, which are depicted below the *x*-axis. With respect to the users, we can at least see that *User4* is a bit far from the others.

Because our goal is to make a prediction for Alice's ratings, we must first find out where Alice would be positioned in this two-dimensional space.

To find out Alice's datapoint, multiply Alice's rating vector [5, 3, 4, 4] by the two-column subset of *U* and the inverse of the two-column singular value matrix Σ.

$$Alice_{2D} = Alice \times U_2 \times \Sigma_2^{-1} = [-0.64, 0.30] \qquad (2.11)$$

Given Alice's datapoint, different strategies can be used to generate a recommendation for her. One option could be to look for neighbors in the compressed two-dimensional space and use their item ratings as predictors for Alice's rating. If we rely again on cosine similarity to determine user similarity, *User1* and *User2* will be the best predictors for Alice in the example. Again, different weighting schemes, similarity thresholds, and strategies for filling missing item ratings (e.g., based on product averages) can be used to fine-tune the prediction. Searching for neighbors in the compressed space is only one of the possible options to make a prediction for Alice. Alternatively, the interaction between user and items in the latent factor space (measured with the cosine similarity metric) can be used to approximate Alice's rating for an item (Koren et al. 2009).

Principal component analysis – Eigentaste. A different approach to dimensionality reduction was proposed by Goldberg et al. (2001) and initially applied to the implementation of a joke recommender. The idea is to preprocess the ratings database using *principal component analysis* (PCA) to filter out the "most important" aspects of the data that account for most of the variance. The authors call their method "Eigentaste," because PCA is a standard statistical analysis method based on the computation of the eigenvalue decomposition of a matrix. After the PCA step, the original rating data are projected along the most relevant of the principal eigenvectors. Then, based on this reduced dataset, users are grouped into clusters of neighbors, and the mean rating for the items is calculated. All these (computationally expensive) steps are done offline. At run time, new users are asked to rate a set of jokes (*gauge set*) on a numerical scale. These ratings are transformed based on the principal components, and the correct cluster is determined. The items with the highest ratings for this cluster are simply retrieved by a look-up in the preprocessed data. Thus, the computational complexity at run time is independent of the number of users, resulting in a "constant time" algorithm. The empirical evaluation and comparison with a basic nearest-neighborhood algorithm show that in some experiments, Eigentaste can provide comparable recommendation accuracy while the computation time can be significantly reduced. The need for a gauge set of, for example, ten ratings is one of the characteristics that may limit the practicality of the approach in some domains.

Discussion. Sarwar et al. (2000a) have analyzed how SVD-based dimensionality reduction affects the quality of the recommendations. Their experiments showed some interesting insights. In some cases, the prediction quality was worse when compared with memory-based prediction techniques, which can be interpreted as a consequence of not taking into account all the available information. On the other hand, in some settings, the recommendation accuracy was better, which can be accounted for by the fact that the dimensionality reduction technique also filtered out some "noise" in the data and, in addition, is capable of detecting nontrivial correlations in the data. To a great extent the quality of the recommendations seems to depend on the right choice of the amount of data reduction – that is, on the choice of the number of singular values to keep in an SVD approach. In many cases, these parameters can, however, be determined and fine-tuned only based on experiments in a certain domain. Koren et al. (2009) talk about 20 to 100 factors that are derived from the rating patterns.

As with all preprocessing approaches, the problem of data updates – how to integrate newly arriving ratings without recomputing the whole "model" again – also must be solved. Sarwar et al. (2002), for instance, proposed a technique that allows for the incremental update for SVD-based approaches. Similarly, George and Merugu (2005) proposed an approach based on co-clustering for building scalable CF recommenders that also support the dynamic update of the rating database.

Since the early experiments with matrix factorization techniques in recommender systems, more elaborate and specialized methods have been developed. For instance, Hofmann (2004; Hofmann and Puzicha 1999) proposed to apply *probabilistic LSA* (pLSA) a method to discover the (otherwise hidden) user communities and interest patterns in the ratings database and showed that good accuracy levels can be achieved based on that method. Hofmann's pLSA method is similar to LSA with respect to the goal of identifying hidden relationships; pLSA is, however, based not on linear algebra but rather on statistics and represents a "more principled approach which has a solid foundation in statistics" (Hofmann 1999).

An overview of recent and advanced topics in matrix factorization for recommender systems can be found in Koren et al. (2009). In this paper, Koren et al. focus particularly on the flexibility of the model and show, for instance, how additional information, such as demographic data, can be incorporated; how temporal aspects, such as changing user preferences, can be dealt with; or how existing rating bias can be taken into account. In addition, they also propose more elaborate methods to deal with missing rating data and report on some insights from applying these techniques in the Netflix prize competition.

2.4.2 Association rule mining

Association rule mining is a common technique used to identify rulelike relationship patterns in large-scale sales transactions. A typical application of this technique is the detection of pairs or groups of products in a supermarket that are often purchased together. A typical rule could be, "If a customer purchases baby food then he or she also buys diapers in 70 percent of the cases". When such relationships are known, this knowledge can, for instance, be exploited for promotional and cross-selling purposes or for design decisions regarding the layout of the shop.

This idea can be transferred to collaborative recommendation – in other words, the goal will be to automatically detect rules such as "If user X liked both *item1* and *item2*, then X will most probably also like *item5*." Recommendations for the active user can be made by evaluating which of the detected rules apply – in the example, checking whether the user liked *item1* and *item2* – and then generating a ranked list of proposed items based on statistics about the co-occurrence of items in the sales transactions.

We can describe the general problem more formally as follows, using the notation from Sarwar et al. (2000b). A (sales) *transaction T* is a subset of the set of available products $P = \{p_1, \ldots, p_m\}$ and describes a set of products that have been purchased together. Association rules are often written in the form $X \Rightarrow Y$, with X and Y being both subsets of P and $X \cap Y = \emptyset$. An association rule $X \Rightarrow Y$ (e.g., *baby-food* $\Rightarrow$ *diapers*) expresses that whenever the elements of X (the rule *body*) are contained in a transaction T, it is very likely that the elements in Y (the rule *head*) are elements of the same transaction.

The goal of rule-mining algorithms such as Apriori (Agrawal and Srikant 1994) is to automatically detect such rules and calculate a measure of quality for those rules. The standard measures for association rules are *support* and *confidence*. The *support* of a rule $X \Rightarrow Y$ is calculated as the percentage of transactions that contain all items of $X \cup Y$ with respect to the number of overall transactions (i.e., the probability of co-occurrence of X and Y in a transaction). *Confidence* is defined as the ratio of transactions that contain all items of $X \cup Y$ to the number of transactions that contain only X – in other words, confidence corresponds to the conditional probability of Y given X.

More formally,

$$support = \frac{number\ of\ transactions\ containing\ X \cup Y}{number\ of\ transactions} \tag{2.12}$$

$$confidence = \frac{number\ of\ transactions\ containing\ X \cup Y}{number\ of\ transactions\ containing\ X} \tag{2.13}$$

Let us consider again our small rating matrix from the previous section to show how recommendations can be made with a rule-mining approach. For demonstration purposes we will simplify the five-point ratings and use only a binary "like/dislike" scale. Table 2.6 shows the corresponding rating matrix; zeros correspond to "dislike" and ones to "like." The matrix was derived from Table 2.2 (showing the mean-adjusted ratings). It contains a 1 if a rating was above a user's average and a 0 otherwise.

Standard rule-mining algorithms can be used to analyze this database and calculate a list of association rules and their corresponding confidence and support values. To focus only on the relevant rules, minimum threshold values for support and confidence are typically defined, for instance, through experimentation.

Table 2.6. *Transformed ratings database for rule mining.*

	Item1	Item2	Item3	Item4	Item5
Alice	1	0	0	0	?
User1	1	0	0	1	1
User2	1	0	1	0	1
User3	0	0	0	1	1
User4	0	1	1	0	0

In the context of collaborative recommendation, a transaction could be viewed as the set of all previous (positive) ratings or purchases of a customer. A typical association that should be analyzed is the question of how likely it is that users who liked *Item1* will also like *Item5* (*Item1* $\Rightarrow$ *Item5*). In the example database, the support value for this rule (without taking Alice's ratings into account) is $2/4$; confidence is $2/2$.

The calculation of the set of interesting association rules with a sufficiently high value for confidence and support can be performed offline. At run time, recommendations for user Alice can be efficiently computed based on the following scheme described by Sarwar et al. (2000b):

(1) Determine the set of $X \Rightarrow Y$ association rules that are relevant for Alice – that is, where Alice has bought (or liked) all elements from X. Because Alice has bought *Item1*, the aforementioned rule is relevant for Alice.
(2) Compute the union of items appearing in the consequent Y of these association rules that have not been purchased by Alice.
(3) Sort the products according to the confidence of the rule that predicted them. If multiple rules suggested one product, take the rule with the highest confidence.
(4) Return the first N elements of this ordered list as a recommendation.

In the approach described by Sarwar et al. (2000b), only the actual purchases ("like" ratings) were taken into account – the system does not explicitly handle "dislike" statements. Consequently, no rules are inferred that express that, for example, whenever a user liked *Item2* he or she would *not* like *Item3*, which could be a plausible rule in our example.

Fortunately, association rule mining can be easily extended to also handle *categorical* attributes so both "like" and "dislike" rules can be derived from the data. Lin et al. (2002; Lin 2000), for instance, propose to transform the usual numerical item ratings into two categories, as shown in the example, and then to map ratings to "transactions" in the sense of standard association rule mining techniques. The detection of rules describing relationships between articles ("whenever *item2* is liked . . .") is only one of the options; the same mechanism can be used to detect like and dislike relationships between users, such as "90 percent of the articles liked by user *A* and user *B* are also liked by user *C*."

For the task of detecting the recommendation rules, Lin et al. (2002) propose a mining algorithm that takes the particularities of the domain into account and specifically searches only for rules that have a certain *target item* (user or article) in the rule head. Focusing the search for rules in that way not only improves the algorithm's efficiency but also allows for the detection of rules for infrequently bought items, which could be filtered out in a global search because of their limited support value. In addition, the algorithm can be parameterized with lower and upper bounds on the number of rules it should try to identify.

Depending on the mining scenario, different strategies for determining the set of recommended items can be used. Let us assume a scenario in which associations between customers instead of items are mined. An example of a detected rule would therefore be "If *User1* likes an item, and *User2* dislikes the item, Alice (the target user) will like the item."

To determine whether an item will be liked by Alice, we can check, for each item, whether the rule "fires" for the item – that is, if *User1* liked it and *User2* disliked it. Based on confidence and support values of these rules, an overall score can be computed for each item as follows (Lin et al. 2002):

$$score_{item_i} = \sum_{rules\ recommending\ item_i} (support_{rule} * confidence_{rule}) \qquad (2.14)$$

If this overall item score surpasses a defined threshold value, the item will be recommended to the target user. The determination of a suitable threshold was done by Lin et al. (2002) based on experimentation.

When item (article) associations are used, an additional cutoff parameter can be determined in advance that describes some minimal support value. This cutoff not only reduces the computational complexity but also allows for the detection of rules for articles that have only a very few ratings.

In the experiments reported by Lin et al. (2002), a mixed strategy was implemented that, as a default, not only relies on the exploitation of user associations but also switches to article associations whenever the support values of the user association rules are below a defined threshold. The evaluation shows that user associations generally yield better results than article associations; article associations can, however, be computed more quickly. It can also be observed from the experiments that a limited number of rules is sufficient for generating good predictions and that increasing the number of rules does not contribute any more to the prediction accuracy. The first observation is interesting because it contrasts the observation made in nearest-neighbor approaches described earlier in which item-to-item correlation approaches have shown to lead to better results.

A comparative evaluation finally shows that in the popular movie domain, the rule-mining approach outperforms other algorithms, such as the one presented by Billsus and Pazzani (1998a), with respect to recommendation quality. In another domain – namely, the recommendation of interesting pages for web users, Fu et al. (2000), also report promising results for using association rules as a mechanism for predicting the relevance of individual pages. In contrast to many other approaches, they do not rely on explicit user ratings for web pages but rather aim to automatically store the navigation behavior of many users in a central repository and then to learn which users are similar with respect to their interests. More recent and elaborate works in that direction, such as those by Mobasher et al. (2001) or Shyu et al. (2005), also rely on web usage data and association rule mining as core mechanisms to predict the relevance of web pages in the context of adaptive user interfaces and web page recommendations.

2.4.3 Probabilistic recommendation approaches

Another way of making a prediction about how a given user will rate a certain item is to exploit existing formalisms of probability theory[4].

A first, and very simple, way to implement collaborative filtering with a probabilistic method is to view the prediction problem as a *classification problem*, which can generally be described as the task of "assigning an object to one of several predefined categories" (Tan et al. 2006). As an example of a classification problem, consider the task of classifying an incoming e-mail message as spam or non-spam. In order to automate this task, a function has to be developed that defines – based, for instance, on the words that occur in the message header or content – whether the message is classified as a spam e-mail or not. The classification task can therefore be seen as the problem of *learning* this mapping function from training examples. Such a function is also informally called the *classification model*.

One standard technique also used in the area of data mining is based on *Bayes classifiers*. We show, with a simplified example, how a basic probabilistic method can be used to calculate rating predictions. Consider a slightly different

---

[4] Although the selection of association rules, based on support and confidence values, as described in the previous section is also based on statistics, association rule mining is usually not classified as a probabilistic recommendation method.

Table 2.7. *Probabilistic models: the rating database.*

	Item1	Item2	Item3	Item4	Item5
Alice	1	3	3	2	?
User1	2	4	2	2	4
User2	1	3	3	5	1
User3	4	5	2	3	3
User4	1	1	5	2	1

ratings database (see Table 2.7). Again, a prediction for Alice's rating of *Item5* is what we are interested in.

In our setting, we formulate the prediction task as the problem of calculating the most probable rating value for *Item5*, given the set of Alice's other ratings and the ratings of the other users. In our method, we will calculate *conditional probabilities* for each possible rating value given Alice's other ratings, and then select the one with the highest probability as a prediction[5].

To predict the probability of rating value 1 for *Item5* we must calculate the conditional probability $P(Item5 = 1|X)$, with X being Alice's other ratings: $X = (Item1 = 1, Item2 = 3, Item3 = 3, Item4 = 2)$.

For the calculation of this probability, the Bayes theorem is used, which allows us to compute this posterior probability $P(Y|X)$ through the *class-conditional* probability $P(X|Y)$, the probability of Y (i.e., the probability of a rating value 1 for *Item5* in the example), and the probability of X, more formally

$$P(Y|X) = \frac{P(X|Y) \times P(Y)}{P(X)} \qquad (2.15)$$

Under the assumption that the attributes (i.e., the ratings users) are *conditionally independent*, we can compute the posterior probability for each value of Y with a *naive* Bayes classifier as follows, d being the number of attributes in each X:

$$P(Y|X) = \frac{\prod_{i=1}^{d} P(X_i|Y) \times P(Y)}{P(X)} \qquad (2.16)$$

In many domains in which naive Bayes classifiers are applied, the assumption of conditional independence actually does not hold, but such classifiers perform well despite this fact.

---

[5] Again, a transformation of the ratings database into "like" and "dislike" statements is possible (Miyahara and Pazzani 2000).

As $P(X)$ is a constant value, we can omit it in our calculations. $P(Y)$ can be estimated for each rating value based on the ratings database: $P(Item5=1) = 2/4$ (as two of four ratings for *Item5* had the value 1), $P(Item5=2)=0$, and so forth. What remains is the calculation of all class-conditional probabilities $P(X_i|Y)$:

$$P(X|Item5=1) = P(Item1=1|Item5=1) \times P(Item2=3|Item5=1)$$
$$\times P(Item3=3|Item5=1) \times P(Item4=2|Item5=1)$$
$$= 2/2 \times 1/2 \times 1/2 \times 1/2$$
$$= 0.125$$

$$P(X|Item5=2) = P(Item1=1|Item5=2) \times P(Item2=3|Item5=2)$$
$$\times P(Item3=3|Item5=2) \times P(Item4=2|Item5=2)$$
$$= 0/0 \times \cdots \times \cdots \times \cdots$$
$$= 0$$

Based on these calculations, given that $P(Item5=1) = 2/4$ and omitting the constant factor $P(X)$ in the Bayes classifier, the posterior probability of a rating value 1 for *Item5* is $P(Item5 = 1|X) = 2/4 \times 0.125 = 0.0625$. In the example ratings database, $P(Item5=1)$ is higher than all other probabilities, which means that the probabilistic rating prediction for Alice will be 1 for *Item5*. One can see in the small example that when using this simple method the estimates of the posterior probabilities are 0 if one of the factors is 0 and that in the worst case, a rating vector cannot be classified. Techniques such as using the *m*-estimate or Laplace smoothing are therefore used to smooth conditional probabilities, in particular for sparse training sets. Of course, one could also – as in the association rule mining approach – use a preprocessed rating database and use "like" and "dislike" ratings and/or assign default ratings only to missing values.

The simple method that we developed for illustration purposes is computationally complex, does not work well with small or sparse rating databases, and will finally lead to probability values for each rating that differ only very slightly from each other. More advanced probabilistic techniques are thus required.

The most popular approaches that rely on a probabilistic model are based on the idea of grouping similar users (or items) into *clusters*, a technique that, in general, also promises to help with the problems of data sparsity and computational complexity.

The naïve Bayes model described by Breese et al. (1998) therefore includes an additional unobserved class variable C that can take a small number of discrete values that correspond to the clusters of the user base. When, again, conditional independence is assumed, the following formula expresses the probability model (mixture model):

$$P(C = c, v_1, \ldots, v_n) = P(C = c) \prod_{i=1}^{n} P(v_i | C = c) \qquad (2.17)$$

where $P(C = c, v_1, \ldots, v_n)$ denotes the probability of observing a full set of values for an individual of class c. What must be derived from the training data are the probabilities of class membership, $P(C = c)$, and the conditional probabilities of seeing a certain rating value when the class is known, $P(v_i | C = c)$. The problem that remains is to determine the parameters for a model and estimate a good number of clusters to use. This information is not directly contained in the ratings database but, fortunately, standard techniques in the field, such as the expectation maximization algorithm (Dempster et al. 1977), can be applied to determine the model parameters.

At run time, a prediction for a certain user u and an item i can be made based on the probability of user u falling into a certain cluster and the probability of liking item i when being in a certain cluster given the user's ratings; see Ungar and Foster (1998) for more details on model estimation for probabilistic clustering for collaborative filtering.

Other methods for clustering can be applied in the recommendation domain to for instance, reduce the complexity problem. Chee et al. (2001), for example, propose to use a modified k-means clustering algorithm to partition the set of users in homogeneous or cohesive groups (clusters) such that there is a high similarity between users with respect to some similarity measure in one cluster and the interpartition similarity is kept at a low level. When a rating prediction has to be made for a user at run time, the system determines the group of the user and then takes the ratings of only the small set of members of this group into account when computing a weighted rating value. The performance of such an algorithm depends, of course, on the number of groups and the respective group size. Smaller groups are better with respect to run-time performance; still, when groups are too small, the recommendation accuracy may degrade. Despite the run-time savings that can be achieved with this technique, such in-memory approaches do not scale for really large databases. A more recent approach that also relies on clustering users with the same tastes into groups with the help of k-means can be found in Xue et al. (2005).

Coming back to the probabilistic approaches, besides such naive Bayes approaches, Breese et al. (1998) also propose another form of implementing a Bayes classifier and modeling the class-conditional probabilities based on *Bayesian belief networks*. These networks allow us to make existing dependencies between variables explicit and encode these relationships in a directed acyclic graph. Model building thus first requires the system to learn the structure of the network (see Chickering et al. 1997) before the required conditional probabilities can be estimated in a second step.

The comparison of the probabilistic methods with other approaches, such as a user-based nearest-neighbor algorithm, shows that the technique based on Bayesian networks slightly outperforms the other algorithms in some test domains, although not in all. In a summary over all datasets and evaluation protocols, the Bayesian network method also exhibits the best overall performance (Breese et al. 1998). For some datasets, however – as in the popular movie domain – the Bayesian approach performs significantly worse than a user-based approach extended with default voting, inverse user frequency, and case amplification.

In general, Bayesian classification methods have the advantages that individual noise points in the data are averaged out and that irrelevant attributes have little or no impact on the calculated posterior probabilities (Tan et al. 2006). Bayesian networks have no strong trend of *overfitting* the model – that is, they can almost always learn appropriately generalized models, which leads to good predictive accuracy. In addition, they can also be used when the data are incomplete.

As a side issue, the run-time performance of the probabilistic approaches described herein is typically much better than that for memory-based approaches, as the model itself is learned offline and in advance. In parallel, Breese et al. (1998) argue that probabilistic approaches are also favorable with respect to memory requirements, partly owing to the fact that the resulting belief networks remain rather small.

An approach similar to the naive Bayes method of Breese et al. (1998) is described by Chen and George (1999), who also provide more details about the treatment of missing ratings and how users can be clustered based on the introduction of a hidden (latent) variable to model group membership. Miyahara and Pazzani (2000), propose a comparably straightforward but effective collaborative filtering technique based on a simple Bayes classifier and, in particular, also discuss the aspect of *feature selection*, a technique commonly used to leave out irrelevant items (features), improve accuracy, and reduce computation time.

A more recent statistical method that uses latent class variables to discover groups of similar users and items is that proposed by Hofmann (2004; Hofmann and Puzicha 1999), and it was shown that further quality improvements can be achieved when compared with the results of Breese et al. (1998). This method is also employed in Google's news recommender, which will be discussed in the next section. A recent comprehensive overview and comparison of different probabilistic approaches and mixture models can be found in Jin et al. (2006).

Further probabilistic approaches are described by Pennock et al. (2000) and Yu et al. (2004), both aiming to combine the ideas and advantages of model-based and memory-based recommendations in a probabilistic framework. Yu et al. (2004) develop what they call a "memory-based probabilistic framework for collaborative filtering". As a framework, it is particularly designed to accommodate extensions for particular challenges such as the new user problem or the problem of selecting a set of peer users from the ratings database; all these extensions are done in a principled, probabilistic way. The new user problem can be addressed in this framework through an *active learning approach* – that is, by asking a new user to rate a set of items as also proposed by Goldberg et al. (2001). The critical task of choosing the items that the new user will hopefully be able to rate is done on a decision-theoretic and probabilistic basis. Moreover, it is also shown how the process of generating and updating the *profile space*, which contains the most "informative" users in the user database and which is constructed to reduce the computational complexity, can be embedded in the same probabilistic framework. Although the main contribution, as the authors state it, is the provision of a framework that allows for extensions on a sound probabilistic basis, their experiments show that with the proposed techniques, comparable or superior prediction accuracy can be achieved when compared, for instance, with the results reported for probabilistic methods described by Breese et al. (1998).

2.5 Recent practical approaches and systems

Our discussion so far has shown the broad spectrum of different techniques that can, in principle, be used to generate predictions and recommendations based on the information from a user–item rating matrix. We can observe that these approaches differ not only with respect to their recommendation quality (which is the main goal of most research efforts) but also in the complexity of the algorithms themselves. Whereas the first memory-based algorithms are also rather straightforward with respect to implementation aspects, others are based on sophisticated (preprocessing and model-updating) techniques. Although mathematical software libraries are available for many methods, their usage requires in-depth mathematical expertise,[6] which may hamper the

---

[6] Such expertise is required, in particular, when the used approach is computationally complex and the algorithms must be applied in an optimized way.

Table 2.8. *Slope One prediction for*
Alice *and* Item5 = 2 + (2 − 1) = 3.

	Item1	Item5
Alice	2	?
User1	1	2

practical usage of these approaches, in particular for small-sized businesses. In a recent paper, Lemire and Maclachlan (2005) therefore proposed a new and rather simple recommendation technique that, despite its simplicity, leads to reasonable recommendation quality. In addition to the goal of easy implementation for "an average engineer", their Slope One prediction schemes should also support on-the-fly data updates and efficient run-time queries. We discuss this method, which is in practical use on several web sites, in the next section.

In general, the number of publicly available reports on real-world commercial recommender systems (large scale or not) is still limited. In a recent paper, Das et al. (2007), report in some detail on the implementation of Google's news personalization engine that was designed to provide personalized recommendations in real time. A summary of this approach concludes the section and sheds light on practical aspects of implementing a large-scale recommender system that has an item catalog consisting of several million news articles and is used by millions of online users.

2.5.1 Slope One predictors

The original idea of "Slope One" predictors is simple and is based on what the authors call a "popularity differential" between items for users. Consider the following example (Table 2.8), which is based on a pairwise comparison of how items are liked by different users.

In the example, *User1* has rated *Item1* with 1 and *Item5* with 2. Alice has rated *Item1* with 2. The goal is to predict Alice's rating for *Item5*. A simple prediction would be to add to Alice's rating for *Item1* the relative difference of *User1*'s ratings for these two items: $p(Alice, Item5) = 2 + (2 − 1) = 3$. The ratings database, of course, consists of many such pairs, and one can take the average of these differences to make the prediction.

In general, the problem consists of finding functions of the form $f(x) = x + b$ (that is why the name is "Slope One") that predict, for a pair of items, the rating for one item from the rating of the other one.

Table 2.9. *Slope One prediction: a more detailed example.*

	Item1	Item2	Item3
Alice	2	5	?
User1	3	2	5
User2	4		3

Let us now look at the following slightly more complex example (Table 2.9) in which we search for a prediction for Alice for *Item3*.[7]

There are two co-ratings of *Item1* and *Item3*. One time *Item3* is rated two points higher ($5 - 3 = 2$), and the other time one point lower, than *Item1* ($3 - 4 = -1$). The average distance between these items is thus $(2 + (-1))/2 = 0.5$. There is only one co-rating of *Item3* and *Item2* and the distance is ($5 - 2) = 3$. The prediction for *Item3* based on *Item1* and Alice's rating of 2 would therefore be $2 + 0.5 = 2.5$. Based on *Item2*, the prediction is $5 + 3 = 8$. An overall prediction can now be made by taking the number of co-ratings into account to give more weight to deviations that are based on more data:

$$pred(Alice, Item3) = \frac{2 \times 2.5 + 1 \times 8}{2 + 1} = 4.33 \qquad (2.18)$$

In detail, the approach can be described as follows, using a slightly different notation (Lemire and Maclachlan 2005) that makes the description of the calculations simpler. The whole ratings database shall be denoted R, as usual. The ratings of a certain user are contained in an incomplete array u, u_i being u's ratings for item i. Lemire and Maclachlan (2005) call such an array an *evaluation*, and it corresponds to a line in the matrix R. Given two items j and i, let $S_{j,i}(R)$ denote the set of evaluations that contain both ratings for i and j – that is, the lines that contain the co-ratings. The average deviation *dev* of two items i and j is then calculated as follows:

$$dev_{j,i} = \sum_{(u_j, u_i) \in S_{j,i}(R)} \frac{u_j - u_i}{|S_{j,i}(R)|} \qquad (2.19)$$

As shown in the example, from every co-rated item i we can make a prediction for item j and user u as $dev_{j,i} + u_i$. A simple combination of these

---

[7] Adapted from Wikipedia (2008).

individual predictions would be to compute the average over all co-rated items:

$$pred(u, j) = \frac{\sum_{i \in Relevant(u, j)}(dev_{j,i} + u_i)}{|Relevant(u, j)|} \qquad (2.20)$$

The function $Relevant(u, j)$ denotes the set of relevant items – those that have at least one co-rating with j by user u. In other words, $Relevant(u, j) = \{i | i \in S(u), i \neq j, |S_{j,i}(R)| > 0\}$, where $S(u)$ denotes the set of entries in u that contain a rating. This formula can be simplified in realistic scenarios and sufficiently dense datasets to $Relevant(u, j) = S(u) - \{j\}$ when $j \in S(u)$.

The intuitive problem of that basic scheme is that it does not take the number of co-rated items into account, although it is obvious that a predictor will be better if there is a high number of co-rated items. Thus, the scheme is extended in such a way that it weights the individual deviations based on the number of co-ratings as follows:

$$pred(u, j) = \frac{\sum_{i \in S(u)-\{j\}}(dev_{j,i} + u_i) * |S_{j,i}(R)|}{\sum_{i \in S(u)-\{j\}} * |S_{j,i}(R)|} \qquad (2.21)$$

Another way of enhancing the basic prediction scheme is to weight the deviations based on the "like and dislike" patterns in the rating database (bipolar scheme). To that purpose, when making a prediction for user j, the relevant item ratings (and deviations) are divided into two groups, one group containing the items that were liked by both users and one group containing items both users disliked. A prediction is made by combining these deviations. The overall effect is that the scheme takes only those ratings into account in which the users agree on a positive or negative rating. Although this might seem problematic with respect to already sparse ratings databases, the desired effect is that the prediction scheme "predicts nothing from the fact that user A likes item K and user B dislikes the same item K" (Lemire and Maclachlan 2005).

When splitting the ratings into like and dislike groups, one should also take the particularities of real-world rating databases into account. In fact, when given a five-point scale (1–5), it can be observed that in typical datasets around 70 percent of the ratings are above the theoretical average of 3. This indicates that, in general, users either (a) tend to provide ratings for items that they like or (b) simply have a tendency to give rather high ratings and interpret a value of 3 to be a rather poor rating value. In the bipolar prediction scheme discussed here, the threshold was thus set to the average rating value of a user instead of using an overall threshold.

An evaluation of the Slope One predictors on popular test databases revealed that the quality of recommendations (measured by the usual metrics; see

Section 7.4.2) is comparable with the performance of existing approaches, such as collaborative filtering based on Pearson correlation and case amplification. The extensions of the basic scheme (weighted predictors, bipolar scheme) also result in a performance improvement, although these improvements remain rather small (1% to 2%) and are thus hardly significant.

Overall, despite its simplicity, the proposed item-based and ratings-based algorithm shows a reasonable performance on popular rating databases. In addition, the technique supports both dynamic updates of the predictions when new ratings arrive and efficient querying at run time (in exchange for increased memory requirements, of course). In the broader context, such rather simple techniques and the availability of small, open-source libraries in different popular programming languages can help to increase the number of real-world implementations of recommender systems.

From an scientific perspective, however, a better understanding and more evaluations on different datasets are required to really understand the particular characteristics of the proposed Slope One algorithms in different applications and settings.

2.5.2 The Google News personalization engine

Google News is an online information portal that aggregates news articles from several thousand sources and displays them (after grouping similar articles) to signed-in users in a personalized way; see Figure 2.4. The recommendation approach is a collaborative one and is based on the click history of the active user and the history of the larger community – that is, a click on a story is interpreted as a positive rating. More elaborate rating acquisition and interpretation techniques are possible, of course; see, for instance, the work of Joachims et al. (2005).

On the news portal, the recommender system is used to fill one specific section with a personalized list of news articles. The main challenges are that (a) the goal is to generate the list in real time, allowing at most one second for generating the content and (b) there are very frequent changes in the "item catalog", as there is a constant stream of new items, whereas at the same time other articles may quickly become out of date. In addition, one of the goals is to immediately react to user interaction and take the latest article reads into account.

Because of the vast number of articles and users and the given response-time requirements, a pure memory-based approach is not applicable and a combination of model-based and memory-based techniques is used. The model-based part is based on two clustering techniques, *probabilistic latent semantic*

Figure 2.4. Google News portal.

indexing (PLSI) as proposed by Hofmann (2004), and – as a new proposal – MinHash as a hashing method used for putting two users in the same cluster (hash bucket) based on the overlap of items that both users have viewed. To make this hashing process scalable, both a particular method for finding the neighbors and Google's own MapReduce technique for distributing the computation over several clusters of machines are employed.

The PLSI method can be seen as a "second generation" probabilistic technique for collaborative filtering that – similar to the idea of the probabilistic clustering technique of Breese et al. (1998) discussed earlier – aims to identify clusters of like-minded users and related items. In contrast to the work of Breese et al. (1998), in which every user belongs to exactly one cluster, in Hofmann's approach hidden variables with a finite set of states for every user–item pair are introduced. Thus, such models can also accommodate the fact that users may have interests in various topics in parallel. The parameters of the resulting mixture model are determined with the standard expectation maximization (EM) method (Dempster et al. 1977). As this process is computationally very expensive, with respect to both the number of operations and the amount of required main memory, an algorithm is proposed for parallelizing the EM computation via MapReduce over several machines. Although this parallelization can significantly speed up the process of learning the probability distributions, it is clearly not sufficient to retrain the network in real time when new users or items appear, because such modifications happen far too often in this domain. Therefore, for new stories, an approximate version of PLSI is applied that can be updated in real time. A recommendation score is computed based on cluster-membership probabilities and per-cluster statistics of the number of clicks for each story. The score is normalized in the interval $[0 \ldots 1]$.

For dealing with new users, the memory-based part of the recommender that analyzes story "co-visits" is important. A co-visit means that an article has been visited by the same user within a defined period of time. The rationale of exploiting such information directly corresponds to an item-based recommendation approach, as described in previous sections. The data, however, are not preprocessed offline, but a special data structure resembling the adjacency of clicks is constantly kept up to date. Predictions are made by iterating over the recent history of the active user and retrieving the neighboring articles from memory. For calculating the actual score, the weights stored in the adjacency matrix are taken into account, and the result is normalized on a 0-to-1 interval.

At run time, the overall recommendation score for each item in a defined set of candidate items is computed as a linear combination of all the scores obtained by the three methods (MinHash, PLSI, and co-visits). The preselection of an appropriate set of recommendation candidates can be done based on different pieces of information, such as language preferences, story freshness, or other user-specific personalization settings. Alternatively, the click history of other users in the same cluster could be used to limit the set of candidate items.

The evaluation of this algorithm on different datasets (movies and news articles) revealed that, when evaluated individually, PLSI performs best, followed by MinHash and the standard similarity-based recommendation. For live data, an experiment was made in which the new technique was compared with a nonpersonalized approach, in which articles were ranked according to their recent popularity. To compare the approaches, recommendation lists were generated that interleaved items from one algorithm with the other. The experiment then measured which items received more clicks by users. Not surprisingly, the personalized approach did significantly better (around 38%) except for the not-so-frequent situations in which there were extraordinarily popular stories. The interesting question of how to weight the scores of the individual algorithms, however, remains open to some extent.

In general, what can be learned from that report is that if we have a combination of massive datasets and frequent changes in the data, significant efforts (with respect to algorithms, engineering, and parallelization) are required such that existing techniques can be employed and real-time recommendations are possible. Pure memory-based approaches are not directly applicable and for model-based approaches, the problem of continuous model updates must be solved.

What is not answered in the study is the question of whether an approach that is not content-agnostic would yield better results. We will see in the next chapter that content-based recommendation techniques – algorithms that base their recommendations on the document content and explicit or learned user interests – are particularly suited for problems of that type.

2.6 Discussion and summary

Of all the different approaches to building recommender systems discussed in this book, CF is the best-researched technique – not only because the first recommender systems built in the mid-1990s were based on user communities that rated items, but also because most of today's most successful online recommenders rely on these techniques. Early systems were built using memory-based neighborhood and correlation-based algorithms. Later, more complex and model-based approaches were developed that, for example, apply techniques from various fields, such as machine learning, information retrieval, and data mining, and often rely on algebraic methods such as SVD.

In recent years, significant research efforts went into the development of more complex probabilistic models as discussed by Adomavicius and Tuzhilin (2005), in particular because the earliest reports of these methods (as in Breese et al. 1998) indicate that they lead to very accurate predictions.

The popularity of the collaborative filtering subfield of recommender systems has different reasons, most importantly the fact that real-world benchmark problems are available and that the data to be analyzed for generating recommendations have a very simple structure: a matrix of item ratings. Thus, the evaluation of whether a newly developed recommendation technique, or the application of existing methods to the recommendation problem, outperforms previous approaches is straightforward, in particular because the evaluation metrics are also more or less standardized. One can easily imagine that comparing different algorithms is not always as easy as with collaborative filtering, in particular if more knowledge is available than just the simple rating matrix. Think, for instance, of conversational recommender applications, in which the user is interactively asked about his or her preferences and in which additional domain knowledge is encoded.

However, the availability of test databases for CF in different domains favored the further development of various and more complex CF techniques. Still, this somehow also narrowed the range of domains on which CF techniques are actually applied. The most popular datasets are about movies and books, and many researchers aim to improve the accuracy of their algorithms only on these datasets. Whether a certain CF technique performs particularly well in one domain or another is unfortunately beyond the scope of many research efforts.

In fact, given the rich number of different proposals, the question of which recommendation algorithm to use under which circumstances is still open, even if we limit our considerations to purely collaborative approaches. Moreover, the accuracy results reported on the well-known test datasets do not convey a clear picture. Many researchers compare their measurements with the already rather old results from Breese et al. (1998) and report that they can achieve better results in one or another setting and experiment. A newer basis of comparison is required, given the dozens of different techniques that have been proposed over the past decade. Based on such a comparison, a new set of "baseline" algorithms could help to get a clearer picture.

Viewed from a practical perspective, one can see that item-based CF, as reported by Linden et al. (2003) and used by Amazon.com, is scalable enough to cope with very large rating databases and also produces recommendations of reasonable quality. The number of reports on other commercial implementations and accompanying technical details (let alone datasets) is unfortunately also limited, so an industry survey in that direction could help the research community validate whether and how new proposals make their way into industrial practice.

In addition, we will see in Chapter 5, which covers hybrid recommendation approaches, that recommendation algorithms that exploit additional information about items or users and combine different techniques can achieve significantly better recommendation results than purely collaborative approaches can. When we observe the trends and developments in the recent past, we can expect that in the next years more information, both about the catalog items and about the users, will be available at very low cost, thus favoring combined or hybrid approaches. The sources of such additional knowledge can be manifold: online users share more and more information about themselves in social networks and online communities; companies exchange item information in electronic form only and increasingly adopt exchange standards including defined product classification schemes. Finally, according to the promise of the "Semantic Web," such item and community information can easily be automatically extracted from existing web sources (see, e.g., Shchekotykhin et al. 2007 for an example of such an approach).

Overall, today's CF techniques are mature enough to be employed in practical applications, provided that moderate requirements are fulfilled. Collaborative recommenders can not be applied in every domain: think of a recommender system for cars, a domain in which no buying history exists or for which the system needs a more detailed picture of the users' preferences. In parallel, CF techniques require the existence of a user community of a certain size, meaning that even for the book or movie domains one cannot apply these techniques if there are not enough users or ratings available.

Alternative approaches to product recommendation that overcome these limitations in one or the other dimension at the price of, for instance, increased development and maintenance efforts will be discussed in the next chapters.

2.7 Bibliographical notes

The earliest reports on what we now call recommender systems were published in the early 1990s. The most cited ones might be the papers on the Tapestry (Goldberg et al. 1992) and the GroupLens (Resnick et al. 1994) systems, both first published in 1992. Tapestry was developed at Xerox Parc for mail filtering and was based on the then rather new idea of exploiting explicit feedback (ratings and annotations) of other users. One of the first uses of the term "collaborative filtering" can be found in this paper. The GroupLens[8] system was also developed for filtering text documents (i.e., news articles), but was designed for use in an open community and introduced the basic idea of automatically finding similar users in the database for making predictions. The Ringo system, presented by Shardanand and Maes (1995) describes a music recommender based on collaborative filtering using Pearson's correlation measure and the mean absolute error (MAE) evaluation metric.

As mentioned earlier, the evaluation of Breese et al. (1998) still serves as an important reference point, in particular as the paper also introduces some special techniques to the compared algorithms.

The first model-based version of the Jester joke recommender (Goldberg et al. 2001) that relied on principal component analysis and clustering was initially proposed around 1999 and is still being developed further (Nathanson et al. 2007). Hofmann and Puzicha published their influential approach based on latent class models for collaborative filtering in 1999. Dimensionality reduction based on SVD was proposed by Sarwar et al. (2000a).

Item-based filtering was analyzed by Sarwar et al. (2001); a short report about Amazon.com's patented implementation and experiences are described by Linden et al. (2003).

---

[8] The homepage of the influential GroupLens research group can be found at http://www.grouplens.org.

Excellent overview papers on CF, which partially inspired the structure of this chapter and which can serve as a starting point for further readings, are those by Schafer et al. (2006) and Anand and Mobasher (2005). Another overview with an impressive list of references to recent techniques for collaborative filtering can be found in the article by Adomavicius and Tuzhilin (2005).

Because recommender systems have their roots in various fields, research papers on collaborative filtering techniques appear in different journals and conferences. Special issues on recommender systems appeared, for instance, in the *Communications of the ACM* (1999), *ACM Transactions on Information Systems* (2004), and more recently in the *Journal of Electronic Commerce* (2007), *IEEE Intelligent Systems* (2007), and *AI Communications* (2008). Many papers also appear first at dedicated workshops series, such as the Intelligent Techniques for Web Personalization Workshop and, more recently, at the ACM Recommender Systems conference.

3
Content-based recommendation

From our discussion so far we see that for applying collaborative filtering techniques, except for the user ratings, nothing has to be known about the items to be recommended. The main advantage of this is, of course, that the costly task of providing detailed and up-to-date item descriptions to the system is avoided. The other side of the coin, however, is that with a pure collaborative filtering approach, a very intuitive way of selecting recommendable products based on their characteristics and the specific preferences of a user is not possible: in the real world, it would be straightforward to recommend the new *Harry Potter* book to Alice, if we know that (a) this book is a fantasy novel and (b) Alice has always liked fantasy novels. An electronic recommender system can accomplish this task only if two pieces of information are available: a description of the item characteristics and a *user profile* that somehow describes the (past) interests of a user, maybe in terms of preferred item characteristics. The recommendation task then consists of determining the items that match the user's preferences best. This process is commonly called *content-based recommendation*. Although such an approach must rely on additional information about items and user preferences, it does not require the existence of a large user community or a rating history – that is, recommendation lists can be generated even if there is only one single user.

In practical settings, *technical* descriptions of the features and characteristics of an item – such as the genre of a book or the list of actors in a movie – are more often available in electronic form, as they are partially already provided by the providers or manufacturers of the goods. What remains challenging, however, is the acquisition of subjective, *qualitative* features. In domains of quality and taste, for example, the reasons that someone likes something are not always related to certain product characteristics and may be based on a sub-jective impression of the item's exterior design. One notable and exceptional endeavor in that context is the "Music Genome Project"[1], whose data are used by the music recommender on the popular Internet radio and music discovery and commercial recommendation site Pandora.com. In that project, songs are manually annotated by musicians with up to several hundred features such as instrumentation, influences, or instruments. Such a manual acquisition process – annotating a song takes about twenty to thirty minutes, as stated by the service providers – is, however, often not affordable.

We will refer to the descriptions of the item characteristics as "content" in this chapter, because most techniques described in the following sections were originally developed to be applied to recommending interesting text documents, such as newsgroup messages or web pages. In addition, in most of these approaches the basic assumption is that the characteristics of the items can be automatically extracted from the document content itself or from unstructured textual descriptions. Typical examples for content-based recommenders are, therefore, systems that recommend news articles by comparing the main keywords of an article in question with the keywords that appeared in other articles that the user has rated highly in the past. Correspondingly, the recommendable items will be often referred to as "documents".

There is no exact border between content-based and knowledge-based systems in the literature; some authors even see content-based approaches as a subset of knowledge-based approaches. In this book, we follow the traditional classification scheme, in which content-based systems are characterized by their focus on exploiting the information in the item descriptions, whereas in knowledge-based systems there typically exists some sort of additional means–end knowledge, such as a utility function, for producing recommendations.

In this chapter we discuss content-based recommendation, focusing particularly on algorithms that have been developed for recommending textually described items and for "learning" the user profile automatically (instead of explicitly asking the user for his or her interests, which is more common in conversational, knowledge-based systems).

3.1 Content representation and content similarity

The simplest way to describe catalog items is to maintain an explicit list of *features* for each item (also often called *attributes, characteristics,* or *item profiles*). For a book recommender, one could, for instance, use the genre, the author's name, the publisher, or anything else that describes the item and store

---

[1] *http://www.pandora.com/mgp.shtml.*

Table 3.1. *Book knowledge base.*

Title	Genre	Author	Type	Price	Keywords
The Night of the Gun	Memoir	David Carr	Paperback	29.90	press and journalism, drug addiction, personal memoirs, New York
The Lace Reader	Fiction, Mystery	Brunonia Barry	Hardcover	49.90	American contemporary fiction, detective, historical
Into the Fire	Romance, Suspense	Suzanne Brockmann	Hardcover	45.90	American fiction, murder, neo-Nazism
. . .					

this information in a relational database system. When the user's preferences are described in terms of his or her interests using exactly this set of features, the recommendation task consists of matching item characteristics and user preferences.

Consider the example in Table 3.1, in which books are described by characteristics such as title, genre, author, type, price, or keywords. Let us further assume that Alice's preferences are captured in exactly the same dimensions (Table 3.2).

A book recommender system can construct Alice's profile in different ways. The straightforward way is to explicitly ask Alice, for instance, for a desired price range or a set of preferred genres. The other way is to ask Alice to rate a set of items, either as a whole or along different dimensions. In the example, the set of preferred genres could be defined manually by Alice, whereas the system may automatically derive a set of keywords from those books that Alice liked, along with their average price. In the simplest case, the set of keywords corresponds to the set of all terms that appear in the document.

Table 3.2. *Preference profile.*

Title	Genre	Author	Type	Price	Keywords
. . .	Fiction, Suspense	Brunonia Barry, Ken Follett	Paperback	25.65	detective, murder, New York

To make recommendations, content-based systems typically work by evaluating how strongly a not-yet-seen item is "similar" to items the active user has liked in the past. Similarity can be measured in different ways in the example. Given an unseen book B, the system could simply check whether the genre of the book at hand is in the list of Alice's preferred genres. Similarity in this case is either 0 or 1. Another option is to calculate the similarity or overlap of the involved keywords. As a typical similarity metric that is suitable for multi-valued characteristics, we could, for example, rely on the Dice coefficient[2] as follows: If every book B_i is described by a set of keywords $keywords(B_i)$, the Dice coefficient measures the similarity between books b_i and b_j as

$$\frac{2 \times |keywords(b_i) \cap keywords(b_j)|}{|keywords(b_i)| + |keywords(b_j)|} \quad (3.1)$$

In principle, depending on the problem at hand, various similarity measures are possible. For instance, in Zanker et al. (2006) an approach is proposed in which several similarity functions for the different item characteristics are used. These functions are combined and weighted to calculate an overall similarity measure for cases in which both structured and unstructured item descriptions are available.

3.1.1 The vector space model and TF-IDF

Strictly speaking, the information about the publisher and the author are actually not the "content" of a book, but rather additional knowledge about it. However, content-based systems have historically been developed to filter and recommend text-based items such as e-mail messages or news. The standard approach in content-based recommendation is, therefore, not to maintain a list of "meta-information" features, as in the previous example, but to use a list of relevant keywords that appear within the document. The main idea, of course, is that such a list can be generated automatically from the document content itself or from a free-text description thereof.

The content of a document can be encoded in such a keyword list in different ways. In a first, and very naïve, approach, one could set up a list of all words that appear in all documents and describe each document by a Boolean vector, where a 1 indicates that a word appears in a document and a 0 that the word does not appear. If the user profile is described by a similar list (1 denoting interest

---

[2] For other measures, see, e.g., Maimon and Rokach (2005) or Baeza-Yaks and Ribaro-Nato (1999).

in a keyword), document matching can be done by measuring the overlap of interest and document content.

The problems with such a simple approach are obvious. First, the simple encoding is based on the assumption that every word has the same importance within a document, although it seems intuitive that a word that appears more often is better suited for characterizing the document. In addition, a larger overlap of the user profile and a document will naturally be found when the documents are longer. As a result, the recommender will tend to propose long documents.

To solve the shortcomings of the simple Boolean approach, documents are typically described using the TF-IDF encoding format (Salton et al. 1975). TF-IDF is an established technique from the field of information retrieval and stands for *term frequency-inverse document frequency*. Text documents can be TF-IDF encoded as vectors in a multidimensional Euclidian space. The space dimensions correspond to the keywords (also called *terms* or *tokens*) appearing in the documents. The coordinates of a given document in each dimension (i.e., for each term) are calculated as a product of two submeasures: term frequency and inverse document frequency.

Term frequency describes how often a certain term appears in a document (assuming that important words appear more often). To take the document length into account and to prevent longer documents from getting a higher relevance weight, some normalization of the document length should be done. Several schemes are possible (see Chakrabarti 2002, Pazzani and Billsus 2007, or Salton and Buckley 1988). A relatively simple one relates the actual number of term occurrences to the maximum frequency of the other keywords of the document as follows (see also Adomavicius and Tuzhilin 2005).

We search for the normalized term frequency value $TF(i, j)$ of keyword i in document j. Let $freq(i, j)$ be the absolute number of occurrences of i in j. Given a keyword i, let $OtherKeywords(i, j)$ denote the set of the other keywords appearing in j. Compute the maximum frequency $maxOthers(i, j)$ as $max(freq(z, j)), z \in OtherKeywords(i, j)$. Finally, calculate $TF(i, j)$ as in Chakrabarti (2002):

$$TF(i, j) = \frac{freq(i, j)}{maxOthers(i, j)} \tag{3.2}$$

Inverse document frequency is the second measure that is combined with term frequency. It aims at reducing the weight of keywords that appear very often in all documents. The idea is that those generally frequent words are not very helpful to discriminate among documents, and more weight should therefore be given to words that appear in only a few documents. Let N be the number of all recommendable documents and $n(i)$ be the number of documents from N in which keyword i appears. The inverse document frequency for i is typically calculated as

$$IDF(i) = log\frac{N}{n(i)} \tag{3.3}$$

The combined TF-IDF weight for a keyword i in document j is computed as the product of these two measures:

$$TF\text{-}IDF(i, j) = TF(i, j) * IDF(i) \tag{3.4}$$

In the TF-IDF model, the document is, therefore, represented not as a vector of Boolean values for each keyword but as a vector of the computed TF-IDF weights.

3.1.2 Improving the vector space model/limitations

TF-IDF vectors are typically large and very sparse. To make them more compact and to remove irrelevant information from the vector, additional techniques can be applied.

Stop words and stemming. A straightforward method is to remove so-called stop words. In the English language these are, for instance, prepositions and articles such as "a", "the", or "on", which can be removed from the document vectors because they will appear in nearly all documents. Another commonly used technique is called *stemming* or *conflation*, which aims to replace variants of the same word by their common stem (root word). The word "stemming" would, for instance, be replaced by "stem", "went" by "go", and so forth.

These techniques further reduce the vector size and at the same time, help to improve the matching process in cases in which the word stems are also used in the user profile. Stemming procedures are commonly implemented as a combination of morphological analysis using, for instance, Porter's suffix-stripping algorithm (Porter 1980) and word lookup in dictionaries such as WordNet.[3] Although this technique is a powerful one in principle, there are some pitfalls, as stemming may also increase the danger of matching the profile with irrelevant documents when purely syntactic suffix stripping is used. For example, both the terms *university* and *universal* are stemmed to *univers*, which may lead to an unintended match of a document with the user profile (Chakrabarti 2002). Other

---

[3] http://wordnet.princeton.edu.

problems arise, in particular, when technical documents with many abbreviations are analyzed or when homonymous words (having multiple meanings) are in the text.

Size cutoffs. Another straightforward method to reduce the size of the document representation and hopefully remove "noise" from the data is to use only the n most informative words. In the Syskill & Webert system (Pazzani and Billsus 1997), for instance, the 128 most informative words (with respect to the expected information gain) were chosen. Similarly, Fab (Balabanović and Shoham 1997) used the top 100 words. The optimal number of words to be used was determined experimentally for the Syskill & Webert system for several domains. The evaluation showed that if too few keywords (say, fewer than 50) were selected, some possibly important document features were not covered. On the other hand, when including too many features (e.g., more than 300), keywords are used in the document model that have only limited importance and therefore represent noise that actually worsens the recommendation accuracy. In principle, complex techniques for "feature selection" can also be applied for determining the most *informative* keywords. However, besides an increase in model complexity, it is argued that learning-based approaches will tend to overfit the example representation to the training data (Pazzani and Billsus 1997). Instead, the usage of external lexical knowledge is proposed to remove words that are not relevant in the domain. Experiments show a consistent accuracy gain when such lexical knowledge is used, in particular when few training examples are available.

Phrases. A further possible improvement with respect to representation accuracy is to use "phrases as terms", which are more descriptive for a text than single words alone. Phrases, or composed words such as "United Nations", can be encoded as additional dimensions in the vector space. Detection of phrases can be done by looking up manually defined lists or by applying statistical analysis techniques (see Chakrabarti 2002 for more details).

Limitations. The described approach of extracting and weighting individual keywords from the text has another important limitation: it does not take into account the context of the keyword and, in some cases, may not capture the "meaning" of the description correctly. Consider the following simple example from Pazzani and Billsus (2007). A free-text description of a steakhouse used in a corresponding recommender system might state that "there is nothing on the menu that a vegetarian would like". In this case, in an automatically generated feature vector, the word *vegetarian* will most probably receive a higher weight than desired and produce an unintended match with a user interested in vegetarian restaurants. Note, however, that in general we may assume that terms appearing in a document are usually well suited for characterizing documents and that a "negative context" – as shown in the example – is less likely to be encountered in documents.

3.2 Similarity-based retrieval

When the item selection problem in collaborative filtering can be described as "recommend items that similar users liked", content-based recommendation is commonly described as "recommend items that are similar to those the user liked in the past". Therefore, the task for a recommender system is again – based on a user profile – to predict, for the items that the current user has not seen, whether he or she will like them. The most common techniques that rely on the vector-space document representation model will be described in this section.

3.2.1 Nearest neighbors

A first, straightforward, method for estimating to what extent a certain document will be of interest to a user is simply to check whether the user liked similar documents in the past. To do this, two pieces of information are required. First, we need some history of "like/dislike" statements made by the user about previous items. Similar to collaborative approaches, these ratings can be provided either explicitly via the user interface or implicitly by monitoring the user's behavior. Second, a measure is needed that captures the similarity of two documents. In most reported approaches, the cosine similarity measure (already described in Section 2.2.1) is used to evaluate whether two document vectors are similar.

The prediction for a not-yet-seen item d is based on letting the k most similar items for which a rating exists "vote" for n (Allan et al. 1998). If, for instance, four out of $k = 5$ of the most similar items were liked by the current user, the system may guess that the chance that d will also be liked is relatively high. Besides varying the neighborhood size k, several other variations are possible, such as binarization of ratings, using a minimum similarity threshold, or weighting of the votes based on the degree of similarity.

Such a k-nearest-neighbor method (kNN) has been implemented, for instance, in the personalized and mobile news access system described by Billsus et al. (2000). In this system, the kNN method was used to model the short-term interests of the users, which is of particular importance in the application domain of news recommendation. On arrival of a new message, the system looks for similar items in the set of stories that were recently rated by the user. By taking into account the last ratings only, the method is thus capable of adapting quickly and focusing on the user's short-term interests, which might be to read follow-up stories to recent events. At the same time, when relying on nearest neighbors, it is also possible to set an upper threshold for item similarity to prevent the system from recommending items that the user most probably has already seen.

In the described system, the kNN method was implemented as part of a multistrategy user profile technique. The system maintained profiles of short-term (ephemeral) and long-term interests. The short-term profile, as described earlier, allows the system to provide the user with information on topics of recent interest. For the long-term model, the system described by Billsus et al. (2000) therefore collects information over a longer period of time (e.g., several months) and also seeks to identify the most *informative* words in the documents by determining the terms that consistently receive high TF-IDF scores in a larger document collection. The prediction of whether an item is of interest with respect to the long-term user model is based on a probabilistic classification technique, which we describe in Section 3.3.1. Details on the threshold values and algorithms used in the experimental systems are described by Billsus and Pazzani (1999).

Given the interest predictions and recommendations for the short-term and the long-term user models, the question remains how to combine them. In the described system, a rather simple strategy is chosen. Neighbors in the short-term model are searched; if no such neighbors exist, the long-term user model is used. Other combinations are also possible. One option would be to acquire the short-term preferences online – for example, by questioning topics of interest and then sorting the matching items based on the information from the long-term preferences.

In summary, kNN-based methods have the advantage of being relatively simple to implement[4], adapt quickly to recent changes, and have the advantage that, when compared with other learning approaches, a relatively small number of ratings is sufficient to make a prediction of reasonable quality. However, as experiments show, the prediction accuracy of pure kNN methods can be lower than those of other more sophisticated techniques.

---

[4] Naive implementations of nearest-neighbor methods may, however, quickly become computationally intensive, so more advanced neighbor search methods may be required; see, e.g., Zezula et al. (2006).

3.2.2 Relevance feedback – Rocchio's method

Another method that is based on the vector-space model and was developed in the context of the pioneering information retrieval (IR) system SMART (Salton 1971) in the late 1960s is Rocchio's *relevance feedback* method. A particular aspect of SMART was that users could not only send (keyword-based) queries to the system but could also give feedback on whether the retrieved items were relevant. With the help of this feedback, the system could then internally extend the query to improve retrieval results in the next round of retrieval.

The SMART system for information retrieval did not exploit such additional information but rather allowed the user to interactively and *explicitly* rate the retrieved documents – that is, to tell the system whether they were relevant. This information is subsequently used to further refine the retrieval results. The rationale of following this approach is that with pure query- and similarity-based methods that do not provide any feedback mechanisms, the retrieval quality depends too strongly on the individual user's capability to formulate queries that contain the right keywords. User-defined queries often consist only of very few and probably polysemous words; a typical query on the web, for instance, consists of only two keywords on average (Chakrabarti 2002).

The relevance feedback loop used in this method will help the system improve and automatically extend the query as follows. The main idea is to first split the already rated documents into two groups, D^+ and D^-, of liked (interesting/relevant) and disliked documents and calculate a *prototype* (or average) vector for these categories. This prototype can also be seen as a sort of centroid of a cluster for relevant and nonrelevant document sets; see Figure 3.1.

The current query Q_i, which is represented as a multidimensional term vector just like the documents, is then repeatedly refined to Q_{i+1} by a weighted addition of the prototype vector of the relevant documents and weighted substraction of the vector representing the nonrelevant documents. As an effect, the query vector should consistently move toward the set of relevant documents as depicted schematically in Figure 3.2.

The proposed formula for computing the modified query Q_{i+1} from Q_i is defined as follows:

$$Q_{i+1} = \alpha * Q_i + \beta \left(\frac{1}{|D^+|} \sum_{d^+ \in D^+} d^+ \right) - \gamma \left(\frac{1}{|D^-|} \sum_{d^- \in D^-} d^- \right) \quad (3.5)$$

The variables α, β, and γ are used to fine-tune the behavior of the "move" toward the more relevant documents. The value of α describes how strongly the last (or original) query should be weighted, and β and γ correspondingly capture how strongly positive and negative feedback should be taken into

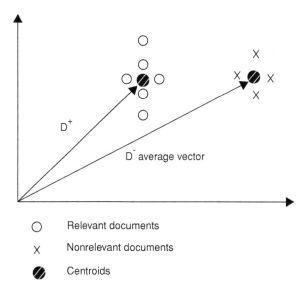

Figure 3.1. Average vectors for relevant and nonrelevant documents.

account in the improvement step. According to the analysis by Buckley et al. (1994), suitable parameter values are, for instance, 8, 16, and 4 (or 1, 2, and 0.25, respectively). These findings indicate that positive feedback is more valuable than negative feedback and it can be even better to take only positive feedback into account.

At first sight, this formula seems intuitive and the algorithm is very simple, but as stated by Pazzani and Billsus (2007), there is no theoretically motivated basis for Formula (3.5), nor can performance or convergence be guaranteed.

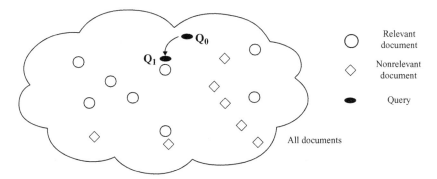

Figure 3.2. Relevance feedback. After feedback, the original query is moved toward the cluster of the relevant documents; see also Manning et al. (2008).

However, empirical evaluations with various document sets showed that useful retrieval performance can be improved, based on the feedback mechanism, already after the first iteration. More feedback iterations show only marginal improvements. An experimental evaluation using variations of this relevance feedback scheme, including an analysis of the effects of different settings, can be found in Salton and Buckley (1997) and Buckley et al. (1994). In practical settings it is also a good idea not to include *all* terms from D^+ and D^- to compute the new query (Chakrabarti 2002), as "one bad word may offset the benefits of many good words", but rather to use only the first 10 or 20 of the most relevant words in terms of the IDF measure.

Overall, the relevance feedback retrieval method and its variations are used in many application domains. It has been shown that the method, despite its simplicity, can lead to good retrieval improvements in real-world settings; see Koenemann and Belkin (1996) for a more detailed study. The main practical challenges – as with most content-based methods – are (a) a certain number of previous item ratings is needed to build a reasonable user model and (b) user interaction is required during the retrieval phase.

The first point can be partially automated by trying to capture the user ratings implicitly – for instance, by interpreting a click on a proposed document as a positive rating. The question of whether such assumptions hold in general – what to do when a user has read an article but was disappointed and what other techniques for gathering implicit feedback can be used – remains open (compare also the discussion on implicit ratings in Section 2.3.1).

Another technique for circumventing the problem of acquiring explicit user feedback is to rely on *pseudorelevance feedback* (blind feedback). The basic idea is to assume that the first n (say, 10) documents that match the query best with respect to the vector similarity measure are considered relevant. The set D^- is not used (γ is set to 0) unless an explicit negative feedback exists.

The second point – that user interaction is required during the proposal generation phase – at first glance appears to be a disadvantage over the fully automated proposal generation process of CF approaches. In fact, interactive query refinement also opens new opportunities for gathering information about the user's *real* preferences and may help the user to "learn" which vocabulary to use to retrieve documents that satisfy his or her information needs. The main assumption, of course, is that the user is capable of formulating a proper initial query, an assumption that might not always hold if we think of terminology aspects, multilanguage translation problems, or simply word misspellings (Manning et al. 2008). Further aspects of interactivity in recommender systems will be covered in more detail in Chapter 4.

Today's web search engines do not provide mechanisms for explicit feedback, although, as various evaluations show, they can lead to improved retrieval performance. Chakrabarti (2002) mentions two reasons for that. First, he argues that today's web users are impatient and are not willing to give explicit feedback on the proposals. Second, second-round queries that include many more terms than the initial query are more problematic from a performance perspective and cannot be answered as quickly as the initial "two-term" queries.

In general, however, query-based retrieval approaches are quite obviously similar to modern web search engines, and the question may arise whether a search engine is also a "content-based recommender". Although until recently popular search engine providers such as Google or Yahoo! did not personalize their search results to particular users, it can be seen also from our news personalization example from the previous section (Das et al. 2007) that a trend toward increased personalization of search results can now be observed. Today we also see that the major players in the field have started to provide more features on their service platforms, which typically include personalized start pages, access to an e-mail service, online document manipulation, document management, and so forth. As users access these features with the same identity, a broad opportunity arises to also personalize the search results more precisely. However, reports on how personalized document rankings can be computed based on these kinds of information and, in particular, how the different relevance metrics, such as PageRank and document-query similarity, can be combined are not yet available.

3.3 Other text classification methods

Another way of deciding whether or not a document will be of interest to a user is to view the problem as a *classification* task, in which the possible classes are "like" and "dislike". Once the content-based recommendation task has been formulated as a classification problem, various standard (supervised) machine learning techniques can, in principle, be applied such that an intelligent system can automatically decide whether a user will be interested in a certain document. *Supervised learning* means that the algorithm relies on the existence of training data, in our case a set of (manually labeled) document-class pairs.

3.3.1 Probabilistic methods

The most prominent classification methods developed in early text classification systems are probabilistic ones. These approaches are based on the naive Bayes

Table 3.3. *Classification based on Boolean feature vector.*

Doc-ID	recommender	intelligent	learning	school	**Label**
1	1	1	1	0	**1**
2	0	0	1	1	**0**
3	1	1	0	0	**1**
4	1	0	1	1	**1**
5	0	0	0	1	**0**
6	1	1	0	0	**?**

assumption of conditional independence (with respect to term occurrences) and have also been successfully deployed in content-based recommenders.

Remember the basic formula to compute the posterior probability for document classification from Section 2.4.3:

$$P(Y|X) = \frac{\prod_{i=1}^{d} P(X_i|Y) \times P(Y)}{P(X)} \tag{3.6}$$

A straightforward application of this model to the classification task is described by Pazzani and Billsus (1997). The possible classes are, of course, "like" and "dislike" (named *hot* and *cold* in some articles). Documents are represented by Boolean feature vectors that describe whether a certain term appeared in a document; the feature vectors are limited to the 128 most informative words, as already mentioned.

Thus, in that model, $P(v_i|C = c)$ expresses the probability of term v_i appearing in a document labeled with class c. The conditional probabilities are again estimated by using the observations in the training data.

Table 3.3 depicts a simple example setting. The training data consist of five manually labeled training documents. Document 6 is a still-unlabeled document. The problem is to decide whether the current user will be interested – that is, whether to recommend the item. To determine the correct class, we can compute the class-conditional probabilities for the feature vector X of Document 6 again as follows:

$$\begin{aligned}
P(X|Label{=}1) = {} & P(recommender{=}1|Label{=}1) \times \\
& P(intelligent{=}1|Label{=}1) \times \\
& P(learning{=}0|Label{=}1) \times P(school{=}0|Label{=}1) \\
= {} & 3/3 \times 2/3 \times 1/3 \times 2/3 \\
\approx {} & 0.149
\end{aligned}$$

The same can be done for the case *Label* = 0, and we see in the simple example that it is more probable that the user is more interested in documents (for instance, web pages) about intelligent recommender systems than in documents about learning in school. In real applications some sort of smoothing must be done for sparse datasets such that individual components of the calculation do not zero out the probability values. Of course, the resulting probability values can be used not only to decide whether a newly arriving document – in, for instance, a news filtering system – is relevant but also to rank a set of not-yet-seen documents. Remember that we also mentioned probabilistic techniques as possible recommendation methods in CF in the previous chapter. In CF, however, the classifier is commonly used to determine the membership of the active user in a cluster of users with similar preferences (by means of a latent class variable), whereas in content-based recommendation the classifier can also be directly used to determine the interestingness of a document.

Obviously, the core assumption of the naive Bayes model that the individual events are conditionally independent does not hold because there exist many term co-occurrences that are far more likely than others – such as the terms *Hong* and *Kong* or *New* and *York*. Nonetheless, the Bayes classifier has been shown to lead to surprisingly good results and is broadly used for text classification. An analysis of the reasons for this somewhat counterintuitive evidence can be found in Domingos and Pazzani (1996, 1997), or Friedman (1997). McCallum and Nigam (1998) summarize the findings as follows: "The paradox is explained by the fact that classification estimation is only a function of the sign (in binary case) of the function estimation; the function approximation can still be poor while classification accuracy remains high."

Besides the good accuracy that can be achieved with the naive Bayes classifier, a further advantage of the method – and, in particular, of the conditional independence assumption – is that the components of the classifier can be easily updated when new data are available and the learning time complexity remains linear to the number of examples; the prediction time is independent of the number of examples (Pazzani and Billsus 1997). However, as with most learning techniques, to provide reasonably precise recommendations, a certain amount of training data (past ratings) is required. The "cold-start" problem also exists for content-based recommenders that require some sort of relevance feedback. Possible ways of dealing with this are, for instance, to let the user manually label a set of documents – although this cannot be done for hundreds of documents – or to ask the user to provide a list of interesting words for each topic category (Pazzani and Billsus 1997).

The Boolean representation of document features has the advantage of simplicity but, of course, the possibly important information on how many times

Table 3.4. *Classification example with term counts.*

DocID	Words	Label
1	recommender intelligent recommender	1
2	recommender recommender learning	1
3	recommender school	1
4	teacher homework recommender	0
5	recommender recommender recommender teacher homework	?

a term occurred in the document is lost at this point. In the Syskill & Webert system, which relies on such a Boolean classifier for each topic category, the relevance of words is taken into account only when the initial set of appropriate keywords is determined. Afterward, the system cannot differentiate anymore whether a keyword appeared only once or very often in the document. In addition, this model also assumes *positional* independence – that is, it does not take into account *where* the term appeared in the document.

Other probabilistic modeling approaches overcome such limitations. Consider for instance, the classification method (example adapted from Manning et al. 2008) in Table 3.4, in which the number of term appearances shall also be taken into account.

The conditional probability of a term v_i appearing in a document of class C shall be estimated by the relative frequency of v_i in all documents of this class:

$$P(v_i|C = c) = \frac{CountTerms(v_i, docs(c))}{AllTerms(docs(c))} \qquad (3.7)$$

where $CountTerms(v_i, docs(c))$ returns the number of appearances of term v_i in documents labeled with c and $AllTerms(docs(c))$ returns the number of all terms in these documents. To prevent zeros in the probabilities, Laplace (add-one) smoothing shall be applied in the example:

$$\hat{P}(v_i|C = c) = \frac{CountTerms(v_i, docs(c)) + 1}{AllTerms(docs(c)) + |V|} \qquad (3.8)$$

where $|V|$ is the number of different terms appearing in all documents (called the "vocabulary"). We calculate the conditional probabilities for the relevant terms appearing in the new document as follows: the total length of the documents classified as "1" is 8, and the length of document 4 classified as "0" is 3. The size of the vocabulary is 6.

$$\hat{P}(recommender|Label = 1) = (5 + 1)/(8 + 6) = 6/14$$

$$\hat{P}(homework|Label = 1) = (0 + 1)/(8 + 6) = 1/14$$

$$\hat{P}(teacher|Label = 1) = (0 + 1)/(8 + 6) = 1/14$$

$$\hat{P}(recommender|Label = 0) = (1 + 1)/(3 + 6) = 2/9$$

$$\hat{P}(homework|Label = 0) = (1 + 1)/(3 + 6) = 2/9$$

$$\hat{P}(teacher|Label = 0) = (1 + 1)/(3 + 6) = 2/9$$

The prior probabilities of a document falling into class 1 or class 0 are 3/4 and 1/4, respectively. The classifier would therefore calculate the posterior probabilities as

$$\hat{P}(Label = 1|v_1 \ldots v_n) = 3/4 \times (3/7)^3 \times 1/14 \times 1/14 \approx 0.0003$$

$$\hat{P}(Label = 0|v_1 \ldots v_n) = 1/4 \times (2/9)^3 \times 2/9 \times 2/9 \approx 0.0001$$

and therefore classify the unlabeled document as being relevant for the user. The classifier has taken the multiple evidence of the term "recommender" into account appropriately. If only the Boolean representation had been used, the classifier would have rejected the document, because two other terms that appear in the document ("homework", "teacher") suggest that it is not relevant, as they also appear in the rejected document 4.

Relation to text classification. The problem of labeling a document as relevant or irrelevant in our document recommendation scenarios can be seen as a special case of the more broader and older *text classification* (text categorization or topic spotting) problem, which consists of assigning a document to a set of predefined classes. Applications of these methods can be found in information retrieval for solving problems such as personal e-mail sorting, detection of spam pages, or sentiment detection (Manning et al. 2008). Different techniques of "supervised learning", such as the probabilistic one described previously, have been proposed. The basis for all the learning techniques is a set of manually annotated training documents and the assumption that the unclassified (new) documents are somehow similar to the manually classified ones. When compared with the described "like/dislike" document recommendation problem, general text classification problems are not limited to only two classes. Moreover, in some applications it is also desirable to assign one document to more than one individual class.

As noted earlier, probabilistic methods that are based on the naive Bayes assumption have been shown to be particularly useful for text classification problems. The idea is that both the training documents and the still unclassified documents are generated by the probability distributions. Basically, two different ways of modeling the documents and their features have been proposed: the multinomial model and the Bernoulli model. The main differences between these models are the "event model" and, accordingly, how the probabilities are estimated from the training data (see McCallum and Nigam 1998, Manning et al. 2008, or Pazzani and Billsus 2007 for a detailed discussion). In the multivariate Bernoulli model, a document is treated as a binary vector that describes whether a certain term is contained in the document. In the multinomial model the number of times a term occurred in a document is also taken into account, as in our earlier example. In both cases, the position of the terms in the document is ignored. Empirical evaluations show that the multinomial model leads to significantly better classification results than does the Bernoulli model (McCallum and Nigam 1998), in particular when it comes to longer documents and classification settings with a higher number of features. An illustrative example for both approaches can be found in Manning et al. (2008).

Finally, another interesting finding in probabilistic text classification is that not only can the manually labeled documents can be used to train the classifier, but still-unlabeled documents can also help to improve classification (Nigam et al. 1998). In the context of content-based recommendation this can be of particular importance, as the training set of manually or implicitly labeled documents is typically very small because every user has his or her personal set of training examples.

3.3.2 Other linear classifiers and machine learning

When viewing the content-based recommendation problem as a classification problem, various other machine learning techniques can be employed. At a more abstract level, most learning methods aim to find coefficients of a linear model to discriminate between relevant and nonrelevant documents.

Figure 3.3 sketches the basic idea in a simplified setting in which the available documents are characterized by only two dimensions. If there are only two dimensions, the classifier can be represented by a line. The idea can, however, also easily be generalized to the multidimensional space in which a two-class classifier then corresponds to a hyperplane that represents the decision boundary.

In two-dimensional space, the line that we search for has the form $w_1 x_1 + w_2 x_2 = b$ where x_1 and x_2 correspond to the vector representation of a document

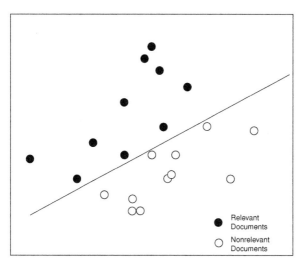

Figure 3.3. A linear classifier in two-dimensional space.

(using, e.g., TF-IDF weights) and w_1, w_2, and b are the parameters to be learned. The classification of an individual document is based on checking whether for a certain document $w_1x_1 + w_2x_2 > b$, which can be done very efficiently. In n-dimensional space, a generalized equation using weight and feature vectors instead of only two values is used, so the classification function is $\vec{w}^T \vec{x} = b$.

Many text classification algorithms are actually linear classifiers, and it can easily be shown that both the naive Bayes classifier and the Rocchio method fall into this category (Manning et al. 2008). Other methods for learning linear classifiers are, for instance, the Widrow-Hoff algorithm (see Widrow and Stearns 1985) or support vector machines (SVM; Joachims 1998). The kNN nearest-neighbor method, on the other hand, is not a linear classifier. In general, infinitely many hyperplanes (or lines in Figure 3.3) exist that can be used to separate the document space. The aforementioned learning methods will typically identify different hyperplanes, which may in turn lead to differences in classification accuracy. In other words, although all classifiers may separate the training data perfectly, they may show differences in their error rates for additional test data. Implementations based on SVM, for instance, try to identify decision boundaries that maximize the distance (called *margin*) to the existing datapoints, which leads to very good classification accuracy when compared with other approaches.

Another challenge when using a linear classifier is to deal with noise in the data. There can be noisy features that mislead the classifier if they are included in the document representation. In addition, there might also be *noise documents* that, for whatever reason, are not near the cluster where they belong. The identification of such noise in the data is, however, not trivial.

A comparative evaluation of different training techniques for text classifiers can be found in Lewis et al. (1996) and in Yang and Liu (1999). Despite the fact that in these experiments some algorithms, and in particular SVM-based ones, performed better than others, there exists no strict guideline as to which technique performs best in every situation. Moreover, it is not always clear whether using a linear classifier is the right choice at all, as there are, of course, many problem settings in which the classification borders cannot be reasonably approximated by a line or hyperplane. Overall, "selecting an appropriate learning method is therefore an unavoidable part of solving a text classification problem" (Manning et al. 2008).

3.3.3 Explicit decision models

Two other learning techniques that have been used for building content-based recommender systems are based on decision trees and rule induction. They differ from the others insofar as they generate an explicit decision model in the training phase.

Decision tree learning based on ID3 or the later C4.5 algorithms (see Quinlan 1993 for an overview) has been successfully applied to many practical problems, such as data mining problems. When applied to the recommendation problem, the inner nodes of the tree are labeled with item features (keywords), and these nodes are used to partition the test examples based, for instance, simply on the existence or nonexistence of a keyword in the document. In a basic setting only two classes, interesting or not, might appear at the leaf nodes. Figure 3.4 depicts an example of such a decision tree.

Determining whether a new document is relevant can be done very efficiently with such a prebuilt classification tree, which can be automatically constructed (learned) from training data without the need for formalizing domain knowledge. Further general advantages of decision trees are that they are well understood, have been successfully applied in many domains, and represent a model that can be interpreted relatively easily.

The main issue in the content-based recommendation problem setting is that we have to work on relatively large feature sets using, for instance, a TF-IDF document representation. Decision tree learners, however, work best when a relatively small number of features exist, which would be the case if we do not use a TF-IDF representation of a document but rather a list of "meta"-features such as author name, genre, and so forth. An experimental evaluation actually shows that decision trees can lead to comparably poor classification

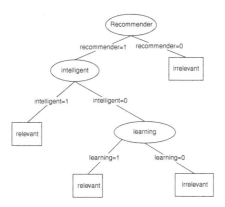

Figure 3.4. Decision tree example.

performance (Pazzani and Billsus 1997). The main reason for this limited performance on large feature sets lies in the typical splitting strategy based on the information gain, which leads to a bias toward small decision trees (Pazzani and Billsus 2007).

For these reasons, decision trees are seldom used for classical content-based recommendation scenarios. One of the few exceptions is the work of Kim et al. (2001), in which decision trees were used for personalizing the set of advertisements appearing on a web page. Still, even though decision trees might not be used directly as the core recommendation technique, they can be used in recommender systems in combination with other techniques to improve recommendation efficiency or accuracy. In Nikovski and Kulev (2006), for example, decision trees are used to compress in-memory data structures for a recommender system based on frequent itemset mining; Bellogín et al. (2010) propose to use decision trees to determine which user model features are the most relevant ones for providing accurate recommendations in a content-based collaborative hybrid news recommender system. Thus, the learning task in this work is to improve the recommendation model itself.

Rule induction is a similar method that is used to extract decision rules from training data. Methods built on the RIPPER algorithm (Cohen 1995, 1996) have been applied with some success for e-mail classification, which is, however, not a core application area of recommender systems. As mentioned by Pazzani and Billsus (2007), the relatively good performance when compared with other classification methods can be partially explained by the elaborate postpruning techniques of RIPPER itself and a particular extension that was made for e-mail classification that takes the specific document structure of e-mails with a subject line and a document body into account. A more recent evaluation and comparison of e-mail classification techniques can be found in Koprinska et al. (2007), which shows that "random forests" (instead of simple trees) perform particularly well on this problem.

In summary, both decision tree learning and rule induction have been successfully applied to specific subproblems such as e-mail classification, advertisement personalization, or cases in which small feature sets are used to describe the items (Bouza et al. 2008), which is a common situation in knowledge-based recommenders. In these settings, two of the main advantages of these learning techniques are that (a) the inferred decision rules can serve as a basis for generating explanations for the system's recommendations and (b) existing prior domain knowledge can be incorporated in the models.

3.3.4 On feature selection

All the techniques described so far rely on the vector representation of documents and on TF-IDF weights. When used in a straightforward way, such document vectors tend to be very long (there are typically thousands of words appearing in the corpus) and very sparse (in every document only a fraction of the words is used), even if stop words are removed and stemming is applied. In practical applications, such long and sparse vectors not only cause problems with respect to performance and memory requirements, but also lead to an effect called *overfitting*. Consider an example in which a very rare word appears by pure chance only in documents that have been labeled as "hot". In the training phase, a classifier could therefore be misled in the direction that this word (which can, in fact, be seen as some sort of noise) is a good indicator of the interestingness of some document. Such overfitting can easily appear when only a limited number of training documents is available.

Therefore, it is desirable to use only a subset of all the terms of the corpus for classification. This process of choosing a subset of the available terms is called *feature selection*. Different strategies for deciding which features to use are possible. Feature selection in the Syskill & Webert recommender system mentioned earlier (Pazzani and Billsus 1997), for instance, is based on domain knowledge and lexical information from WordNet. The evaluation reported by Pazzani and Billsus (1997) shows not only that the recommendation accuracy is improved when irrelevant features are removed, but also that using around 100 "informative" features leads to the best results.

Another option is to apply frequency-based feature selection and use domain- or task-specific heuristics to remove words that are "too rare" or appear "too often" based on empirically chosen thresholds (Chakrabarti 2002).

Table 3.5. χ^2 *contingency table.*

	Term t appeared	Term t missing
Class "relevant"	A	B
Class "irrelevant"	C	D

For larger text corpora, such heuristics may not be appropriate, however, and more elaborate, statistics-based methods are typically employed. In theory, one could find the optimal feature subset by training a classifier on every possible subset of features and evaluate its accuracy. Because such an approach is computationally infeasible, the value of individual features (keywords) is rather evaluated independently and a ranked list of "good" keywords, according to some utility function, is constructed. The typical measures for determining the utility of a keyword are the χ^2 test, the mutual information measure, or Fisher's discrimination index (see Chakrabarti 2002).

Consider, for example, the χ^2 test, which is a standard statistical method to check whether two events are independent. The idea in the context of feature selection is to analyze, based on training data, whether certain classification outcomes are connected to a specific term occurrence. When such a statistically significant dependency for a term can be identified, we should include this term in the feature vector used for classification.

In our problem setting, a 2×2 contingency table of classification outcomes and occurrence of term t can be set up for every term as in Table 3.5 when we assume a binary document model in which the actual number of occurrences of a term in a document is not relevant.

The symbols A to D in the table can be directly taken from the training data: Symbol A stands for the number of documents that contained term t and were classified as relevant, and B is the number of documents that were classified as relevant but did not contain the term. Symmetrically, C and D count the documents that were classified as irrelevant. Based on these numbers, the χ^2 test measures the deviation of the given counts from those that we would statistically expect when conditional independence is given. The χ^2 value is calculated as follows:

$$\chi^2 = \frac{(A + B + C + D)(AD - BC)^2}{(A + B)(A + C)(B + D)(C + D)} \tag{3.9}$$

Higher values for χ^2 indicate that the events of term occurrence and membership in a class are not independent.

To select features based on the χ^2 test, the terms are first ranked by decreasing order of their χ^2 values. The logic behind that is that we want to include those features that help us to determine class membership (or nonmembership) first – that is, those for which class membership and term occurrence are correlated. After sorting the terms, according to the proposal by Chakrabarti (2002), a number of experiments should be made to determine the optimal number of features to use for the classifier.

As mentioned previously, other techniques for feature selection, such as mutual information or Fisher's discriminant, have also been proposed for use in information retrieval scenarios. In many cases, however, these techniques result more or less in the same set of keywords (maybe in different order) as long as different document lengths are taken into account (Chakrabarti 2002, Manning et al. 2008).

3.4 Discussion

3.4.1 Comparative evaluation

Pazzani and Billsus (1997) present a comparison of several learning-based techniques for content-based recommendation. Experiments were made for several relatively small and manually annotated document collections in different domains. The experiments made with the Syskill & Webert system were set up in a way in which a subset of documents was used to learn the user profile, which was then used to predict whether the user would be interested in the unseen documents.

The percentage of correctly classified documents was taken as an accuracy measure. The accuracy of the different recommenders varied relatively strongly in these experiments (from 60 percent to 80 percent). As with most learning algorithms, the most important factor was the size of the training set (up to fifty documents in these tests). For some example sets, the improvements were substantial and an accuracy of more than 80 percent could be achieved. In some domains, however, the classifier never significantly exceeded chance levels.

Overall, the detailed comparison of the algorithms (using twenty training examples in each method) brought no clear winner. What could be seen is that decision-tree learning algorithms, which we did not cover in detail, did not perform particularly well in the given setting and that the "nearest neighbors" method performed poorly in some domains. The Bayesian and Rocchio methods performed consistently well in all domains, and no significant differences could be found. In the experiments, a neural net method with a nonlinear activation function was also evaluated but did not lead to improvements in classification accuracy.

In the Syskill & Webert system, a decision for a Bayes classifier was finally chosen, as it not only worked well in all tested domains (even if the assumption of conditional independence does not hold) but it also is relatively fast with respect to learning and predicting. It also seemed that using only Boolean document representation in the classifier – as opposed to TF-IDF weights – does not significantly affect the recommendation accuracy (Pazzani and Billsus 1997).

Finally, Manning et al. (2008) also mention that Bayes classifiers seem to work well in many domains and summarize several techniques that have been developed to improve classifier performance, such as *feature engineering* (the manual or automatic selection of "good" features), *hierarchical classification* for large category taxonomies, or taking into account that different feature sets could be used for the different zones of a document.

3.4.2 Limitations

Pure content-based recommender systems have known limitations, which rather soon led to the development of hybrid systems that combine the advantages of different recommendation techniques. The Fab system is an early example of such a hybrid system; Balabanović and Shoham (1997) mention the following limitations of content-based recommenders.

Shallow content analysis. Particularly when web pages are the items to be recommended, capturing the quality or interestingness of a web page by looking at the textual contents alone may not be enough. Other aspects, such as aesthetics, usability, timeliness, or correctness of hyperlinks, also determine the quality of a page. Shardanand and Maes (1995) also mention that when keywords are used to characterize documents, a recommender cannot differentiate between well-written articles and comparably poor papers that, naturally, use the same set of keywords. Furthermore, in some application domains the text items to be recommended may not be long enough to extract a good set of discriminating features. A typical example is the recommendation of jokes (Pazzani and Billsus 2007): Learning a good preference profile from a very small set of features may be difficult by itself; at the same time it is nearly impossible to distinguish, for instance, good lawyer jokes from bad ones.

Information in hypertext documents is also more and more contained in multimedia elements, such as images, as well as audio and video sequences. These contents are also not taken into account when only a shallow text analysis is done. Although some recent advances have been made in the area of feature extraction from text documents, research in the extraction of features from multimedia content is still at an early stage. Early results in the music domain have been reported, for instance, by (Li et al. 2003; automated genre detection) or (Shen et al. 2006; singer identification). More research can be expected in that direction in the near future, as the web already now is established as a major distribution channel for digital music, in which personalized music recommendations play an important role. Similar things happen in the video domain, where, in particular, the new opportunities of semantic annotation based on the MPEG-7 standard (ISO/IEC 15938) also allow enhanced annotation capabilities.

If no automatic extraction of descriptive features is possible, manual annotation is a theoretical option. Many authors agree that in most domains manual annotation is too costly. However, new opportunities arise in light of what is called Web 2.0, in which Internet users more and more play the role of content providers. It can already be observed that today's web users actively and voluntarily annotate content such as images or videos on popular web portals (collaborative tagging). Although these tags are mostly not taken from limited-size ontologies and may be inconsistent, they could serve as a valuable resource for determining further features of a resource. How such user-provided tags can be exploited to recommend resources to users in social web platforms will be discussed in more detail in Chapter 11.

Overspecialization. Learning-based methods quickly tend to propose *more of the same* – that is, such recommenders can propose only items that are somehow similar to the ones the current user has already (positively) rated. This can lead to the undesirable effect that *obvious* recommendations are made and the system, for instance, recommends items that are too similar to those the user already knows. A typical example is a news filtering recommender that proposes a newspaper article that covers the same story that the user has already seen in another context. The system described by Billsus and Pazzani (1999) therefore defines a threshold to filter out not only items that are too different from the profile but also those that are too similar. A set of more elaborate metrics for measuring novelty and redundancy has been analyzed by Zhang et al. (2002).

A general goal therefore is to increase the *serendipity* of the recommendation lists – that is, to include "unexpected" items in which the user might be interested, because expected items are of little value for the user. A simple way of avoiding monotonous lists is to "inject a note of randomness" (Shardanand and Maes 1995).

A discussion of this additional aspect of recommender system quality, which also applies to systems that are based on other prediction techniques, can be found, for instance, in McNee et al. (2006). Ziegler et al. (2005) propose a technique for generating more diverse recommendation lists ("topic diversification"). A recent proposal for a metric to measure the serendipity of the recommendation lists can be found in Satoh et al. (2007).

Acquiring ratings. The cold-start problem, which we discussed for collaborative systems, also exists in a slightly different form for content-based recommendation methods. Although content-based techniques do not require a large user community, they require at least an initial set of ratings from the user, typically a set of explicit "like" and "dislike" statements. In all described filtering techniques, recommendation accuracy improves with the number of ratings; significant performance increases for the learning algorithms were reported by Pazzani and Billsus (1997) when the number of ratings was between twenty and fifty. However, in many domains, users might not be willing to rate so many items before the recommender service can be used. In the initial phase, it could be an option to ask the user to provide a list of keywords, either by selecting from a list of topics or by entering free-text input.

Again, in the context of Web 2.0, it might be an option to "reuse" information that the user may have provided or that was collected in the context of another personalized (web) application and take such information as a starting point to incrementally improve the user profile.

3.5 Summary

In this chapter we have discussed different methods that are commonly referred to as content-based recommendation techniques. The roots of most approaches can be found in the field of information retrieval (IR), as the typical IR tasks of information filtering or text classification can be seen as a sort of recommendation exercise. The presented approaches have in common that they aim to learn a model of the user's interest preferences based on explicit or implicit feedback. Practical evaluations show that a good recommendation accuracy can be achieved with the help of various machine learning techniques. In contrast to collaborative approaches, these techniques do not require a user community in order to work.

However, challenges exist. The first one concerns user preference elicitation and new users. Giving explicit feedback is onerous for the user, and deriving implicit feedback from user behavior (such as viewing item details for a certain period of time) can be problematic. All learning techniques require a certain amount of training data to achieve good results; some learning methods tend to overfit the training data so the danger exists that the recommendation lists contain too many similar items.

The border between content-based recommenders and other systems is not strictly defined. Automatic text classification or information filtering are classical IR methods. In the context of recommender systems, perhaps the main difference is that these classical IR tasks are personalized – in other words, a general spam e-mail detector or a web search engine should not be viewed as a recommender system. If we think of personal e-mail sorting (according to different automatically detected document categories) or personalization of search results, however, the border is no longer clear.

Another fuzzy border is between content-based recommenders and knowledge-based ones. A typical difference between them is that content-based recommenders generally work on text documents or other items for which features can be automatically extracted and for which some sort of learning technique is employed. In contrast, knowledge-based systems rely mainly on externally provided information about the available items.

With respect to industrial adoption of content-based recommendation, one can observe that pure content-based systems are rarely found in commercial environments. Among the academic systems that were developed in the mid-1990s, the following works are commonly cited as successful examples demonstrating the general applicability of the proposed techniques.

Syskill & Webert (Pazzani et al. 1996, Pazzani and Billsus 1997) is probably the most-cited system here and falls into the category of web browsing assistants that use past ratings to predict whether the user will be interested in the links on a web page (using "thumbs up" and "thumbs down" annotations). Personal Web Watcher (Mladenic 1996) is a similar system and browsing assistant, which, however, exploits document information in a slightly different way than does Syskill & Webert; see also Mladenic 1999. The Information Finder system (Krulwich and Burkey 1997) aims to achieve similar goals but is based on a special phrase extraction technique and Bayesian networks. Newsweeder (Lang 1995) is an early news filtering system based on a probabilistic method. NewsRec (Bomhardt 2004) is a more recent system that is not limited to a specific document type, such as news or web pages, and is based on SVM as a learning technique.

As the aforementioned limitations of pure content-based recommenders are critical in many domains, researchers relatively quickly began to combine them with other techniques into hybrid systems. Fab (Balabanović and Shoham

1997) is an early example of a collaborative/content-based hybrid that tries to combine the advantages of both techniques. Many other hybrid approaches have been proposed since then and will be discussed in Chapter 5. Reports on pure content-based systems are relatively rare today. Examples of newer systems can be found in domains in which recent advances have been made with respect to automated feature extraction, such as music recommendation. Even there, however, hybrids using collaborative filtering techniques are also common; see, for instance, Logan (2004) or Yoshii et al. (2006).

3.6 Bibliographical notes

The roots of several techniques that are used in content-based recommenders are in the fields of information retrieval and information filtering. An up-to-date introduction on IR and its methods is given in the text book by Manning et al. (2008). The work covers several techniques discussed in this chapter, such as TF-IDF weighting, the vector-space document model, feature selection, relevance feedback, and naive Bayes text classification. It describes additional classification techniques based on SVM, linear regression, and clustering; it also covers further specific information retrieval techniques, such as LSI, which were also applied in the recommendation domain as described in Chapter 2, and, finally, it discusses aspects of performance evaluation for retrieval and filtering systems.

A similar array of methods is discussed in the textbook by Chakrabarti (2002), which has a strong focus on web mining and practical aspects of developing the technical infrastructure that is needed to, for instance, crawl the web.

IR methods have a long history. A 1992 review of information filtering techniques and, in particular, on the then newly developed LSI method can be found in Foltz and Dumais (1992). Housman and Kaskela (1970) is an overview paper on methods of "selective dissemination of information", a concept that early on implemented some of the features of modern filtering methods.

A recent overview of content-based recommendation techniques (in the context of adaptive web applications and personalization) is given by Pazzani and Billsus (2007). More details about the influential Syskill & Webert recommender system can be found in the original paper (Pazzani et al. 1996) by the same authors. In Pazzani and Billsus (1997), a comparative evaluation of different classification techniques from nearest neighbors over decision trees, Bayes classifiers, and neural nets is given.

Adomavicius and Tuzhilin (2005) give a compact overview on content-based methods and show how such approaches fit into a more general framework of recommendation methods. They also provide an extensive literature review that can serve as a good starting point for further reading. The structure of this chapter mainly follows the standard schemes developed by Adomavicius and Tuzhilin (2005) and Pazzani and Billsus (2007).

4

Knowledge-based recommendation

4.1 Introduction

Most commercial recommender systems in practice are based on collaborative filtering (CF) techniques, as described in Chapter 2. CF systems rely solely on the user ratings (and sometimes on demographic information) as the only knowledge sources for generating item proposals for their users. Thus, no additional knowledge – such as information about the available movies and their characteristics – has to be entered and maintained in the system.

Content-based recommendation techniques, as described in Chapter 3, use different knowledge sources to make predictions whether a user will like an item. The major knowledge sources exploited by content-based systems include category and genre information, as well as keywords that can often be automatically extracted from textual item descriptions. Similar to CF, a major advantage of content-based recommendation methods is the comparably low cost for knowledge acquisition and maintenance.

Both collaborative and content-based recommender algorithms have their advantages and strengths. However, there are many situations for which these approaches are not the best choice. Typically, we do not buy a house, a car, or a computer very frequently. In such a scenario, a pure CF system will not perform well because of the low number of available ratings (Burke 2000). Furthermore, time spans play an important role. For example, five-year-old ratings for computers might be rather inappropriate for content-based recommendation. The same is true for items such as cars or houses, as user preferences evolve over time because of, for example, changes in lifestyles or family situations. Finally, in more complex product domains such as cars, customers often want to define their requirements explicitly – for example, "the maximum price of the car is x and the color should be black". The formulation of such requirements is not typical for pure collaborative and content-based recommendation frameworks.

Knowledge-based recommender systems help us tackle the aforementioned challenges. The advantage of these systems is that no ramp-up problems exist, because no rating data are needed for the calculation of recommendations. Recommendations are calculated independently of individual user ratings: either in the form of *similarities* between customer requirements and items or on the basis of explicit *recommendation rules*. Traditional interpretations of what a recommender system is focus on the *information filtering* aspect (Konstan et al. 1997, Pazzani 1999a), in which items that are likely to be of interest for a certain customer are filtered out. In contrast, the recommendation process of knowledge-based recommender applications is highly interactive, a foundational property that is a reason for their characterization as *conversational systems* (Burke 2000). This interactivity aspect triggered a slight shift from the interpretation as a filtering system toward a wider interpretation where recommenders are defined as systems that "guide a user in a personalized way to interesting or useful objects in a large space of possible options or that produce such objects as output" (Burke 2000). Recommenders that rely on knowledge sources not exploited by collaborative and content-based approaches are by default defined as knowledge-based recommenders by Burke (2000) and Felfernig and Burke (2008).

Two basic types of knowledge-based recommender systems are *constraint-based* (Felfernig and Burke 2008, Felfernig et al. 2006–07, Zanker et al. 2010) and *case-based* systems (Bridge et al. 2005, Burke 2000). Both approaches are similar in terms of the recommendation process: the user must specify the requirements, and the system tries to identify a solution. If no solution can be found, the user must change the requirements. The system may also provide explanations for the recommended items. These recommenders, however, differ in the way they use the provided knowledge: case-based recommenders focus on the retrieval of similar items on the basis of different types of similarity measures, whereas constraint-based recommenders rely on an explicitly defined set of recommendation rules. In constraint-based systems, the set of recommended items is determined by, for instance, searching for a set of items that fulfill the recommendation rules. Case-based systems, on the other hand, use similarity metrics to retrieve items that are similar (within a predefined threshold) to the specified customer requirements. Constraint-based and case-based knowledge representations will be discussed in the following subsections.

4.2 Knowledge representation and reasoning

In general, knowledge-based systems rely on detailed knowledge about item characteristics. A snapshot of such an item catalog is shown in Table 4.1 for

Table 4.1. *Example product assortment: digital cameras (Felfernig et al. 2009).*

id	price(€)	mpix	opt-zoom	LCD-size	movies	sound	waterproof
p_1	148	8.0	4×	2.5	no	no	yes
p_2	182	8.0	5×	2.7	yes	yes	no
p_3	189	8.0	10×	2.5	yes	yes	no
p_4	196	10.0	12×	2.7	yes	no	yes
p_5	151	7.1	3×	3.0	yes	yes	no
p_6	199	9.0	3×	3.0	yes	yes	no
p_7	259	10.0	3×	3.0	yes	yes	no
p_8	278	9.1	10×	3.0	yes	yes	yes

the digital camera domain. Roughly speaking, the recommendation problem consists of selecting items from this catalog that match the user's needs, preferences, or hard requirements. The user's requirements can, for instance, be expressed in terms of desired values or value ranges for an item feature, such as "the price should be lower than 300€" or in terms of desired functionality, such as "the camera should be suited for sports photography".

Following the categorization from the previous section, we now discuss how the required domain knowledge is encoded in typical knowledge-based recommender systems. A constraint-based recommendation problem can, in general, be represented as a *constraint satisfaction problem* (Felfernig and Burke 2008, Zanker et al. 2010) that can be solved by a constraint solver or in the form of a *conjunctive query* (Jannach 2006a) that is executed and solved by a database engine. Case-based recommendation systems mostly exploit similarity metrics for the retrieval of items from a catalog.

4.2.1 Constraints

A classical constraint satisfaction problem (CSP)[1] can be described by a-tuple (V, D, C) where

- V is a set of variables,
- D is a set of finite domains for these variables, and
- C is a set of constraints that describes the combinations of values the variables can simultaneously take (Tsang 1993).

A solution to a CSP corresponds to an assignment of a value to each variable in V in a way that all constraints are satisfied.

---

[1] A discussion of different CSP algorithms can be found in Tsang (1993).

Table 4.2. *Example recommendation task (V_C, V_{PROD}, C_R, C_F, C_{PROD}, REQ) and the corresponding recommendation result (RES).*

V_C	{*max-price*(0 . . . 1000), *usage*(*digital, small-print, large-print*), *photography* (*sports, landscape, portrait, macro*)}
V_{PROD}	{*price*(0 . . . 1000), *mpix*(3.0 . . . 12.0), *opt-zoom*(4× . . . 12×), *lcd-size* (2.5 . . . 3.0), *movies*(*yes, no*), *sound*(*yes, no*), *waterproof*(*yes, no*)}
C_F	{*usage = large-print → mpix > 5.0*} (*usage* is a customer property and *mpix* is a product property)
C_R	{*usage = large-print → max-price > 200*} (*usage* and *max-price* are customer properties)
C_{PROD}	{(*id*=p1 ∧ *price*=148 ∧ *mpix*=8.0 ∧ *opt-zoom*=4× ∧ *lcd-size*=2.5 ∧ *movies*=*no* ∧ *sound*=*no* ∧ *waterproof*=no) ∨ · · · ∨ (*id*=p8 ∧ *price*=278 ∧ *mpix*=9.1 ∧ *opt-zoom*=10× ∧ *lcd-size*=3.0 ∧ *movies*=*yes* ∧ *sound*=*yes* ∧ *waterproof*=yes)}
REQ	{*max-price = 300, usage = large-print, photography = sports*}
RES	{*max-price = 300, usage = large-print, photography = sports, id = p8, price*=278, *mpix*=9.1, *opt-zoom*=10×, *lcd-size*=3.0, *movies*=*yes*, *sound*=*yes*, *waterproof*=yes}

Constraint-based recommender systems (Felfernig and Burke 2008, Felfernig et al. 2006–07, Zanker et al. 2010) can build on this formalism and exploit a *recommender knowledge base* that typically includes two different sets of variables ($V = V_C \cup V_{PROD}$), one describing potential customer requirements and the other describing product properties. Three different sets of constraints ($C = C_R \cup C_F \cup C_{PROD}$) define which items should be recommended to a customer in which situation. Examples for such variables and constraints for a digital camera recommender, as described by Jannach (2004), and Felfernig et al. (2006–07), are shown in Table 4.2.

- *Customer properties (V_C)* describe the possible customer requirements (see Table 4.2). The customer property *max-price* denotes the maximum price acceptable for the customer, the property *usage* denotes the planned usage of photos (print versus digital organization), and *photography* denotes the predominant type of photos to be taken; categories are, for example, sports or portrait photos.
- *Product properties (V_{PROD})* describe the properties of products in an assortment (see Table 4.2); for example, *mpix* denotes possible resolutions of a digital camera.

- *Compatibility constraints (C_R)* define allowed instantiations of customer properties – for example, *if large-size photoprints are required, the maximal accepted price must be higher than 200* (see Table 4.2).
- *Filter conditions (C_F)* define under which conditions which products should be selected – in other words, filter conditions define the relationships between customer properties and product properties. An example filter condition is *large-size photoprints require resolutions greater than 5 mpix* (see Table 4.2).
- *Product constraints (C_{PROD})* define the currently available product assortment. An example constraint defining such a product assortment is depicted in Table 4.2. Each conjunction in this constraint completely defines a product (item) – all product properties have a defined value.

The task of identifying a set of products matching a customer's wishes and needs is denoted as a *recommendation task*. The customer requirements *REQ* can be encoded as unary constraints over the variables in V_C and V_{PROD} – for example, *max-price* $= 300$.

Formally, each solution to the CSP ($V = V_C \cup V_{PROD}$, D, $C = C_R \cup C_F \cup C_{PROD} \cup REQ$) corresponds to a consistent recommendation. In many practical settings, the variables in V_C do not have to be instantiated, as the relevant variables are already bound to values through the constraints in *REQ*. The task of finding such valid instantiations for a given constraint problem can be accomplished by every standard constraint solver. A consistent recommendation *RES* for our example recommendation task is depicted in Table 4.2.

Conjunctive queries. A slightly different way of constraint-based item retrieval for a given catalog, as shown in Table 4.1, is to view the item selection problem as a data filtering task. The main task in such an approach, therefore, is not to find valid variable instantiations for a CSP but rather to construct a conjunctive database query that is executed against the item catalog. A *conjunctive query* is a database query with a set of selection criteria that are connected conjunctively.

For example, $\sigma_{[mpix \geq 10, price < 300]}(P)$ is such a conjunctive query on the database table P, where σ represents the selection operator and $[mpix \geq 10, price < 300]$ the corresponding selection criteria. If we exploit conjunctive queries (database queries) for item selection purposes, V_{PROD} and C_{PROD} are represented by a database table P. Table attributes represent the elements of V_{PROD} and the table entries represent the constraint(s) in C_{PROD}. In our working example, the set of available items is $P = \{p_1, p_2, p_3, p_4, p_5, p_6, p_7, p_8\}$ (see Table 4.1).

Queries can be defined that select different item subsets from P depending on the requirements in REQ. Such queries are directly derived from the filter conditions (C_F) that define the relationship between customer requirements and the corresponding item properties. For example, the filter condition $usage = large\text{-}print \rightarrow mpix > 5.0$ denotes the fact that if customers want to have large photoprints, the resolution of the corresponding camera ($mpix$) must be > 5.0. If a customer defines the requirement $usage = large\text{-}print$, the corresponding filter condition is active, and the consequent part of the condition will be integrated in a corresponding conjunctive query. The existence of a recommendation for a given set REQ and a product assortment P is checked by querying P with the derived conditions (consequents of filter conditions). Such queries are defined in terms of selections on P formulated as $\sigma_{[criteria]}(P)$, for example, $\sigma_{[mpix \geq 10]}(P) = \{p_4, p_7\}$.[2]

4.2.2 Cases and similarities

In case-based recommendation approaches, items are retrieved using similarity measures that describe to which extent item properties match some given user's requirements. The so-called distance similarity (McSherry 2003a) of an item p to the requirements $r \in REQ$ is often defined as shown in Formula 4.1. In this context, $sim(p, r)$ expresses for each *item attribute value* $\phi_r(p)$ its distance to the customer requirement $r \in REQ$ – for example, $\phi_{mpix}(p_1) = 8.0$. Furthermore, w_r is the importance weight for requirement r.[3]

$$similarity(p, REQ) = \frac{\sum_{r \in REQ} w_r * sim(p, r)}{\sum_{r \in REQ} w_r} \qquad (4.1)$$

In real-world scenarios, there are properties a customer would like to maximize – for example, the resolution of a digital camera. There are also properties that customers want to minimize – for example, the price of a digital camera or the risk level of a financial service. In the first case we are talking about "more-is-better" (MIB) properties; in the second case the corresponding properties are denoted with "less-is-better" (LIB).

To take those basic properties into account in our similarity calculations, we introduce the following formulae for calculating local similarities

---

[2] For reasons of simplicity in the following sections we assume $V_C = V_{PROD}$ – that is, customer requirements are directly defined on the technical product properties. Queries on a product table P will be then written as $\sigma_{[REQ]}(P)$.

[3] A detailed overview of different types of similarity measures can be found in Wilson and Martinez 1997. Basic approaches to determine the importance of requirements (w) are discussed in Section 4.3.4.

(McSherry 2003a). First, in the case of MIB properties, the local similarity between p and r is calculated as follows:

$$sim(p, r) = \frac{\phi_r(p) - min(r)}{max(r) - min(r)} \tag{4.2}$$

The local similarity between p and r in the case of LIB properties is calculated as follows:

$$sim(p, r) = \frac{max(r) - \phi_r(p)}{max(r) - min(r)} \tag{4.3}$$

Finally, there are situations in which the similarity should be based solely on the distance to the originally defined requirements. For example, if the user has a certain run time of a financial service in mind or requires a certain monitor size, the shortest run time as well as the largest monitor will not represent an optimal solution. For such cases we have to introduce a third type of local similarity function:

$$sim(p, r) = 1 - \frac{|\phi_r(p) - r|}{max(r) - min(r)} \tag{4.4}$$

The similarity measures discussed in this section are often the basis for different case-based recommendation systems, which will be discussed in detail in Section 4.4. *Utility-based recommendation* – as, for instance, mentioned by Burke (2000) – can be interpreted as a specific type of knowledge-based recommendation. However, this approach is typically applied in combination with constraint-based recommendation (Felfernig et al. 2006–07) and sometimes as well, in combination with case-based recommenders (Reilly et al. 2007b). Therefore, this approach will be discussed in Section 4.3.4 as a specific functionality in the context of constraint-based recommendation.

4.3 Interacting with constraint-based recommenders

The general interaction flow of a knowledge-based, conversational recommender can be summarized as follows.

- The user specifies his or her initial preferences – for example, by using a web-based form. Such forms can be identical for all users or personalized to the specific situation of the current user. Some systems use a question/answer preference elicitation process, in which the questions can be asked either all at once or incrementally in a wizard-style, interactive dialog, as described by Felfernig et al. (2006–07).

- When enough information about the user's requirements and preferences has been collected, the user is presented with a set of matching items. Optionally, the user can ask for an explanation as to why a certain item was recommended.
- The user might revise his or her requirements, for instance, to see alternative solutions or narrow down the number of matching items.

Although this general user interaction scheme appears to be rather simple in the first place, practical applications are typically required to implement more elaborate interaction patterns to support the end user in the recommendation process. Think, for instance, of situations in which none of the items in the catalog satisfies all user requirements. In such situations, a conversational recommender should intelligently support the end user in resolving the problem and, for example, proactively propose some action alternatives.

In this section we analyze in detail different techniques to support users in the interaction with constraint-based recommender applications. These techniques help improve the usability of these applications and achieve higher user acceptance in dimensions such as trust or satisfaction with the recommendation process and the output quality (Felfernig et al. 2006–07).

4.3.1 Defaults

Proposing default values. Defaults are an important means to support customers in the requirements specification process, especially in situations in which they are unsure about which option to select or simply do not know technical details (Huffman and Kahn 1998). Defaults can support customers in choosing a reasonable alternative (an alternative that realistically fits the current preferences). For example, if a customer is interested in printing large-format pictures from digital images, the camera should support a resolution of more than 5.0 megapixels (default). The negative side of the coin is that defaults can also be abused to manipulate consumers to choose certain options. For example, users can be stimulated to buy a park distance control functionality in a car by presenting the corresponding default value (Herrmann et al. 2007). Defaults can be specified in various ways:

- *Static defaults*: In this case, one default is specified per customer property – for example, *default(usage)=large-print*, because typically users want to generate posters from high-quality pictures.
- *Dependent defaults*: In this case a default is defined on different combinations of potential customer requirements – for example, *default(usage=small-print, max-price)* = 300.

Table 4.3. *Example of customer interaction data.*

customer (user)	price	opt-zoom	lcd-size
cu_1	400	10×	3.0
cu_2	300	10×	3.0
cu_3	150	4×	2.5
cu_4	200	5×	2.7
cu_5	200	5×	2.7

- *Derived defaults*: When the first two default types are strictly based on a declarative approach, this third type exploits existing interaction logs for the automated derivation of default values.

The following example sketches the main idea and a basic scheme for derived default values. Assume we are given the sample interaction log in Table 4.3. The only currently known requirement of a new user should be *price=400*; the task is to find a suitable default value for the customer requirement on the optical zoom (*opt-zoom*). From the interaction log we see that there exists a customer (cu_1) who had similar requirements (*price=400*). Thus, we could take cu_1's choice for the optical zoom as a default also for the new user.

Derived defaults can be determined based on various schemes; basic example approaches to the determination of suitable default values are, for example, *1-nearest neighbor* and *weighted majority voter*.

- *1-Nearest neighbor*: The 1-nearest neighbor approach can be used for the prediction of values for one or a set of properties in V_C. The basic idea is to determine the entry of the interaction log that is as close as possible to the set of requirements (*REQ*) specified by the customer. The 1-nearest neighbor is the entry in the example log in Table 4.3 that is most similar to the customer requirements in *REQ* (see Formula 4.1). In our working example, the nearest neighbor for the set of requirements $REQ = \{r_1 : price = 400, r_2 : opt\text{-}zoom = 10\times\}$ would be the interaction log entry for customer cu_1. If, for example, the variable *lcd-size* is not specified by the current customer, the recommender application could propose the value 3.0.
- *Weighted majority voter*: The weighted majority voter proposes customer property values that are based on the voting of a set of neighbor items for a specific property. It operates on a set of n-nearest neighbors, which can be calculated on the basis of Formula 4.1. Let us assume that the three-nearest neighbors for the requirements $REQ = \{r_1 : price = 400\}$ are the interaction

log entries for the customers $\{cu_1, cu_2, cu_4\}$ and we want to determine a default for the property *opt-zoom*. The majority value for *opt-zoom* would then be $10\times$, which, in this context, can be recommended as the default.

For weighted majority voters as well as for simple 1-nearest-neighbor-based default recommendations, it is not possible to guarantee that the requirements (including the defaults) allow the derivation of a recommendation. For example, if $REQ = \{r_1 : opt\text{-}zoom = 3\times\}$ then the weighted majority voter approach would recommend *lcd-size* $= 2.7$, assuming that the three-nearest neighbors are $\{cu_3, cu_4, cu_5\}$; the corresponding query $\sigma_{[opt\text{-}zoom=3x, lcd\text{-}size=2.7]}(P)$ would result in the empty set $\emptyset$. The handling of such situations will be discussed in Subsection 4.3.2.

Selecting the next question. Besides using defaults to support the user in the requirements specification process, the interaction log and the default mechanism can also be applied for identifying properties that may be interesting for the user within the scope of a recommendation session. For example, if a user has already specified requirements regarding the properties *price* and *opt-zoom*, defaults could propose properties that the user could be interested to specify next. Concepts supporting such a functionality are discussed in the following paragraphs.

Proposing defaults for properties to be presented next is an important functionality, as most users are not interested in specifying values for all properties – they rather want to specify the conditions that are important for them, but then immediately move on to see the recommended items. Different approaches to the selection of interesting questions are discussed by Mahmood and Ricci (2007). The precondition for such approaches is the availability of user interaction logs (see Table 4.4). One basic approach to the determination of defaults for the presentation of selectable customer properties is discussed by Mahmood and Ricci (2007), in which question recommendation is based on the principle of frequent usage (popularity). Such a popularity value can be calculated using Formula 4.5, in which the recommendation of a question depends strictly on the number of previous selections of other users – see, for example, Table 4.4. By analyzing the interaction log of Table 4.4, *popularity*(*price*, *pos* : 1) $= 0.6$, whereas *popularity*(*mpix*, *pos* : 1) $= 0.4$. Consequently, for the first question, the property *price* would be selected.

$$popularity(attribute, pos) = \frac{\#selections(attribute, pos)}{\#sessions} \qquad (4.5)$$

Another approach for supporting question selection is to apply weighted-majority voters (Felfernig and Burke 2008). If, for example, a user has already

Table 4.4. *Order of selected customer properties; for example, in session 4*
(ID = 4) mpix *has been selected as first customer property to be specified.*

ID	pos:1	pos:2	pos:3	pos:4	pos:5	pos:6	$\cdots$
1	price	opt-zoom	mpix	movies	LCD-size	sound	$\cdots$
2	price	opt-zoom	mpix	movies	LCD-size	–	$\cdots$
3	price	mpix	opt-zoom	lcd-size	movies	sound	$\cdots$
4	mpix	price	opt-zoom	lcd-size	movies	–	$\cdots$
5	mpix	price	lcd-size	opt-zoom	movies	sound	$\cdots$

selected the properties *price* and *opt-zoom*, the weighted majority voter would identify the sessions with ID $\{1, 2\}$ as nearest neighbors (see Table 4.4) for the given set of requirements and then propose *mpix* as the next interesting question.

4.3.2 Dealing with unsatisfiable requirements and empty result sets

In our example, a given set of requirements $REQ = \{r_1 : price <= 150, r_2 : opt\text{-}zoom = 5x, r_3 : sound = yes, r_4 : waterproof = yes\}$ cannot be fulfilled by any of the products in $P = \{p_1, p_2, p_3, p_4, p_5, p_6, p_7, p_8\}$ because $\sigma_{[price<=150, opt\text{-}zoom=5x, sound=yes, waterproof=yes]}(P) = \emptyset$.

Many recommender systems are not able to propose a way out of such a "no solution could be found" dilemma. One option to help the user out is to incrementally and automatically relax constraints of the recommendation problem until a corresponding solution has been found. Different approaches to deal with this problem have been proposed in the literature. All of them share the same basic goal of identifying relaxations to the original set of constraints (Jannach 2006a, O'Sullivan et al. 2007, Felfernig et al. 2004, Felfernig et al. 2009). For the sake of better understandability, we assume that the user's requirements are directly related to item properties V_{PROD}.

In this section we discuss one basic approach in more detail. This approach is based on the idea of identifying and resolving requirements-immanent conflicts induced by the set of products in P. In such situations users ask for help that can be provided, for example, by the indication of a *minimal* set of requirements that should be changed in order to find a solution. In addition to a point to such unsatisfiable requirements, users could also be interested in repair proposals – that is, in adaptations of the initial requirements in such a way that the recommender is able to calculate a solution (Felfernig et al. 2009).

The calculation of such repairs can be based on the concepts of model-based diagnosis (MBD; Reiter 1987) – the basis for the automated identification and repair of minimal sets of faulty requirements (Felfernig et al. 2004). MBD starts with a description of a system that is, in the case of recommender applications, a predefined set of products $p_i \in P$. If the actual system behavior is in contradiction to the intended system behavior (the unintended behavior is reflected by the fact that no solution could be found), the diagnosis task is to identify the system components (in our context represented by the user requirements in *REQ*) that, when we assume that they function abnormally, explain the discrepancy between the actual and the intended behavior of the system under consideration.

In the context of our problem setting, a diagnosis is a minimal set of user requirements whose repair (adaptation) will allow the retrieval of a recommendation. Given $P = \{p_1, p_2, \ldots, p_n\}$ and $REQ = \{r_1, r_2, \ldots, r_m\}$ where $\sigma_{[REQ]}(P) = \emptyset$, a knowledge-based recommender system would calculate a set of diagnoses $\Delta = \{d_1, d_2, \ldots, d_k\}$ where $\sigma_{[REQ-d_i]}(P) \neq \emptyset \; \forall d_i \in \Delta$. A diagnosis is a minimal set of elements $\{r_1, r_2, \ldots, r_k\} = d \subseteq REQ$ that have to be repaired in order to restore consistency with the given product assortment so at least one solution can be found: $\sigma_{[REQ-d]}(P) \neq \emptyset$. Following the basic principles of MBD, the calculation of diagnoses $d_i \in \Delta$ is based on the determination and resolution of conflict sets. A conflict set CS (Junker 2004) is defined as a subset $\{r_1, r_2, \ldots, r_l\} \subseteq REQ$, such that $\sigma_{[CS]}(P) = \emptyset$. A conflict set CS is minimal if and only if (iff) there does not exist a CS' with $CS' \subset CS$.

As mentioned, no item in P completely fulfills the requirements $REQ = \{r_1 : price <= 150, r_2 : opt\text{-}zoom = 5\times, r_3 : sound=yes, r_4 : waterproof=yes\}$: $\sigma_{[price<=150,opt\text{-}zoom=5x,sound=yes,waterproof=yes]}(P) = \emptyset$. The corresponding conflict sets are $CS_1 = \{r_1, r_2\}$, $CS_2 = \{r_2, r_4\}$, and $CS_3 = \{r_1, r_3\}$, as $\sigma_{[CS1]}(P) = \emptyset$, $\sigma_{[CS2]}(P) = \emptyset$, and $\sigma_{[CS3]}(P) = \emptyset$. The identified conflict sets are minimal, as $\neg \exists CS_1': CS_1' \subset CS_1 \wedge \sigma_{[CS1']}(P) = \emptyset$, $\neg \exists CS_2': CS_2' \subset CS_2 \wedge \sigma_{[CS2']}(P) = \emptyset$, and $\neg \exists CS_3': CS_3' \subset CS_3 \wedge \sigma_{[CS3']}(P) = \emptyset$.

Diagnoses $d_i \in \Delta$ can be calculated by resolving conflicts in the given set of requirements. Because of its minimality, one conflict can be easily resolved by deleting one of the elements from the conflict set. After having deleted at least one element from each of the identified conflict sets, we are able to present a corresponding diagnosis. The diagnoses derived from the conflict sets $\{CS_1, CS_2, CS_3\}$ in our working example are $\Delta = \{d_1:\{r_1, r_2\}, d_2:\{r_1, r_4\}, d_3:\{r_2, r_3\}\}$. The calculation of such diagnoses (see Figure 4.1) starts with the first identified conflict set (CS_1) (1). CS_1 can be resolved in two alternative ways: by deleting either r_1 or r_2. Both of these alternatives are explored following a breadth-first search regime. After deleting r_1 from *REQ*, the next conflict

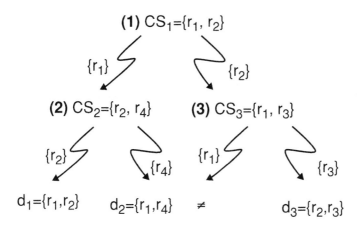

Figure 4.1. Calculating diagnoses for unsatisfiable requirements.

set is CS_2 (2), which also allows two different relaxations, namely r_2 and r_4. Deleting the elements of CS_2 leads to the diagnoses d_1 and d_2. After deleting r_2 from CS_1, the next returned conflict set is CS_3 (3). Both alternative deletions for CS_3, in principle, lead to a diagnosis. However, the diagnosis $\{r_1, r_2\}$ is already contained in d_1; consequently, this path is not expanded further, and the third and final diagnosis is d_3.

Calculating conflict sets. A recent and general method for the calculation of conflict sets is QUICKXPLAIN (Algorithm 4.1), an algorithm that calculates one conflict set at a time for a given set of constraints. Its divide-and-conquer strategy helps to significantly accelerate the performance compared to other approaches (for details see, e.g., Junker 2004).

QUICKXPLAIN has two input parameters: first, P is the given product assortment $P = \{p_1, p_2, \ldots, p_m\}$. Second, $REQ = \{r_1, r_2, \ldots, r_n\}$ is a set of requirements analyzed by the conflict detection algorithm.

QUICKXPLAIN is based on a recursive divide-and-conquer strategy that divides the set of requirements into the subsets REQ_1 and REQ_2. If both subsets contain about 50 percent of the requirements (the splitting factor is $\frac{n}{2}$), all the requirements contained in REQ_2 can be deleted (ignored) after a single consistency check if $\sigma_{[REQ_1]}(P) = \emptyset$. The splitting factor of $\frac{n}{2}$ is generally recommended; however, other factors can be defined. In the best case (e.g., all elements of the conflict belong to subset REQ_1) the algorithm requires $\log_2 \frac{n}{u} + 2u$ consistency checks; in the worst case, the number of consistency checks is $2u(\log_2 \frac{n}{u} + 1)$, where u is the number of elements contained in the conflict set.

Algorithm 4.1 QUICKXPLAIN(P, REQ)

Input: trusted knowledge (items) P; Set of requirements REQ
Output: minimal conflict set CS
if $\sigma_{[REQ]}(P) \neq \emptyset$ or $REQ = \emptyset$ **then** return $\emptyset$
else return QX$'$ $(P, \emptyset, \emptyset, REQ)$;

Function QX$'$(P, B, Δ, REQ)
if $\Delta \neq \emptyset$ and $\sigma_{[B]}(P) = \emptyset$ **then** return $\emptyset$;
if $REQ = \{r\}$ **then** return $\{r\}$;
let $\{r_1, \ldots, r_n\} = REQ$;
let k be $\frac{n}{2}$;
$REQ_1 \leftarrow r_1, \ldots, r_k$ and $REQ_2 \leftarrow r_{k+1}, \ldots, r_n$;
$\Delta_2 \leftarrow$ QX$'$(P, $B \cup REQ_1$, REQ_1, REQ_2);
$\Delta_1 \leftarrow$ QX$'$(P, $B \cup \Delta_2$, Δ_2, REQ_1);
return $\Delta_1 \cup \Delta_2$;

To show how the algorithm QUICKXPLAIN works, we will exemplify the calculation of a conflict set on the basis of our working example (see Figure 4.2) – that is, P $= \{p_1, p_2, \ldots, p_8\}$ and REQ $= \{r_1$:price$\leq$150, r_2:opt-zoom$=$5x, r_3:sound$=$yes, r_4:waterproof$=$yes$\}$. First, the main routine is activated (1), which checks whether $\sigma_{[REQ]}(P) \neq \emptyset$. As this is not the case, the recursive routine QX$'$ is activated (2). This call results in call (3) (to obtain Δ_2), which itself results in $\emptyset$, as $\Delta \neq \emptyset$ and $\sigma_{[B]}(P) = \emptyset$. To obtain Δ_1, call (4) directly activates call (5) and call (6), and each those last calls identifies a corresponding conflict element (r_2 and r_1). Thus, CS_1:$\{r_1, r_2\}$ is returned as the first conflict set.

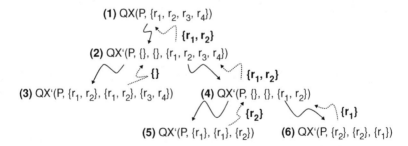

(1) QX(P, $\{r_1, r_2, r_3, r_4\}$)
                    $\{r_1, r_2\}$
(2) QX$'$(P, $\{\}$, $\{\}$, $\{r_1, r_2, r_3, r_4\}$)
                $\{\}$ $\{r_1, r_2\}$
(3) QX$'$(P, $\{r_1, r_2\}$, $\{r_1, r_2\}$, $\{r_3, r_4\}$)   **(4)** QX$'$(P, $\{\}$, $\{\}$, $\{r_1, r_2\}$)
                                $\{r_2\}$ $\{r_1\}$
                    (5) QX$'$(P, $\{r_1\}$, $\{r_1\}$, $\{r_2\}$)   **(6)** QX$'$(P, $\{r_2\}$, $\{r_2\}$, $\{r_1\}$)

Figure 4.2. Example: calculation of conflict sets using QUICKXPLAIN.

Algorithm 4.2 MINRELAX(P, REQ)

Input: Product assortment P; set of requirements REQ
Output: Complete set of all minimal diagnoses Δ
$\Delta \leftarrow \emptyset$;
forall $p_i \in P$ **do**
   PSX $\leftarrow$ product-specific-relaxation(p_i, REQ);
   SUB $\leftarrow \{r \in \Delta - r \subset PSX\}$;
   if SUB $\neq \emptyset$ **then** continue with next p_i;
   SUPER $\leftarrow \{r \in \Delta - PSX \subset r\}$;
   if SUPER $\neq \emptyset$ **then** $\Delta \leftarrow \Delta - $ SUPER;
   $\Delta \leftarrow \Delta \cup \{PSX\}$;
return Δ;

Besides the usage within an MBD procedure, the conflicts computed with QUICKXPLAIN can also be used in interactive relaxation scenarios as described by McSherry (2004), in which the user is presented with one or more remaining conflicts and asked to choose one of the conflict elements to retract. For an example of such an algorithm, see Jannach 2006b.

Fast in-memory computation of relaxations with MINRELAX. As long as the set of items is specified explicitly (as in Table 4.1), the calculation of diagnoses can be achieved without the explicit determination and resolution of conflict sets (Jannach 2006a). MINRELAX (Algorithm 4.2) is such an algorithm to determine the complete set of diagnoses. The previously discussed approach based on the resolution of conflict sets is still indispensable in interactive settings, in which users should be able to manually resolve conflicts, and in settings in which items are not enumerated but described in the form of generic product structures (Felfernig et al. 2004).

The MINRELAX algorithm for determining the complete set of minimal diagnoses has been introduced by Jannach (2006a). This algorithm calculates, for each item $p_i \in P$ and the requirements in REQ, a corresponding product-specific relaxation PSX. PSX is a minimal diagnosis $d \in \Delta$ (the set of all minimal diagnoses) if there is no set r such that $r \subset PSX$. For example, the PSX for item p_1 and the requirements $\{r_1, r_2, r_3, r_4\}$ is the ordered set $\{1, 0, 0, 1\}$, which corresponds to the first column of Table 4.5 (only the requirements r_1 and r_4 are satisfied by item p_1).

The performance of MINRELAX is $(n * (n + 1))/2$ subset checks in the worst case, which can be conducted efficiently with in-memory bitset

Table 4.5. *Intermediate representation: item-specific relaxations PSX for* $p_i \in P$.

	p_1	p_2	p_3	p_4	p_5	p_6	p_7	p_8
$r_1 : price \leq 150$	1	0	0	0	0	0	0	0
$r_2 : opt\text{-}zoom = 5\times$	0	1	0	0	0	0	0	0
$r_3 : sound = yes$	0	1	1	0	1	1	1	1
$r_4 : waterproof = yes$	1	0	0	1	0	0	0	1

operations (Jannach 2006a). Table 4.5 depicts the relationship between our example requirements and each $p_i \in P$. Because of the explicit enumeration of all possible items in P, we can determine for each requirement/item combination whether the requirement is supported by the corresponding item. Each column of Table 4.5 represents a diagnosis; our goal is to identify the diagnoses that are minimal.

4.3.3 Proposing repairs for unsatisfiable requirements

After having identified the set of possible diagnoses ($\triangle$), we must propose repair actions for each of those diagnoses – in other words, we must identify possible adaptations for the existing set of requirements such that the user is able to find a solution (Felfernig et al. 2009). Alternative repair actions can be derived by querying the product table P with $\pi_{[attributes(d)]}\sigma_{[REQ-d]}(P)$. This query identifies all possible repair alternatives for a single diagnosis $d \in \triangle$ where $\pi_{[attributes(d)]}$ is a projection and $\sigma_{[REQ-d]}(P)$ is a selection of -tuples from P that satisfy the criteria in *REQ–d*. Executing this query for each of the identified diagnoses produces a complete set of possible repair alternatives. For reasons of simplicity we restrict our example to three different repair alternatives, each belonging to exactly one diagnosis. Table 4.6 depicts the complete set of repair alternatives $REP = \{rep_1, rep_2, rep_3\}$ for our working example, where

- $\pi_{[attributes(d1)]}\sigma_{[REQ-d1]}(P) = \pi_{[price,opt\text{-}zoom]}\sigma_{[r3:sound=yes,r4:waterproof=yes]}(P) = \{price=278, opt\text{-}zoom=10\times\}$
- $\pi_{[attributes(d2)]}\sigma_{[REQ-d2]}(P) = \pi_{[price,waterproof]}\sigma_{[r2:opt\text{-}zoom=5x,r3:sound=yes]}(P) = \{price=182, waterproof=no\}$
- $\pi_{[attributes(d3)]}\sigma_{[REQ-d3]}(P) = \pi_{[opt\text{-}zoom,sound]}\sigma_{[r1:price<=150,r4:waterproof=yes]}(P) = \{opt\text{-}zoom=4\times, sound=no\}$

Table 4.6. *Repair alternatives for requirements in REQ.*

repair	price	opt-zoom	sound	waterproof
rep_1	278	10×	✓	✓
rep_2	182	✓	✓	no
rep_3	✓	4×	no	✓

4.3.4 Ranking the items/utility-based recommendation

It is important to rank recommended items according to their utility for the customer. Because of primacy effects that induce customers to preferably look at and select items at the beginning of a list, such rankings can significantly increase the trust in the recommender application as well as the willingness to buy (Chen and Pu 2005, Felfernig et al. 2007)

In knowledge-based conversational recommenders, the ranking of items can be based on the multi-attribute utility theory (MAUT), which evaluates each item with regard to its utility for the customer. Each item is evaluated according to a predefined set of dimensions that provide an aggregated view on the basic item properties. For example, *quality* and *economy* are dimensions in the domain of digital cameras; *availability*, *risk*, and *profit* are such dimensions in the financial services domain. Table 4.7 exemplifies the definition of scoring rules that define the relationship between item properties and dimensions. For example, a digital camera with a *price* lower than or equal to 250 is evaluated, with Q score of 5 regarding the dimension *quality* and 10 regarding the dimension *economy*.

We can determine the utility of each item p in P for a specific customer (Table 4.8). The customer-specific item utility is calculated on the basis of Formula 4.6, in which the index j iterates over the number of predefined dimensions (in our example, #(dimensions)=2: *quality* and *economy*), *interest*(j) denotes a user's interest in dimension j, and *contribution*(p, j) denotes the contribution of item p to the interest dimension j. The value for *contribution*(p, j) can be calculated by the scoring rules defined in Table 4.7 – for example, the contribution of item p_1 to the dimension *quality* is $5 + 4 + 6 + 6 + 3 + 7 + 10 = 41$, whereas its contribution to the dimension *economy* is $10 + 10 + 9 + 10 + 10 + 10 + 6 = 65$.

To determine the overall utility of item p_1 for a specific customer, we must take into account the customer-specific interest in each of the given dimensions – *interest*(j). For our example we assume the customer preferences depicted in Table 4.9. Following Formula 4.6, for customer cu_1, the

Table 4.7. *Example scoring rules regarding the dimensions* quality *and* economy.

	value	quality	economy
price	≤250	5	10
	>250	10	5
mpix	≤8	4	10
	>8	10	6
opt-zoom	≤9	6	9
	>9	10	6
LCD-size	≤2.7	6	10
	>2.7	9	5
movies	yes	10	7
	no	3	10
sound	yes	10	8
	no	7	10
waterproof	yes	10	6
	no	8	10

utility of item p_2 is 49*0.8 + 64*0.2 = 52.0 and the overall utility of item p_8 would be 69*0.8 + 43*0.2 = 63.8. For customer cu_2, item p_2 has the utility 49*0.4 + 64*0.6 = 58.0 and item p_8 has the utility 69*0.4 + 43*0.6 = 53.4. Consequently, item p_8 has a higher utility (and the highest utility) for cu_1, whereas item p_2 has a higher utility (and the highest utility) for cu_2. Formula 4.6 follows the principle of the similarity metrics introduced in Section 4.2: *interest(j)* corresponds to the weighting of requirement r_j and *contribution(p, j)* corresponds to the local similarity function *sim(p, r)* (McSherry 2003a).

Table 4.8. *Item utilities for customer* cu_1 *and customer* cu_2.

	quality	economy	cu_1	cu_2
p_1	$\sum(5, 4, 6, 6, 3, 7, 10) = 41$	$\sum(10, 10, 9, 10, 10, 10, 6) = 65$	45.8 [8]	55.4 [6]
p_2	$\sum(5, 4, 6, 6, 10, 10, 8) = 49$	$\sum(10, 10, 9, 10, 7, 8, 10) = 64$	52.0 [7]	58.0 [1]
p_3	$\sum(5, 4, 10, 6, 10, 10, 8) = 53$	$\sum(10, 10, 6, 10, 7, 8, 10) = 61$	54.6 [5]	57.8 [2]
p_4	$\sum(5, 10, 10, 6, 10, 7, 10) = 58$	$\sum(10, 6, 6, 10, 7, 10, 6) = 55$	57.4 [4]	56.2 [4]
p_5	$\sum(5, 4, 6, 10, 10, 10, 8) = 53$	$\sum(10, 10, 9, 6, 7, 8, 10) = 60$	54.4 [6]	57.2 [3]
p_6	$\sum(5, 10, 6, 9, 10, 10, 8) = 58$	$\sum(10, 6, 9, 5, 7, 8, 10) = 55$	57.4 [3]	56.2 [5]
p_7	$\sum(10, 10, 6, 9, 10, 10, 8) = 63$	$\sum(5, 6, 9, 5, 7, 8, 10) = 50$	60.4 [2]	55.2 [7]
p_8	$\sum(10, 10, 10, 9, 10, 10, 10) = 69$	$\sum(5, 6, 6, 5, 7, 8, 6) = 43$	63.8 [1]	53.4 [8]

Table 4.9. *Customer-specific preferences*
represent the values for interest(j) *in Formula 4.6.*

customer (user)	quality	economy
cu_1	80%	20%
cu_2	40%	60%

The concepts discussed here support the calculation of personalized rankings for a given set of items. However, such utility-based approaches can be applied in other contexts as well – for example, the calculation of utilities of specific repair alternatives (personalized repairs; Felfernig et al. 2006) or the calculation of utilities of explanations (Felfernig et al. 2008b).

$$utility(p) = \sum_{j=1}^{\#(dimensions)} interest(j) * contribution(p, j) \qquad (4.6)$$

There exist different approaches to determining a customer's degree of interest in a certain dimension (*interest(j)* in Formula 4.6). Such preferences can be explicitly defined by the user (user-defined preferences). Preferences can also be predefined in the form of scoring rules (utility-based preferences) derived by analyzing logs of previous user interactions (e.g., conjoint analysis). These basic approaches will be exemplified in the following paragraphs.

User-defined preferences. The first and most straightforward approach is to directly ask the customer for his or her preferences within the scope of a recommendation session. Clearly, this approach has the main disadvantage that the overall interaction effort for the user is nearly doubled, as for many of the customer properties the corresponding importance values must be specified. A second problem with this basic approach is that the recommender user interface is obtrusive in the sense that customers are interrupted in their preference construction process and are forced to explicitly specify their preferences beforehand.

Utility-based preferences. A second possible approach to determining customer preferences is to apply the scoring rules of Table 4.7. If we assume, for example, that a customer has specified the requirements $REQ = \{r_1 : price <= 200, r_2 : mpix = 8.0, r_3 : opt\text{-}zoom = 10\times, r_4 : lcd\text{-}size <= 2.7\}$, we can directly derive the instantiations of the corresponding dimensions by applying the scoring rules in Table 4.7. In our case, the dimension *quality*

Table 4.10. *Ranking of price/mpix stimuli: $price_1[100–159]$, $price_2[160–199]$, $price_3[200–300]$, $mpix_1[5.0–8.0]$, and $mpix_2[8.1–11.0]$.*

	$price_1$	$price_2$	$price_3$	$avg(mpix_x)$
$mpix_1$	4	5	6	5
$mpix_2$	2	1	3	2
$avg(price_x)$	3	3	4.5	3.5

would be $5 + 4 + 10 + 6 = 25$ and the dimension *economy* would be $10 + 10 + 6 + 10 = 36$. This would result in a relative importance for *quality* with a value of $\frac{25}{25+36} = 0.41$ and a relative importance for *economy* with the value $\frac{36}{25+36} = 0.59$.

Conjoint analysis. The following simple example should characterize the basic principle of conjoint analysis (Belanger 2005). In this example, a user (test person) is confronted with different *price/mpix* value combinations (stimuli). The user's task is to rank those combinations; for example, the combination $mpix_2[8.1–11.0]$ / $price_2[160–199]$ gets the highest ranking (see Tables 4.10 and 4.11). The average values for the columns inform us about the average ranking for the corresponding *price* interval ($avg(price_x)$). The average values for the rows inform us about the average rankings for the corresponding *mpix* interval $avg(mpix_x)$. The average value over all rankings is $avg(ranking) = 3.5$.

The information we can extract from Tables 4.10 and 4.11 is the deviation from the average ranking for specific property values – for

Table 4.11. *Effects of customer property changes on overall utility: changes in* mpix *have a higher impact on the overall utility than changes in price.*

	$avg(ranking) − avg(mpix_x)$
$avg(ranking) − avg(mpix_1)$	−1.5
$avg(ranking) − avg(mpix_2)$	1.5
$avg(ranking) − avg(price_1)$	0.5
$avg(ranking) − avg(price_2)$	0.5
$avg(ranking) − avg(price_3)$	−1.0

example, $avg(price_x)$ from $avg(ranking)$: $avg(ranking) - avg(price_1) = 0.5$, $avg(ranking) - avg(price_2) = 0.5, avg(ranking) - avg(price_3) = -1$. Furthermore, $avg(ranking) - avg(mpix_1) = -1.5$ and $avg(ranking) - avg(mpix_2) = 1.5$. The $avg(price_x)$ span is 1.5 ($-1 \ldots 0.5$) whereas the span for $avg(mpix_x)$ is 3.0 ($-1.5 \ldots 1.5$). Following the ideas of conjoint analysis (Belanger 2005) we are able to conclude that *price* changes have a lower effect on the overall utility (for this customer) than changes in terms of *megapixels*. This result is consistent with the idea behind the ranking in our example (Tables 4.10 and 4.11), as the highest ranking was given to the combination $price_2/mpix_2$, where a higher price was accepted in order to ensure high quality of technical features (here: *mpix*). Consequently, we can assign a higher importance to the technical property *mpix* compared with the property *price*.

In this section we provided an overview of concepts that typically support users in the interaction with a constraint-based recommender application. *Diagnosis and repair* concepts support users in situations in which no solution could be found. *Defaults* provide support in the requirements specification process by proposing reasonable alternatives – a negative connotation is that defaults can be abused to manipulate users. *Utility-based ranking* mechanisms support the ordering of information units such as items on a result page, repair alternatives provided by a diagnosis and repair component, and the ranking of explanations for recommended items. These concepts form a toolset useful for the implementation of constraint-based recommender applications. A commercial application built on the basis of those concepts is presented in Section 4.5.

4.4 Interacting with case-based recommenders

Similar to constraint-based recommenders, earlier versions of case-based recommenders followed a pure *query-based approach*, in which users had to specify (and often respecify) their requirements until a target item (an item that fits the user's wishes and needs) has been identified (Burke 2002a). Especially for nonexperts in the product domain, this type of requirement elicitation process can lead to tedious recommendation sessions, as the interdependent properties of items require a substantial domain knowledge to perform well (Burke 2002a). This drawback of pure query-based approaches motivated the development of *browsing-based approaches* to item retrieval, in which users – maybe not knowing what they are seeking – are navigating in the item space with the goal to find useful alternatives. *Critiquing* is an effective way to

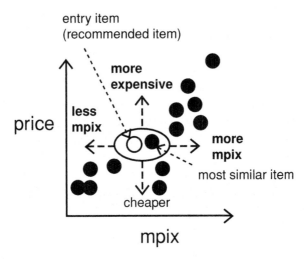

Figure 4.3. Critique-based navigation: items recommended to the user can be critiqued regarding different item properties (e.g., *price* or *mpix*).

support such navigations and, in the meantime, it is one of the key concepts of case-based recommendation; this concept will be discussed in detail in the following subsections.

4.4.1 Critiquing

The idea of critiquing (Burke 2000, Burke et al. 1997) is that users specify their change requests in the form of goals that are not satisfied by the item currently under consideration (*entry item* or *recommended item*). If, for example, the *price* of the currently displayed digital camera is too high, a critique *cheaper* can be activated; if the user wants to have a camera with a higher resolution (*mpix*), a corresponding critique *more mpix* can be selected (see Figure 4.3).

Further examples for critiques are "the hotel location should be nearer to the sea" or "the apartment should be more modern-looking". Thus, critiques can be specified on the level of technical properties as well as on the level of abstract dimensions.

State-of-the-art case-based recommenders are integrating query-based with browsing-based item retrieval (Burke 2002a). On one hand, critiquing supports an effective navigation in the item space; on the other hand, similarity-based case retrieval supports the identification of the *most similar items* – that is, items similar to those currently under consideration. Critiquing-based recommender

---

Algorithm 4.3 SIMPLECRITIQUING(q, CI)

Input: Initial user query q; Candidate items CI
procedure SIMPLECRITIQUING(q, CI)
  repeat
    $r \leftarrow$ ITEMRECOMMEND(q, CI);
    $q \leftarrow$ USERREVIEW(r, CI);
  until empty(q)
end procedure

procedure ITEMRECOMMEND(q, CI)
  $CI \leftarrow \{ci \in CI$: satisfies($ci$, q)$\}$;
  $r \leftarrow$ mostsimilar(CI, q);
  return r;
end procedure

procedure USERREVIEW(r, CI)
  $q \leftarrow$ critique(r);
  $CI \leftarrow CI - r$;
  return q;
end procedure

---

systems allow users to easily articulate preferences without being forced to specify concrete values for item properties (see the previous example). The goal of critiquing is to achieve time savings in the item selection process and, at the same time, achieve at least the same recommendation quality as standard query-based approaches. The major steps of a critiquing-based recommender application are the following (see Algorithm 4.3, SIMPLECRITIQUING).

Item recommendation. The inputs for the algorithm SIMPLECRITIQUING[4] are an initial *user query q*, which specifies an initial set of requirements, and a set of *candidate items CI* that initially consists of all the available items (the product assortment). The algorithm first activates the procedure ITEMRECOMMEND, which is responsible for selecting an item r to be presented to the user. We denote the item that is displayed in the first critiquing cycle as *entry item* and all other items displayed thereafter as *recommended items*. In the first critiquing

---

[4] The notation used in the algorithm is geared to Reilly et al. (2005a).

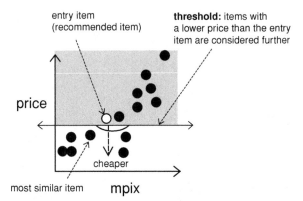

Figure 4.4. Critique-based navigation: remaining candidate items (items with bright background) after a critique on the entry item property *price*.

cycle, the retrieval of such items is based on a user query q that represents a set of initial requirements. Entry items are typically determined by calculating the similarity between the requirements and the candidate items. After the first critiquing cycle has been completed, recommended items are determined by the procedure ITEMRECOMMEND on the basis of the similarity between the currently recommended item and those items that fulfill the criteria of the critique specified by the user.

Item reviewing. The user reviews the recommended (entry) item and either accepts the recommendation or selects another critique, which triggers a new critiquing cycle (procedure USERREVIEW). If a critique has been triggered, only the items (the *candidate items*) that fulfill the criteria defined in the critique are further taken into account – this reduction of CI is done in procedure ITEMRECOMMEND. For example, if a user activates the critique *cheaper*, and the *price* of the recommended (entry) camera is 300, the recommender excludes cameras with a *price* greater than or equal to 300 in the following critiquing cycle.

4.4.2 Compound critiquing

In our examples so far we primarily considered the concept of *unit critiques*; such critiques allow the definition of change requests that are related to a single item property. Unit critiques have a limited capability to effectively narrow down the search space. For example, the unit critique on *price* in Figure 4.4 eliminates only about half the items.

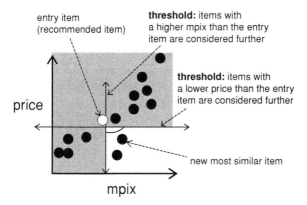

entry item
(recommended item)

threshold: items with
a higher mpix than the entry
item are considered further

threshold: items with
a lower price than the entry
item are considered further

price

new most similar item

mpix

Figure 4.5. Critique-based navigation: remaining candidate items (items with bright background) after a compound critique on *price* and *mpix*.

Allowing the specification of critiques that operate over multiple properties can significantly improve the efficiency of recommendation dialogs, for example, in terms of a reduced number of critiquing cycles. Such critiques are denoted as *compound critiques*. The effect of compound critiques on the number of eliminated items (items not fulfilling the criteria of the critique) is shown in Figure 4.5. The compound critique *cheaper and more mpix* defines additional goals on two properties that should be fulfilled by the next proposed recommendation.

An important advantage of compound critiques is that they allow a faster progression through the item space. However, compound critiques still have disadvantages as long as they are formulated *statically*, as all critique alternatives are available for every item displayed. For example, in the context of a high-end computer with the *fastest CPU available* on the market and the *maximum available storage capacity*, a corresponding critique *faster CPU and more storage capacity* (or *more efficient*) would be still proposed by a static compound critiquing approach. In the following subsection we will present the *dynamic critiquing* approach that helps to solve this problem.

4.4.3 Dynamic critiquing

Dynamic critiquing exploits *patterns*, which are generic descriptions of differences between the recommended (entry) item and the candidate items – these patterns are used for the derivation of compound critiques. Critiques are denoted as dynamic because they are derived on the fly in each critiquing cycle. Dynamic critiques (Reilly et al. 2007b) are calculated using the concept

---

Algorithm 4.4 DYNAMICCRITIQUING(q, CI)

Input: Initial user query q; Candidate items CI;
       number of compound critiques per cycle k;
       minimum support for identified association rules σ_{min}

procedure DYNAMICCRITIQUING(q, CI, k, σ_{min})
  repeat
    $r \leftarrow$ ITEMRECOMMEND(q, CI);
    $CC \leftarrow$ COMPOUNDCRITIQUES(r, CI, k, σ_{min});
    $q \leftarrow$ USERREVIEW(r, CI, CC);
  until empty(q)
end procedure

procedure ITEMRECOMMEND(q, CI)
  $CI \leftarrow \{ci \in CI: \text{satisfies}(ci, q)\}$;
  $r \leftarrow \text{mostsimilar}(CI, q)$;
  return r;
end procedure

procedure USERREVIEW(r, CI, CC)
  $q \leftarrow \text{critique}(r, CC)$;
  $CI \leftarrow CI - r$;
  return q;
end procedure

procedure COMPOUNDCRITIQUES(r, CI, k, σ_{min})
  $CP \leftarrow$ CRITIQUEPATTERNS(r, CI);
  $CC \leftarrow$ APRIORI(CP, σ_{min});
  $SC \leftarrow$ SELECTCRITIQUES(CC, k);
  return SC;
end procedure

---

of association rule mining (Agrawal and Srikant 1994). Such a rule can be, for example, "42.9% of the remaining digital cameras have a higher zoom and a lower price". The critique that corresponds to this property combination is "more zoom and lower price". A dynamic critiquing cycle consists of the following basic steps (see Algorithm 4.4, DYNAMICCRITIQUING[5]).

---

[5] The algorithm has been developed by Reilly et al. (2005a).

The inputs for the algorithm are an initial user query q, which specifies the initial set of requirements, a set of candidate items CI that initially consists of all the available items, k as the maximum number of compound critiques to be shown to the user in one critiquing cycle, and σ_{min} as the minimum support value for calculated association rules.

Item recommendation. Similar to the SIMPLECRITIQUING algorithm discussed in Section 4.4.1, the DYNAMICCRITIQUING algorithm first activates the procedure ITEMRECOMMEND, which is responsible for returning one *recommended item r* (respectively, *entry item* in the first critiquing cycle). On the basis of this item, the algorithm starts the calculation of compound critiques $cc_i \in CC$ by activating the procedure COMPOUNDCRITIQUES, which itself activates the procedures CRITIQUEPATTERNS (identification of critique patterns), APRIORI (mining compound critiques from critique patterns), and SELECTCRITIQUES (ranking compound critiques). These functionalities will be discussed and exemplified in the following paragraphs. The identified compound critiques in CC are then shown to the user in USERREVIEW. If the user selects a critique – which could be a unit critique on a specific item property as well as a compound critique – this forms the criterion of the new user query q. If the resulting query q is empty, the critiquing cycle can be stopped.

Identification of critique patterns. Critique patterns are a generic representation of the differences between the currently recommended item (entry item) and the candidate items. Table 4.12 depicts a simple example for the derivation of critique patterns, where item ei_8 is assumed to be the entry item and the items $\{ci_1, \ldots, ci_7\}$ are the candidate items. On the basis of this example, critique patterns can be easily generated by comparing the properties of item ei_8 with the properties of $\{ci_1, \ldots, ci_7\}$. For example, compared with item ei_8, item ci_1 is cheaper, has less *mpix*, a lower *opt-zoom*, a smaller *lcd-size*, and does not have a *movie* functionality. The corresponding critique pattern for item ci_1 is $(<, <, <, <, \neq)$. A complete set of critiquing patterns in our example setting is shown in Table 4.12. These patterns are the result of calculating the type of difference for each combination of *recommended* (*entry*) and *candidate item*. In the algorithm DYNAMICCRITIQUING, critique patterns are determined on the basis of the procedure CRITIQUINGPATTERNS.

Mining compound critiques from critique patterns. The next step is to identify compound critiques that frequently co-occur in the set of critique patterns. This approach is based on the assumption that critiques correspond to feature combinations of interest to the user – that is, a user would like to adapt the

Table 4.12. *Critique patterns (C P) are generated by analyzing the differences between the* recommended item *(the* entry item, *E I) and the* candidate items *(C I). In this example, item ei_8 is assumed to be the* entry item, $\{ci_1, \ldots, ci_7\}$ *are assumed to be the candidate items, and* $\{cp_1, \ldots, cp_7\}$ *are the critique patterns.*

	id	price	mpix	opt-zoom	LCD-size	movies
entry item (EI)	ei_8	278	9.1	9×	3.0	yes
	ci_1	148	8.0	4×	2.5	no
	ci_2	182	8.0	5×	2.7	yes
	ci_3	189	8.0	10×	2.5	yes
candidate items (CI)	ci_4	196	10.0	12×	2.7	yes
	ci_5	151	7.1	3×	3.0	yes
	ci_6	199	9.0	3×	3.0	yes
	ci_7	259	10.0	10×	3.0	yes
	cp_1	<	<	<	<	≠
	cp_2	<	<	<	<	=
	cp_3	<	<	>	<	=
critique patterns (CP)	cp_4	<	>	>	<	=
	cp_5	<	<	<	=	=
	cp_6	<	<	<	=	=
	cp_7	<	>	>	=	=

requirements in exactly the proposed combination (Reilly et al. 2007b). For critique calculation, Reilly et al. (2007b) propose applying the APRIORI algorithm (Agrawal and Srikant 1994). The output of this algorithm is a set of *association rules* $p \Rightarrow q$, which describe relationships between elements in the set of critique patterns. An example is $>_{zoom} \Rightarrow <_{price}$, which can be derived from the critique patterns of Table 4.12. This rule denotes the fact that given $>_{zoom}$ as part of a critique pattern, $<_{price}$ is contained in the same critique pattern. Examples for association rules and the corresponding compound critiques that can be derived from the critique patterns in Table 4.12 are depicted in Table 4.13.

Each association rule is additionally characterized by *support* and *confidence* values. Support (SUPP) denotes the number of critique patterns that include all the elements of the antecedent and consequent of the association rule (expressed in terms of the percentage of the number of critique patterns). For example, the support of association rule ar_1 in Table 4.13 is 28.6 percent; of the seven critique patterns, exactly two include the antecedent and consequent part of association rule ar_1. Confidence (CONF) denotes the ratio between critique

Table 4.13. *Example association rules (AR) and the compound
critiques (CC) derived from C P in Table 4.12.*

association rules (AR)	compound critiques (CC)	SUPP	CONF
ar_1: $>_{mpix}$ $\Rightarrow$ $>_{zoom}$	cc_1: $>_{mpix(9.1)}$, $>_{zoom(9x)}$	28.6	100.0
ar_2: $>_{zoom}$ $\Rightarrow$ $<_{price}$	cc_2: $>_{zoom(9x)}$, $<_{price(278)}$	42.9	100.0
ar_3: $=_{movies}$ $\Rightarrow$ $<_{price}$	cc_3: $=_{movie(yes)}$, $<_{price(278)}$	85.7	100.0

patterns containing all the elements of the antecedent and consequent of the
association rule and those containing only the antecedent part. For all the
association rules in Table 4.13 the confidence level is 100.0 percent – that is,
if the antecedent part of the association rule is confirmed by the pattern, the
consequent part is confirmed as well. In the algorithm DYNAMICCRITIQUING,
compound critiques are determined on the basis of the procedure APRIORI
that represents a basic implementation of the APRIORI algorithm (Agrawal and
Srikant 1994).

Ranking of compound critiques. The number of compound critiques can
become very large, which makes it important to filter out the most relevant
critiques for the user in each critiquing cycle. Critiques with *low support* have
the advantage of significantly reducing the set of candidate items, but at the
same time they decrease the probability of identifying the target item. Critiques
with *high support* can significantly increase the probability of finding the target
item. However, these critiques eliminate a low number of candidate cases, which
leads to a larger number of critiquing cycles in recommendation sessions. Many
existing recommendation approaches rank compound critiques according to
the support values of association rules, because the lower the support of the
corresponding association rules, the more candidate items can be eliminated. In
our working example, such a ranking of compound critiques $\{cc_1, cc_2, cc_3\}$ is
cc_1, cc_2, and cc_3. Alternative approaches to the ranking of compound critiques
are discussed, for example, by Reilly et al. (2004), where low support, high
support, and random critique selection have been compared. This study reports
a lower number of interaction cycles in the case that compound critiques are
sorted ascending based on their support value. The issue of critique selection is
in need of additional empirical studies focusing on the optimal balance between
a low number of interaction cycles and the number of excluded candidate items.
In the algorithm DYNAMICCRITIQUING, compound critiques are selected on the
basis of the procedure SELECTCRITIQUES.

Item reviewing. At this stage of a recommendation cycle all the relevant information for deciding about the next action is available for the user: the recommended (entry) item and the corresponding set of compound critiques. The user reviews the recommended item and either accepts the recommendation or selects a critique (unit or compound), in which case a new critiquing cycle is started. In the algorithm DYNAMICCRITIQUING, item reviews are conducted by the user in USERREVIEW.

4.4.4 Advanced item recommendation

After a critique has been selected by the user, the next item must be proposed (*recommended item* for the next critiquing cycle). An approach to doing this – besides the application of simple similarity measures (see Section 4.2) – is described by Reilly et al. (2007b), where a *compatibility score* is introduced that represents the percentage of compound critiques $cc_i \in CC_U$ that already have been selected by the user and are consistent with the candidate item ci. This compatibility-based approach to item selection is implemented in Formula 4.7.

$$compatibility(ci, CC_U) = \frac{|\{cc_i \in CC_U : satisfies(cc_i, ci)\}|}{|CC_U|} \quad (4.7)$$

CC_U represents a set of (compound) critiques already selected by the user; $satisfies(cc_i, ci) = 1$ indicates that critique cc_i is *consistent* with candidate item ci and $satisfies(cc_i, ci) = 0$ indicates that critique cc_i is *inconsistent* with ci. On the basis of this compatibility measure, Reilly et al. (2007b) introduce a new quality measure for a certain candidate item ci (see Formula 4.8). This formula assigns the highest values to candidate items ci that are as compatible as possible with the already selected compound critiques and also as similar as possible to the currently recommended item ri.

$$quality(ci, ri, CC_U) = compatibility(ci, CC_U) * similarity(ci, ri) \quad (4.8)$$

Table 4.14 exemplifies the application of Formula 4.8. Let us assume that $CC_U = \{cc_2: >_{zoom(9x)}, <_{price(278)}\}$ is the set of critiques that have been selected by the user in USERREVIEW – in other words, only one critique has been selected up to now. Furthermore, we assume that $ri = ei_8$. Then the resulting set of new candidate items $CI = \{ci_3, ci_4, ci_7\}$. In this case, item ci_4 has by far the highest quality value and thus would be the item r returned by ITEMRECOMMEND $- quality(ci_4, ei_8, cc_2:>_{zoom(9x)}, <_{price(278)} = 0.61)$. This item selection approach helps take into account already selected critiques – that is, preferences already specified are not ignored in future critiquing cycles. Further

Table 4.14. *Dynamic critiquing:* quality *of candidate items* $CI = \{ci_3, ci_4,$ $ci_7\}$ *with regard to* $ri = ei_8$ *and* $CC_U = \{cc_2 :>_{zoom(9x)}, <_{price(278)}\}$.

candidate item ci	compatibility(ci, CC_U)	similarity(ci, ri)	*quality*
ci_3	1.0	0.40	0.40
ci_4	1.0	0.61	0.61
ci_7	1.0	0.41	0.41

related item recommendation approaches are discussed by Reilly et al. (2004), who focus especially on the issue of consistency in histories of already selected critiques. For example, if the user has initially specified the upper bound for the price with $<_{price(150)}$ and later specifies $>_{price(300)}$, one of those critiques must be removed from the critique history to still have available candidate items for the next critiquing cycle.

4.4.5 Critique diversity

Compound critiques are a powerful mechanism for effective item search in large assortments – especially for users who are nonexperts in the corresponding product domain. All the aforementioned critiquing approaches perform well as long as there are no "hot spots," in which many similar items are concentrated in one area of the item space. In such a situation a navigation to other areas of the item space can be very slow. Figure 4.6 depicts such a situation, in which a compound critique on *price* and *mpix* leads to recommended items that are quite similar to the current one.

An approach to avoid such situations is presented by McCarthy et al. (2005), who introduce a quality function (see Formula 4.9) that prefers compound critiques (cc) with low support values (many items can be eliminated) that are at the same time diversified from critiques CC_{Curr} already selected for presentation in the current critiquing cycle.

$$quality(cc, CC_{Curr}) = support(cc) * overlap(cc, CC_{Curr}) \qquad (4.9)$$

The support for a compound critique cc corresponds to the support of the corresponding association rule – for example, support(ar_3) = support(cc_3) = 85.7 (see Table 4.13). The overlap between the currently investigated compound critique cc and CC_{Curr} can be calculated on the basis of Formula 4.10. This formula determines the overlap between items supported by cc and items supported by critiques of CC_{Curr} – that is, *items*(CC_{Curr}) denotes the items

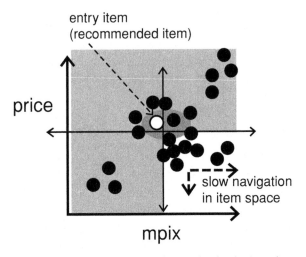

Figure 4.6. Critique-based navigation: slow navigation in dense item spaces.

accepted by the critiques in CC_{Curr}.

$$overlap(c, CP) = \frac{|items(\{cc\}) \cap items(CC_{Curr})|}{|items(\{cc\}) \cup items(CC_{Curr})|} \qquad (4.10)$$

The lower the value of the function *quality* (preferred are low support and low overlap to already presented critiques), the higher the probability for a certain critique to be presented in the next critiquing cycle. Table 4.15 depicts the result of applying Formula 4.9 to $CC_{Curr} = \{cc_2: >_{zoom(9\times)}, <_{price(278)}\}$ and two candidate association rules $\{ar_2, ar_3\}$ (see Table 4.13). Conforming to this formula, the quality of compound critique cc_1 is higher than the quality of cc_3.

Table 4.15. *Dynamic critiquing: quality of compound critiques* $CC = \{cc_1, cc_3\}$ *derived from association rules* $AR = \{ar_1, ar_3\}$ *assuming that* $CC_{Curr} = \{cc_2: >_{zoom(9\times)}, <_{price(278)}\}$.

compound critiques (CC)	support(cc)	overlap(cc, CP)	quality
cc_1: $>_{mpix(9.1)}$, $>_{zoom(9\times)}$	28.6	66.7	0.19
cc_3: $=_{movies(yes)}$, $<_{price(278)}$	85.7	50.0	0.43

4.5 Example applications

In the final part of this chapter we take a more detailed look at two commercial recommender applications: a constraint-based recommender application developed for a Hungarian financial service provider and a case-based recommendation environment developed for recommending restaurants located in Chicago.

4.5.1 The VITA constraint-based recommender

We now move from our working example (digital camera recommender) to the domain of financial services. Concretely, we take a detailed look at the VITA financial services recommender application, which was built for the Fundamenta loan association in Hungary (Felfernig et al. 2007b). VITA supports sales representatives in sales dialogs with customers. It has been developed on the basis of the CWAdvisor recommender environment presented by Felfernig et al. (2006)

Scenario. Sales representatives in the financial services domain are challenged by the increased complexity of service solutions. In many cases, representatives do not know which services should be recommended in which contexts, and how those services should be explained. In this context, the major goal of financial service providers is to improve the overall productivity of sales representatives (e.g., in terms of the number of sales dialogs within a certain time period or number of products sold within a certain time period) and to increase the advisory quality in sales dialogs. Achieving these goals is strongly correlated to both an increase in the overall productivity and a customer's interest in long-term business connections with the financial service provider.

Software developers must deal with highly complex and frequently changing recommendation knowledge bases. Knowledge-based recommender technologies can improve this situation because they allow effective knowledge base development and maintenance processes.

The Fundamenta loan association in Hungary decided to establish knowledge-based recommender technologies to improve the performance of sales representatives and to reduce the overall costs of developing and maintaining related software components. In line with this decision, Fundamenta defined the following major goals:

- *Improved sales performance*: within the same time period, sales representatives should be able to increase the number of products sold.

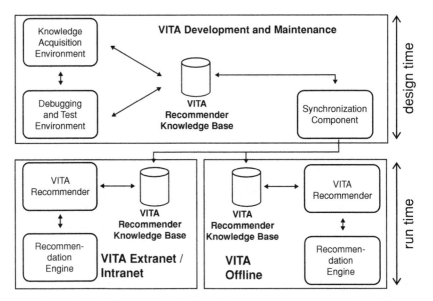

Figure 4.7. Architecture of the VITA sales support environment (Felfernig et al. 2007).

- *Effective software development and maintenance*: the new technologies should ease the development of sales knowledge bases.

Application description. The resulting VITA sales support environment (Figure 4.7) supports two basic (and similar) advisory scenarios. On one hand, VITA is a web server application used by Fundamenta sales representatives and external sales agents for the preparation and conducting of sales dialogs. On the other hand, the same functionality is provided for sales representatives using their own laptops. In this case, new versions of sales dialogs and knowledge bases are automatically installed when the sales representative is connected with the Fundamenta intranet.

For both scenarios, a knowledge acquisition environment supports the automated testing and debugging of knowledge bases. Such a knowledge base consists of the following elements (see Section 4.3):

- *Customer properties*: each customer must articulate his or her requirements, which are the elementary precondition for reasonable recommendations. Examples for customer properties in the financial services domain are *age*, *intended run time of the service*, *existing loans in the portfolio*, and the like.

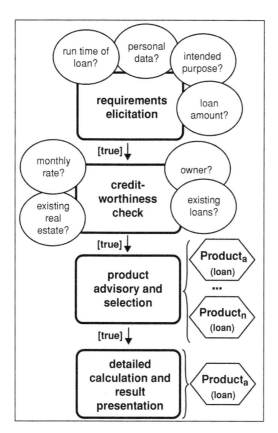

Figure 4.8. Example of advisory process definition (loan advisory) (Felfernig et al. 2007b).

- *Product properties and instances*: each product is described in terms of a set of predefined properties such as *recommended run time, predicted performance, expected risk,* and so on.
- *Constraints*: restrictions that define which products should be recommended in which context. A simple example of such a constraint is *customers with a low preparedness to take risks should receive recommendations that do not include high-risk products.*
- *Advisory process definition*: explicit definitions of sales dialogs are represented in the form of state charts (Felfernig and Shchekotykhin 2006) that basically define the context in which questions should be posed to the user (an example of a simple advisory process definition is depicted in Figure 4.8).

In Fundamenta applications, recommendation calculations are based on the execution of conjunctive queries (for an example, see Section 4.3). Conjunctive queries are generated directly from requirements elicited within the scope of a recommendation session. The recommendation process follows the advisory process definition.

A loan recommendation process is structured in different phases (*requirements elicitation, creditworthiness check, product advisory/selection,* and *detailed calculation/result presentation*). In the first phase, basic information regarding the customer (*personal data*) and the major purpose of and requirements regarding the loan (e.g., *loan amount, run time of loan*) are elicited. The next task in the recommendation process is to check the customer's creditworthiness on the basis of detailed information regarding the customer's current financial situation and available financial securities. At this time, the application checks whether a solution can be found for the current requirements. If no such solution is available (e.g., too high an amount of requested money for the available financial securities), the application tries to determine alternatives that restore the consistency between the requirements and the available set of products. After the successful completion of the phase *creditworthiness check*, the recommender application proposes different available loan alternatives (redemption alternatives are also taken into account). After selecting one of those alternatives, the recommendation process continues with a detailed calculation of specific product properties, such as the monthly redemption rates of the currently selected alternative. A screen shot of the VITA environment is depicted in Figure 4.9.

Knowledge acquisition. In many commercial recommender projects, generalists who possess deep domain knowledge as well as technical knowledge about recommender technologies are lacking. On one hand, knowledge engineers know how to create recommender applications; on the other hand, domain experts know the details of the product domain but do not have detailed technical knowledge of recommenders. This results in a situation in which technical experts have the responsibility for application development and domain experts are solely responsible for providing the relevant product, marketing, and sales knowledge. This type of process is error-prone and creates unsatisfactory results for all project members.

Consequently, the overall goal is to further improve knowledge-based recommender technologies by providing tools that allow a shift of knowledge base development competencies from knowledge engineers to domain experts. The knowledge acquisition environment (CWAdvisor [Felfernig et al. (2006)]) that is used in the context of VITA supports the development of recommender

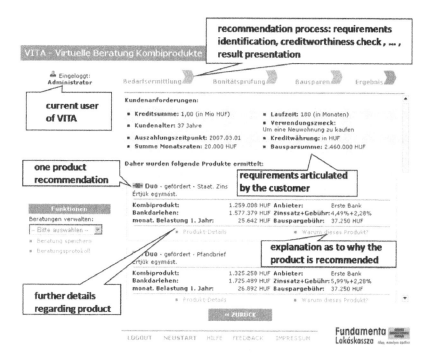

Figure 4.9. Screen shot of VITA sales support environment (Felfernig et al. 2007b).

knowledge bases and recommender process definitions on a graphic level. A glimpse of the basic functionalities of this acquisition is given in Figure 4.10. Different recommender applications can be maintained in parallel – for example, investment and financing recommenders in the financial services domain. Each of those recommenders is defined by a number of product properties, customer properties, and constraints that are responsible for detecting inconsistent customer requirements and for calculating recommendations.

A simple example of the definition of constraints in the financial services domain is given in Figure 4.11. This constraint indicates that high rates of return require a willingness to take risks. The CWAdvisor environment (Felfernig et al. 2006) supports rapid prototyping processes by automatically translating recommender knowledge bases and process definitions into a corresponding executable application. Thus customers and engineers are able to immediately detect the consequences of changes introduced into the knowledge base and the corresponding process definition.

Ninety percent of the changes in the VITA knowledge base are associated with the underlying product assortment because of new products and changing

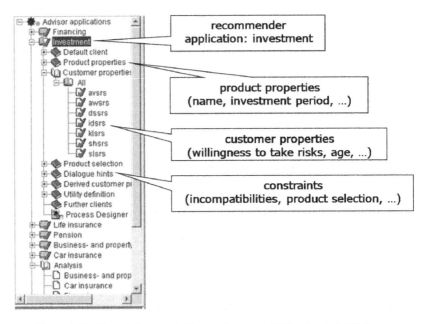

Figure 4.10. Knowledge acquisition environment (Felfernig et al. 2006).

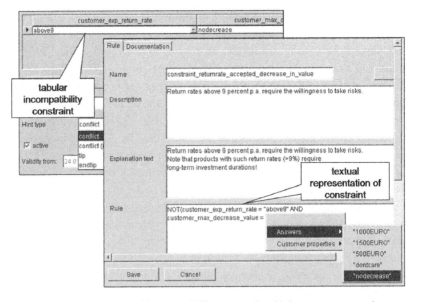

Figure 4.11. Example of incompatibility constraint: high return rates are incompatible with low preparedness to take risks (Felfernig et al. 2006).

interest rates (Felfernig et al. 2007b). Change requests are collected centrally and integrated into the VITA recommender knowledge base once a month. The remaining 10 percent of the changes are related to the graphical user interface and to the explanation and visualization of products. These changes are taken into account in new versions of the recommender application, which are published quarterly.

4.5.2 The Entree case-based recommender

A well-known example of a critiquing-based commercial recommender application is Entree, a system developed for the recommendation of restaurants in Chicago (Burke 2000, Burke et al. 1997). The initial goal was to guide participants in the 1996 Democratic National Convention in Chicago, but its success prolonged its usage for several years.

Scenario. Restaurant recommendation is a domain with a potentially large set of items that are described by a predefined set of properties. The domain is complex because users are often unable to fully define their requirements. This provides a clear justification for the application of recommendation technologies (Burke et al. 1997). Users interact with Entree via a web-based interface with the goal of identifying a restaurant that fits their wishes and needs; as opposed to the financial services scenario discussed in Section 4.5.1, no experts are available in this context who support users in the item retrieval process. FindMe technologies introduced by Burke et al. (1997) were the major technological basis for the Entree recommender. These technologies implement the idea of critique-based recommendation that allows an intuitive navigation in complex item spaces, especially for users who are not experts in the application domain.

Application description. A screenshot of an Entree-type system is shown in Figure 4.12.

There are two entry points to the system: on one hand, the recommender can use a specific reference restaurant as a starting point (preselected on the basis of textual input and a text-based retrieval (Burke et al. 1997)); on the other hand, the user is able to specify the requirements in terms of typical restaurant properties such as *price, cuisine type,* or *noise level* or in terms of high-level properties such as *restaurant with a nice atmosphere* (Burke et al. 1997). High-level properties (e.g., *nice atmosphere*) are translated to low-level item properties – for example, *restaurant with wine cellar and quiet location.* High-level properties

Figure 4.12. Example critiquing-based restaurant recommender.

can be interpreted as a specific type of compound critique, as they refer to a collection of basic properties. The Entree system strictly follows a static critiquing approach, in which a predefined set of critiques is available in each critique cycle. In each cycle, Entree retrieves a set of candidate items from the item database that fulfill the criteria defined by the user (Burke 2002a). Those items are then sorted according to their similarity to the currently recommended item, and the most similar items are returned. Entree does not maintain profiles of users; consequently, a recommendation is determined solely on the basis of the

Figure 4.13. Example of critiquing-based restaurant recommender: result display after one critiquing cycle.

currently displayed item and the critique specified by the user. A simple scenario of interacting with Entree-type recommender applications follows (see Figure 4.13).

The user starts the interaction with searching for a known restaurant, for example, the Biergasthof in Vienna. As shown in Figure 4.12, the recommender manages to identify a similar restaurant named *Brauhof* that is located in the city of Graz. The user, in principle, likes the recommended restaurant but would prefer a less expensive one and triggers the *Less $$* critique. The result of this query is the *Brau Stüberl* restaurant in city of Graz, which has similar characteristics to *Brauhof* but is less expensive and is now acceptable for the user.

Knowledge acquisition. The quality of a recommender application depends on the quality of the underlying knowledge base. When implementing a case-based recommender application, different types of knowledge must be taken into account. A detailed description of the cases in terms of a high number of item attributes requires investing more time into the development of the underlying similarity measures (Burke 2002a). In Entree, for each item a corresponding local similarity measure is defined that explains item similarity in the context of one specific attribute. For example, two restaurants may be very similar in the dimension *cuisine* (e.g., both are Italian restaurants) but may be completely different in the dimension *price*. The global similarity metric is then the result of combining the different local similarity metrics. An important aspect in this context is that similarity metrics must reflect a user's understanding of the item space, because otherwise the application will not be successful (Burke 2002a). Another important aspect to be taken into account is the quality of the underlying item database. It must be correct, complete, and up to date to be able to generate recommendations of high quality. In the restaurant domain, item knowledge changes frequently, and the information in many cases has to be kept up to date by humans, which can be costly and error-prone.

4.6 Bibliographical notes

Applications of knowledge-based recommendation technologies have been developed by a number of groups. For example, Ricci and Nguyen (2007) demonstrate the application of critique-based recommender technologies in mobile environments, and Felfernig and Burke (2008) and Felfernig et al. (2006–07) present successfully deployed applications in the domains of financial services and consumer electronics. Burke (2000) and Burke et al. (1997) provide a detailed overview of knowledge-based recommendation approaches in application domains such as restaurants, cars, movies, and consumer electronics. Further well-known scientific contributions to the field of critiquing-based recommender applications can be found in Lorenzi and Ricci (2005), McGinty and Smyth (2003), Salamo et al. (2005), Reilly et al. (2007a), and Pu et al. (2008). Felfernig and Burke (2008) introduce a categorization of principal recommendation approaches and provide a detailed overview of constraint-based recommendation technologies and their applications. Zanker et al. (2010) formalize different variants of constraint-based recommendation problems and empirically compare the performance of the solving mechanisms. Jiang et al. (2005) introduce an approach to multimedia-enhanced recommendation of digital cameras, in which changes in customer requirements not only result in a changed set of recommendations, but those changes are also animated. For example, a change in the personal goal from portrait pictures to sports photography would result in a lens exchange from a standard lens to a fast lens designed especially for the high-speed movements typical in sports scenes. Thompson et al. (2004) present a knowledge-based recommender based on a combination of knowledge-based approaches with a natural language interface that helps reduce the overall interaction effort.

5

Hybrid recommendation approaches

The three most prominent recommendation approaches discussed in the previous chapters exploit different sources of information and follow different paradigms to make recommendations. Although they produce results that are considered to be personalized based on the assumed interests of their recipients, they perform with varying degrees of success in different application domains. Collaborative filtering exploits a specific type of information (i.e., item ratings) from a user model together with community data to derive recommendations, whereas content-based approaches rely on product features and textual descriptions. Knowledge-based algorithms, on the other hand, reason on explicit knowledge models from the domain. Each of these basic approaches has its pros and cons – for instance, the ability to handle data sparsity and cold-start problems or considerable ramp-up efforts for knowledge acquisition and engineering. These have been discussed in the previous chapters. Figure 5.1 sketches a recommendation system as a black box that transforms input data into a ranked list of items as output. User models and contextual information, community and product data, and knowledge models constitute the potential types of recommendation input. However, none of the basic approaches is able to fully exploit all of these. Consequently, building hybrid systems that combine the strengths of different algorithms and models to overcome some of the aforementioned shortcomings and problems has become the target of recent research. From a linguistic point of view, the term *hybrid* derives from the Latin noun *hybrida* (of mixed origin) and denotes an object made by combining two different elements. Analogously, hybrid recommender systems are technical approaches that combine several algorithm implementations or recommendation components. The following section introduces opportunities for hybridizing algorithm variants and illustrates them with several examples.

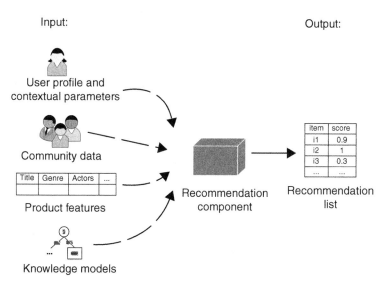

Figure 5.1. Recommender system as a black box.

5.1 Opportunities for hybridization

Although many recommender applications are actually hybrids, little theoretical work has focused on how to hybridize algorithms and in which situations one can expect to benefit from hybridization. An excellent example for combining different recommendation algorithm variants is the Netflix Prize competition[1], in which hundreds of students and researchers teamed up to improve a collaborative movie recommender by hybridizing hundreds of different collaborative filtering techniques and approaches to improve the overall accuracy. Robin Burke's article, "Hybrid Recommender Systems: Survey and Experiments" (2002b) is a well-known survey of the design space of different hybrid recommendation algorithms. It proposes a taxonomy of different classes of recommendation algorithms. Collaborative filtering and content-based and knowledge-based recommender systems are the three base approaches covered in this chapter. Furthermore, Burke (2002b) investigates utility-based recommendation that can be considered as a specific subset of knowledge-based recommender systems, because a utility scheme can be seen as another specific encoding of explicit personalization knowledge. Demographic recommender systems make collaborative propositions based on demographic user profiles.

---

[1] See http://www.netflixprize.com for reference.

As mentioned in Chapter 2, demographic information can be seen as just an additional piece of user knowledge that can be exploited to determine similar peers on the web and is therefore a variant of a collaborative approach. For instance, in the case that few user ratings are available, demographic data can be used to bootstrap a recommender system as demonstrated by Pazzani (1999b). Thus, the basic recommendation paradigm is one dimension of the problem space and will be discussed further in the following subsection.

The second characterizing dimension is the system's hybridization design – the method used to combine two or more algorithms. Recommendation components can work in parallel before combining their results, or two or more single recommender systems may be connected in a pipelining architecture in which the output of one recommender serves as input for the next one.

5.1.1 Recommendation paradigms

In general, a recommendation problem can be treated as a utility function *rec* that predicts the usefulness of an item *i* in a set of items *I* for a specific user *u* from the universe of all users *U*. Adomavicius and Tuzhilin (2005) thus formalized *rec* as a function $U \times I \mapsto R$. In most applications, *R* is the interval $[0 \ldots 1]$ representing the possible utility scores of a recommended item. An item's utility must always be seen in the microeconomic context of a specific user and the sales situation he or she is currently in. Utility therefore denotes an item's capability to fulfill an abstract goal, such as best satisfying the assumed needs of the user or maximizing the retailer's conversion rate. Consequently, the prediction task of a recommender algorithm is to presume this utility score for a given user and item. Utility-based or knowledge-based recommendation systems, for instance, derive the score values directly from a priori known utility schemes; collaborative filtering methods estimate them from community ratings. In contrast, the selection task of any recommender system *RS* is to identify those *n* items from a catalog *I* that achieve the highest utility scores for a given user *u*:

$$RS(u, n) = \{i_1, \ldots, i_k, \ldots, i_n\}, \text{ where} \tag{5.1}$$

$$i_1, \ldots, i_n \in I \text{ and}$$

$$\forall k \ rec(u, i_k) > 0 \wedge rec(u, i_k) > rec(u, i_{k+1})$$

Thus, a recommendation system *RS* will output a ranked list of the top *n* items that are presumably of highest utility for the given user. We will come back to this formalization in the following sections to specify the different hybridization designs.

As already mentioned, we focus on the three base recommendation paradigms: collaborative, content-based and knowledge-based. The collaborative principle assumes that there are clusters of users who behave in a similar way and have comparable needs and preferences. The task of a collaborative recommender is thus to determine similar peers and derive recommendations from the set of their favorite items. Whereas the content-based paradigm follows a "more of the same" approach by recommending items that are similar to those the user liked in the past, knowledge-based recommendation assumes an additional source of information: explicit personalization knowledge. As described in Chapter 4, this knowledge can, for instance, take the form of logical constraints that map the user's requirements onto item properties. Multiattribute utility schemes and specific similarity measures are alternate knowledge representation mechanisms. These knowledge models can be acquired from a third party, such as domain experts; be learned from past transaction data; or use a combination of both. Depending on the representation of the personalization knowledge, some form of reasoning must take place to identify the items to recommend.

Consequently, the choice of the recommendation paradigm determines the type of input data that is required. As outlined in Figure 5.1, four different types exist. The user model and contextual parameters represent the user and the specific situation he or she is currently in. For instance, the items the user has rated so far; the answers the user has given in a requirements elicitation dialogue; demographic background information such as address, age, or education; and contextual parameters such as the season of the year, the people who will accompany the user when he or she buys or consumes the item (e.g., watching a movie), or the current location of the user. The latter contextual parameters are of particular interest and are therefore extensively studied in mobile or more generally pervasive application domains; in Chapter 12 we will specifically focus on personalization strategies applied in physical environments.

It can be assumed that all these user- and situation-specific data are stored in the user's profile. Consequently, all recommendation paradigms require access to this user model to personalize the recommendations. However, depending on the application domain and the usage scenario, only limited parts of the user model may be available. Therefore, not all hybridization variants are possible or advisable in every field of application.

As depicted in Table 5.1, recommendation paradigms selectively require community data, product features, or knowledge models. Collaborative filtering, for example, works solely on community data and the current user profile.

Table 5.1. *Input data requirements of recommendation algorithms.*

Paradigm	User profile and contextual parameters	Community data	Product features	Knowledge models
Collaborative	Yes	Yes	No	No
Content-based	Yes	No	Yes	No
Knowledge-based	Yes	No	Yes	Yes

5.1.2 Hybridization designs

The second dimension that characterizes hybrid algorithms is their design. Burke's taxonomy (2002b) distinguishes among seven different hybridization strategies that we will refer to in this book. Seen from a more general perspective, however, the seven variants can be abstracted into only three base designs: monolithic, parallelized, and pipelined hybrids. *Monolithic* denotes a hybridization design that incorporates aspects of several recommendation strategies in one algorithm implementation. As depicted in Figure 5.2, several recommenders contribute virtually because the hybrid uses additional input data that are specific to another recommendation algorithm, or the input data are augmented by one technique and factually exploited by the other. For instance, a content-based recommender that also exploits community data to determine item similarities falls into this category.

The two remaining hybridization designs require at least two separate recommender implementations, which are consequently combined. Based on their input, parallelized hybrid recommender systems operate independently of one another and produce separate recommendation lists, as sketched in Figure 5.3. In a subsequent hybridization step, their output is combined into a final set of recommendations. Following Burke's taxonomy, the *weighted*, *mixed*, and *switching* strategies require recommendation components to work in parallel.

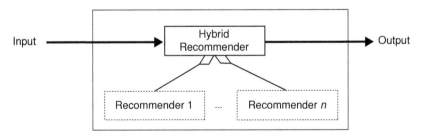

Figure 5.2. Monolithic hybridization design.

When several recommender systems are joined together in a pipeline architecture, as depicted in Figure 5.4, the output of one recommender becomes part
of the input of the subsequent one. Optionally, the subsequent recommender
components may use parts of the original input data, too. The *Cascade* and
meta-level hybrids, as defined by Burke (2002b), are examples of such pipeline
architectures.

The following sections examine each of the three base hybridization designs
in more detail.

5.2 Monolithic hybridization design

Whereas the other two designs for hybrid recommender systems consist of
two or more components whose results are combined, monolithic hybrids consist of a single recommender component that integrates multiple approaches
by preprocessing and combining several knowledge sources. Hybridization is
thus achieved by a built-in modification of the algorithm behavior to exploit
different types of input data. Typically, data-specific preprocessing steps are
used to transform the input data into a representation that can be exploited
by a specific algorithm paradigm. Following Burke's taxonomy (2002b), both
feature combination and *feature augmentation* strategies can be assigned to this
category.

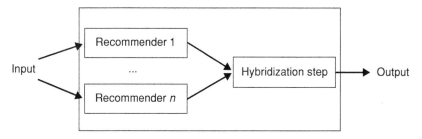

Figure 5.3. Parallelized hybridization design.

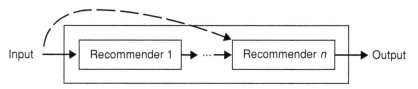

Figure 5.4. Pipelined hybridization design.

Table 5.2. *Community and product knowledge.*

User	Item1	Item2	Item3	Item4	Item5		Item	Genre
Alice		1		1			Item1	romance
User1		1	1		1		Item2	mystery
User2	1	1			1		Item3	mystery
User3	1		1				Item4	mystery
User4					1		Item5	fiction

5.2.1 Feature combination hybrids

A feature combination hybrid is a monolithic recommendation component that uses a diverse range of input data. For instance, Basu et al. (1998) proposed a feature combination hybrid that combines collaborative features, such as a user's likes and dislikes, with content features of catalog items. The following example illustrates this technique in the book domain.

Table 5.2 presents several users' unary ratings of a product catalog. Ratings constitute implicitly observed user feedback, such as purchases. Product knowledge is restricted to an item's genre. Obviously, in a pure collaborative approach, without considering any product features, both User1 and User2 would be judged as being equally similar to Alice. However, the feature-combination approach of Basu et al. identifies new hybrid features based on community and product data. This reflects "the common human-engineering effort that involves inventing good features to enable successful learning" (Basu et al. 1998).

Table 5.3 thus provides a hybrid encoding of the information contained in Table 5.2. These features are derived by the following rules. If a user bought mainly books of genre *X* (i.e., two-thirds of the total purchases and at least two books) we set the user characteristic *User likes many X books* to true.

Table 5.3. *Hybrid input features.*

Feature	Alice	User1	User2	User3	User4
User likes many *mystery* books	true	true			
User likes some *mystery* books			true	true	
User likes many *romance* books					
User likes some *romance* books			true	true	
User likes many *fiction* books					
User likes some *fiction* books		true	true		true

Table 5.4. *Different types of user feedback.*

User	R_{nav}	R_{view}	R_{ctx}	R_{buy}
Alice	n_3, n_4	i_5	k_5	$\emptyset$
User1	n_1, n_5	i_3, i_5	k_5	i_1
User2	n_3, n_4	i_3, i_5, i_7	$\emptyset$	i_3
User3	n_2, n_3, n_4	i_2, i_4, i_5	k_2, k_4	i_4

Analogously, we also set *User likes some X books* to true, requiring one-third of the user's purchases, or at least one item in absolute numbers. Although the transformation seems to be quite trivial at first glance, it nevertheless reflects that some knowledge, such as an item's genre, can lead to considerable improvements. Initially Alice seems to possess similar interests to User1 and User2, but the picture changes after transforming the user/item matrix. It actually turns out that User2 behaves in a very indeterminate manner by buying *some* books of all three genres. In contrast, User1 seems to focus specifically on *mystery* books, as Alice does. For this example we determine similar peers by simply matching user characteristics, whereas Basu et al. (1998) used set-valued user characteristics and the inductive rule learner Ripper (Cohen 1996) in their original experiments.

Another approach for feature combination was proposed by Zanker and Jessenitschnig (2009b), who exploit different types of rating feedback based on their predictive accuracy and availability.

Table 5.4 depicts a scenario in which several types of implicitly or explicitly collected user feedback are available as unary ratings, such as navigation actions R_{nav}, click-throughs on items' detail pages R_{view}, contextual user requirements R_{ctx}, or actual purchases R_{buy}. These categories differentiate themselves by their availability and their aptness for making predictions. For instance, navigation actions and page views of online users occur frequently, whereas purchases are less common. In addition, information about the user's context, such as the keywords used for searching or input to a conversational requirements elicitation dialog, is typically a very useful source for computing recommendations (Zanker and Jessenitschnig 2009a). In contrast, navigation actions typically contain more noise, particularly when users randomly surf and explore a shop. Therefore, rating categories may be prioritized – for example $(R_{buy}, R_{ctx}) \prec R_{view} \prec R_{nav}$, where priority decreases from left to right.

Thus, when interpreting all rating data in Table 5.4 in a uniform manner, User2 and User3 seem to have the most in common with Alice, as both share at least three similar ratings. However, the algorithm for neighborhood determination by Zanker and Jessenitschnig (2009b) uses the given precedence rules and thus initially uses R_{buy} and R_{ctx} as rating inputs to find similar peers. In this highly simplified example, User1 would be identified as being most similar to Alice. In the case that not enough similar peers can be determined with satisfactory confidence, the feature combination algorithm includes additional, lower-ranked rating input. A similar principle for feature combination was also exploited by Pazzani (1999b), who used demographic user characteristics to bootstrap a collaborative recommender system when not enough item ratings were known.

Because of the simplicity of feature combination approaches, many current recommender systems combine collaborative and/or content-based features in one way or another. However, the combination of input features using knowledge-based approaches, such as constraints, with content-based or collaborative knowledge sources has remained largely unexplored. We suggest that critique-based systems that elicit user feedback in the form of constraints on a specific item, such as *price should be less than the price for item a*, could be a suitable starting point for future research.

5.2.2 Feature augmentation hybrids

Feature augmentation is another monolithic hybridization design that may be used to integrate several recommendation algorithms. In contrast with feature combination, this hybrid does not simply combine and preprocess several types of input, but rather applies more complex transformation steps. In fact, the output of a contributing recommender system augments the feature space of the actual recommender by preprocessing its knowledge sources. However, this must not be mistaken for a pipelined design, as we will discuss in the following section, because the implementation of the contributing recommender is strongly interwoven with the main component for reasons of performance and functionality.

Content-boosted collaborative filtering is an actual example of this variant (Melville et al. 2002). It predicts a user's assumed rating based on a collaborative mechanism that includes content-based predictions.

Table 5.5 presents an example user/item matrix, together with rating values for Item5 ($v_{User,Item5}$) as well as the Pearson correlation coefficient $P_{Alice,User}$ signifying the similarity between Alice and the respective users. Furthermore, it denotes the number of user ratings (n_{User}) and the number of overlapping ratings between Alice and the other users ($n_{Alice,User}$). The rating matrix for this example is complete, because it consists not only of users' actual ratings $r_{u,i}$, but also of content-based predictions $c_{u,i}$ in the case that a user rating

Table 5.5. *Hybrid input features.*

User	$v_{User,Item5}$	$P_{Alice,User}$	n_{User}	$n_{Alice,User}$
Alice	?		40	
User1	4	0.8	14	6
User2	2.2	0.7	55	28

is missing. Melville et al. (2002) therefore first create a pseudo-user-ratings vector $v_{u,i}$:

$$v_{u,i} = \begin{cases} r_{u,i} & : \text{if user } u \text{ rated item } i \\ c_{u,i} & : \text{else content-based prediction} \end{cases}$$

Based on these pseudo ratings, the algorithm computes predictions in a second step. However, depending on the number of rated items and the number of co-rated items between two users, weighting factors may be used to adjust the predicted rating value for a specific user–item pair a, i, as illustrated by the following equation for a content-boosted collaborative recommender.

$$rec_{cbcf}(a, i) = \left(sw_a c_{a,i} + \sum_{\substack{u=1 \\ u \neq a}}^{n} hw_{a,u} P_{a,u} v_{u,i} \right) \Big/ \left(sw_a + \sum_{\substack{u=1 \\ u \neq a}}^{n} hw_{a,u} P_{a,u} \right)$$

$$(5.2)$$

$hw_{a,u}$ is the hybrid correlation weight that is defined as follows:

$$hw_{a,u} = sg_{a,u} + hm_{a,u}, \text{ where}$$

$$sg_{a,u} = \begin{cases} \frac{n_{a,u}}{50} & : \text{if } n_{a,u} < 50 \\ 1 & : \text{else} \end{cases}$$

$$hm_{a,u} = \frac{2m_a m_u}{m_a + m_u} \text{ with}$$

$$m_i = \begin{cases} \frac{n_i}{50} & : \text{if } n_i < 50 \\ 1 & : \text{else} \end{cases}$$

The hybrid correlation weight ($hw_{a,u}$) adjusts the computed Pearson correlation based on a significance weighting factor $sg_{a,u}$ (Herlocker et al. 1999) that favors peers with more co-rated items. When the number of co-rated items is greater than 50, the effect tends to level off. In addition, the harmonic mean weighting factor ($hm_{a,u}$) reduces the influence of peers with a low number of user ratings, as content-based pseudo-ratings are less reliable if derived from a small number of user ratings.

Similarly, the self-weighting factor sw_i reflects the confidence in the algorithm's content-based prediction, which obviously also depends on the number of original rating values of any user i. The constant max was set to 2 by Melville et al. (2002).

$$sw_i = \begin{cases} \frac{n_i}{50} \times max & : \text{if } n_i < 50 \\ max & : \text{else} \end{cases}$$

Thus, assuming that the content-based prediction for Item5 is $c_{Alice,\ Item5} = 3$, $rec_{cbcf}(Alice, Item5)$ is computed as follows:

$$\frac{1.6 \times 3 + (0.535 \times 0.8 \times 4 + 1.45 \times 0.7 \times 2.2)}{1.6 + (0.535 \times 0.8 + 1.45 \times 0.7)} = \frac{8.745}{3.043} = 2.87$$

Consequently, the predicted value (on a 1–5 Likert scale) indicates that Alice will not be euphoric about Item5, although the most similar peer, User1, rated it with a 4. However, the weighting and adjustment factors place more emphasis on User 2 and the content-based prediction. Also note that we ignored rating adjustments based on users' rating averages as outlined in Chapter 2 and defined in Formula 2.3 for reasons of simplicity.

Additional applications of feature augmentation hybrids are presented in Mooney and Roy's (1999) discussion of a content-based book recommender and in Torres et al.'s (2004) recommendation of research papers. The latter employs, among other hybrid algorithm variants, a feature augmentation algorithm that interprets article citations as collaborative recommendations.

5.3 Parallelized hybridization design

Parallelized hybridization designs employ several recommenders side by side and employ a specific hybridization mechanism to aggregate their outputs. Burke (2002b) elaborates on the *mixed*, *weighted*, and *switching* strategies. However, additional combination strategies for multiple recommendation lists, such as majority voting schemes, may also be applicable.

5.3.1 Mixed hybrids

A mixed hybridization strategy combines the results of different recommender systems at the level of the user interface, in which results from different techniques are presented together. Therefore the recommendation result for user u

and item i of a mixed hybrid strategy is the set of -tuples $\langle score, k \rangle$ for each of its n constituting recommenders rec_k:

$$rec_{mixed}(u, i) = \bigcup_{k=1}^{n} \langle rec_k(u, i), k \rangle \qquad (5.3)$$

The top-scoring items for each recommender are then displayed to the user next to each other, as in Burke et al. (1997) and Wasfi (1999). However, when composing the different results into a single entity, such as a television viewing schedule, some form of conflict resolution is required. In the personalized television application domain, Cotter and Smyth (2000) apply predefined precedence rules between different recommender functions. Zanker et al. (2007) describe another form of a mixed hybrid, which merges the results of several recommendation systems. It proposes bundles of recommendations from different product categories in the tourism domain, in which for each category a separate recommender is employed. A recommended bundle consists, for instance, of accommodations, as well as sport and leisure activities, that are derived by separate recommender systems. A Constraint Satisfaction Problem (CSP) solver is employed to resolve conflicts, thus ensuring that only consistent sets of items are bundled together according to domain constraints such as "activities and accommodations must be within 50 km of each other".

5.3.2 Weighted hybrids

A weighted hybridization strategy combines the recommendations of two or more recommendation systems by computing weighted sums of their scores. Thus, given n different recommendation functions rec_k with associated relative weights β_k:

$$rec_{weighted}(u, i) = \sum_{k=1}^{n} \beta_k \times rec_k(u, i) \qquad (5.4)$$

where item scores need to be restricted to the same range for all recommenders and $\sum_{k=1}^{n} \beta_k = 1$. Obviously, this technique is quite straightforward and is thus a popular strategy for combining the predictive power of different recommendation techniques in a weighted manner. Consider an example in which two recommender systems are used to suggest one of five items for a user *Alice*. As can be easily seen from Table 5.6, these recommendation lists are hybridized by using a uniform weighting scheme with $\beta_1 = \beta_2 = 0.5$. The weighted hybrid recommender rec_w thus produces a new ranking by combining the scores from rec_1 and rec_2. Items that are recommended by only one of the two users, such as Item2, may still be ranked highly after the hybridization step.

Table 5.6. *Recommendations of weighted hybrid.*

item	rec_1 score	rec_1 rank	rec_2 score	rec_2 rank	rec_w score	rec_w rank
Item1	0.5	1	0.8	2	0.65	1
Item2	0		0.9	1	0.45	2
Item3	0.3	2	0.4	3	0.35	3
Item4	0.1	3	0		0.05	
Item5	0		0		0	

If the weighting scheme is to remain static, the involved recommenders must also produce recommendations of the same relative quality for all user and item combinations. To estimate weights, an empirical bootstrapping approach can be taken. For instance, Zanker and Jessenitschnig (2009a) conducted a sensitivity analysis between a collaborative and a knowledge-based recommender in the cigar domain to identify the optimum weighting scheme. The P-Tango system (Claypool et al. 1999) is another example of such a system, blending the output of a content-based and a collaborative recommender in the news domain. This approach uses a dynamic weighting scheme and will be explained in detail in the following section. Starting from a uniform distribution, it dynamically adjusts the relative weights for each user to minimize the predictive error in cases in which user ratings are available. Furthermore, it adapts the weights on a per-item basis to correctly reflect the relative strengths of each prediction algorithm. For instance, the collaborative filtering gains weight if the item has been rated by a higher number of users.

We explain such a dynamic weighting approach by returning to the initial example. Let us assume that Alice purchased Item1 and Item4, which we will interpret as positive unary ratings. We thus require a weighting that minimizes a goal metric such as the mean absolute error (MAE) of predictions of a user for her rated items R (see Chapter 2).

$$MAE = \frac{\sum_{r_i \in R} \sum_{k=1}^n \beta_k \times |rec_k(u, i) - r_i|}{|R|} \tag{5.5}$$

If we interpret Alice's purchases as very strong relevance feedback for her interest in an item, then the set of Alice's actual ratings R will contain $r_1 = r_4 = 1$. Table 5.7 summarizes the absolute errors of rec_1 and rec_2's predictions (denoted in Table 5.6) for different weighting parameters. Table 5.7 shows that the MAE improves as rec_2 is weighted more strongly. However, when examining the situation more closely, it is evident that the weight assigned to

Table 5.7. *Dynamic weighting parameters, absolute errors, and MAEs for user Alice.*

β_1	β_2	item	r_i	rec_1	rec_2	error	MAE
0.1	0.9	Item1	1	0.5	0.8	0.23	
		Item4	1	0.1	0	0.99	**0.61**
0.3	0.7	Item1	1	0.5	0.8	0.29	
		Item4	1	0.1	0	0.97	0.63
0.5	0.5	Item1	1	0.5	0.8	0.35	
		Item4	1	0.1	0	0.95	0.65
0.7	0.3	Item1	1	0.5	0.8	0.41	
		Item4	1	0.1	0	0.93	0.67
0.9	0.1	Item1	1	0.5	0.8	0.47	
		Item4	1	0.1	0	0.91	0.69

rec_1 should be strengthened. Both rated items are ranked higher by rec_1 than by rec_2 – Item1 is first instead of second and Item4 is third, whereas rec_2 does not recommend Item4 at all. Thus, when applying a weighted strategy, one must ensure that the involved recommenders assign scores on comparable scales or apply a transformation function beforehand. Obviously, the assignment of dynamic weighting parameters stabilizes as more rated items are made available by users. In addition, alternative error metrics, such as mean squared error or rank metrics, can be explored. Mean squared error puts more emphasis on large errors, whereas rank metrics focus not on recommendation scores but on ranks.

In the extreme case, dynamic weight adjustment could be implemented as a switching hybrid. There, the weights of all but one dynamically selected recommenders are set to 0, and the output of a single remaining recommender is assigned the weight of 1.

5.3.3 Switching hybrids

Switching hybrids require an oracle that decides which recommender should be used in a specific situation, depending on the user profile and/or the quality of recommendation results. Such an evaluation could be carried out as follows:

$$\exists_1 k : 1 \ldots n \; rec_{switching}(u, i) = rec_k(u, i) \tag{5.6}$$

where k is determined by the switching condition. For instance, to overcome the cold-start problem, a knowledge-based and collaborative switching hybrid could initially make knowledge-based recommendations until enough rating data are available. When the collaborative filtering component can deliver recommendations with sufficient confidence, the recommendation strategy could be switched. Furthermore, a switching strategy can be applied to optimize results in a similar fashion to the NewsDude system (Billsus and Pazzani 2000). There, two content-based variants and a collaborative strategy are employed in an ordered manner to recommend news articles. First, a content-based nearest neighbor recommender is used. If it does not find any closely related articles, a collaborative filtering system is invoked to make cross-genre propositions; finally, a naive Bayes classifier finds articles matching the long-term interest profile of the user. Zanker and Jessenitschnig (2009a) proposed a switching strategy that actually switches between two hybrid variants of collaborative filtering and knowledge-based recommendation. If the first algorithm, a cascade hybrid, delivers fewer than n recommendations, the hybridization component switches to a weighted variant as a fallback strategy. Even more adaptive switching criteria can be thought of that could even take contextual parameters such as users' intentions or expectations into consideration for algorithm selection. For instance, van Setten (2005) proposed the domain-independent Duine framework that generalizes the selection task of a prediction strategy and discusses several machine learning techniques in that context. To summarize, the quality of the switching mechanism is the most crucial aspect of this hybridization variant.

5.4 Pipelined hybridization design

Pipelined hybrids implement a staged process in which several techniques sequentially build on each other before the final one produces recommendations for the user. The pipelined hybrid variants differentiate themselves mainly according to the type of output they produce for the next stage. In other words, a preceding component may either preprocess input data to build a model that is exploited by the subsequent stage or deliver a recommendation list for further refinement.

5.4.1 Cascade hybrids

Cascade hybrids are based on a sequenced order of techniques, in which each succeeding recommender only refines the recommendations of its predecessor. The recommendation list of the successor technique is thus restricted to items that were also recommended by the preceding technique.

Formally, assume a sequence of n techniques, where rec_1 represents the recommendation function of the first technique and rec_n the last one. Consequently, the final recommendation score for an item is computed by the nth technique. However, an item will be suggested by the kth technique only if the $(k-1)$th technique also assigned a nonzero score to it. This applies to all $k \geq 2$ by induction as defined in Formula (5.7).

$$rec_{cascade}(u, i) = rec_n(u, i) \tag{5.7}$$

where $\forall k \geq 2$ must hold:

$$rec_k(u, i) = \begin{cases} rec_k(u, i) & : rec_{k-1}(u, i) \neq 0 \\ 0 & : \text{else} \end{cases}$$

Thus in a cascade hybrid all techniques, except the first one, can only change the ordering of the list of recommended items from their predecessor or exclude an item by setting its utility to 0. However, they may not introduce new items – items that have already been excluded by one of the higher-priority techniques – to the recommendation list. Thus cascading strategies do have the unfavorable property of potentially reducing the size of the recommendation set as each additional technique is applied. As a consequence, situations can arise in which cascade algorithms do not deliver the required number of propositions, thus decreasing the system's usefulness. Therefore cascade hybrids may be combined with a switching strategy to handle the case in which the cascade strategy does not produce enough recommendations. One such hybridization step that switches to weighted strategy was proposed by Zanker and Jessenitschnig (2009a).

As knowledge-based recommenders produce recommendation lists that are either unsorted or contain many ties among items' scores, cascading them with another technique to sort the results is a natural choice. EntreeC is a knowledge-based restaurant recommender that is cascaded with a collaborative filtering algorithm to recommend restaurants (Burke 2002b). However, in contrast to the definition of cascade hybrid given here, EntreeC uses only the second recommender to break ties. Advisor Suite is a domain independent knowledge-based recommender shell discussed in Chapter 4 (Felfernig et al. 2006–07). It includes an optional utility-based sorting scheme that can further refine recommendations in a cascade design.

5.4.2 Meta-level hybrids

In a meta-level hybridization design, one recommender builds a model that is exploited by the principal recommender to make recommendations. Formula

(5.8) formalizes this behavior, wherein the nth recommender exploits a model Δ that has been built by its predecessor. However, in all reported systems so far, n has always been 2.

$$rec_{meta-level}(u, i) = rec_n(u, i, \Delta_{rec_{n-1}}) \qquad (5.8)$$

For instance, the Fab system (Balabanović and Shoham 1997) exploits a collaborative approach that builds on user models that have been built by a content-based recommender. The application domain of Fab is online news. Fab employs a content-based recommender that builds user models based on a vector of term categories and the users' degrees of interest in them. The recommendation step, however, does not propose items that are similar to the user model, but employs a collaborative technique. The latter determines the user's nearest neighbors based on content models and recommends items that similar peers have liked. Pazzani (1999b) referred to this approach as *collaboration via content* and presented a small user study based on restaurant recommendation. It showed that the hybrid variant performs better than base techniques, especially when users have only a few items in common. Zanker (2008) evaluated a further variant of meta-level hybridization that combines collaborative filtering with knowledge-based recommendation. The hybrid generates binary user preferences of the form $a \rightarrow b$, where a represents a user requirement and b a product feature. If, for example, a user of the cigar advisor described by Zanker (2008) were looking for a gift and finally bought a cigar of the brand Montecristo, then the constraint *for_whom* $=$ *"gift"* $\rightarrow$ *brand* $=$ *"Montecristo"* becomes part of the user's profile. When computing a recommendation, a collaborative filtering step retrieves all such constraints from a user's peers. A knowledge-based recommender finally applies these restrictions to the product database and derives item propositions. An evaluation showed that this approach produced more successful predictions than a manually crafted knowledge base or an impersonalized application of all generated constraints.

Golovin and Rahm (2004) applied a reinforcement learning (RL) approach for exploiting context-aware recommendation rules. The authors exploited different top-N recommendation procedures as well as sequence patterns and frequent-item sets to generate weighted recommendation rules that are used by a reinforcement learning component to make predictions. Rules specify which item should be presented in which situation, where the latter is characterized by the product content and the user model, including contextual parameters such as daytime or season of the year. Thus, RL acts as a principal recommender that adjusts the weights of the recommendation rule database based on user feedback.

5.5 Discussion and summary

In this chapter we discussed the opportunities for combining different algorithm variants and presented a taxonomy for hybridization designs. In summary, no single hybridization variant is applicable in all circumstances, but it is well accepted that all base algorithms can be improved by being hybridized with other techniques. For instance, in the Netflix Prize competition, the winners employed a weighted hybridization strategy in which weights were determined by regression analysis (Bell et al. 2007). Furthermore, they adapted the weights based on particular user and item features, such as number of rated items, that can be classified as a switching hybrid that changes between different weighted hybrids according to the presented taxonomy of hybridization variants.

One of the main reasons that little research focuses on comparing different recommendation strategies and especially their hybrids is the lack of appropriate datasets. Although collaborative movie recommendations or content-based news recommenders are comparably well researched application domains, other application domains for recommender systems and algorithm paradigms receive less attention. Therefore, no empirically backed conclusions about the advantages and disadvantages of different hybridization variants can be drawn, but, depending on the application domain and problem type, different variants should be explored and compared. Nevertheless, enhancing an existing recommendation application by exploiting additional knowledge sources will nearly always pay off. With respect to the required engineering effort, the following can be said.

Monolithic designs are advantageous if little additional knowledge is available for inclusion on the feature level. They typically require only some additional preprocessing steps or minor modifications in the principal algorithm and its data structures.

Parallelized designs are the least invasive to existing implementations, as they act as an additional postprocessing step. Nevertheless, they add some additional runtime complexity and require careful matching of the recommendation scores computed by the different parallelized algorithms.

Pipelined designs are the most ambitious hybridization designs, because they require deeper insight into algorithm's functioning to ensure efficient runtime computations. However, they typically perform well when two antithetic recommendation paradigms, such as collaborative and knowledge-based, are combined.

5.6 Bibliographical notes

Only a few articles focus specifically on the hybridization of recommendation algorithms in general. The most comprehensive work in this regard is Burke's article, "Hybrid recommender systems: Survey and experiments", which appeared in *User Modeling and User-Adapted Interaction* in 2002. It developed the taxonomy of recommendation paradigms and hybridization designs that guided this chapter, and is the most referenced article in this respect. A revised version appeared as a chapter in the Springer state-of-the-art survey *The Adaptive Web* by the same author in 2007. It not only constitutes a comprehensive source of reference for published works on hybrid algorithms, but also includes the most extensive comparative evaluation of different hybrid algorithm variants. Burke (2007) compared forty one different algorithms based on the Entree dataset (Burke 1999). In contrast to many earlier comparative studies on the movie domain (Balabanović and Shoham 1997, Pazzani 1999b, Sarwar et al. 2000b), the Entree dataset also allows the exploration of knowledge-based algorithm variants.

Adomavicius and Tuzhilin (2005) provide an extensive state-of-the-art survey on current recommender systems' literature that includes a taxonomy that differentiates between collaborative and content-based recommenders and hybrid variants thereof; however, it lacks a discussion about knowledge-based algorithms.

Zanker et al. (2007) present a recent evaluation of several algorithm variants that compares knowledge-based variants with collaborative ones on a commercial dataset from the cigar domain. These experiments were further developed by Zanker and Jessenitschnig (2009a) with the focus being placed on explicit user requirements, such as keywords and input, to conversational requirements elicitation dialogs as the sole type of user feedback. They explored weighted, switching, and cascade hybridization variants of knowledge-based and collaborative recommendation paradigms.

6

Explanations in recommender systems

6.1 Introduction

"The digital camera *Profishot* is a must-buy for you because . . ." "In fact, for your requirements as a semiprofessional photographer, you should not use digital cameras of type *Lowcheap* because . . ." Such information is commonly exchanged between a salesperson and a customer during in-store recommendation processes and is usually termed an *explanation* (Brewer et al. 1998).

The concept of explanation is frequently exploited in human communication and reasoning tasks. Consequently, research within artificial intelligence – in particular, into the development of systems that mimic human behavior – has shown great interest in the nature of explanations. Starting with the question, "What is an explanation?", we are confronted with an almost unlimited number of possibilities.

Explanations such as (1) "The car type Jumbo-Family-Van of brand Rising-Sun would be well suited to your family because you have four children and the car has seven seats"; (2) "The light bulb shines because you turned it on"; (3) "I washed the dishes because my brother did it last time"; or simply (4) "You have to do your homework because your dad said so", are examples of explanations depending on circumstances and make the construction of a generic approach for producing explanations difficult. The work of Brewer et al. (1998) distinguishes among functional, causal, intentional, and scientific explanations. Functional explanations (such as explanation 1) deal with the functions of systems. Causal explanations (such as explanation 2) provide causal relationships between events. Intentional explanations (such as explanations 3 and 4) give reasons for human behavior. Scientific explanations are exploited to express relations between the concepts formulated in various scientific fields and are typically based on refutable theories. Unfortunately, there is no accepted unified theory describing the concept of explanation. Consequently, it is unclear how to design a general method for generating explanations.

Facing such fundamental challenges, one might ask why recommender systems should deal with explanations at all. The answer is related to the two parties providing and receiving recommendations. For example, a selling agent may be interested in promoting particular products, whereas a buying agent is concerned about making the right buying decision. Explanations are important pieces of information that can be exploited by both agents throughout the communication process to increase their performance. Different agents will formulate explanations with different intentions – for example, a buying agent looks for bargains and explanations that justify decisions, whereas a selling agent tries to improve profits by providing convincing arguments to the buying agent. We choose to analyze the phenomenon of explanations from a pragmatic viewpoint. Despite the diversity of proposals for characterizing the concept of explanation, almost all sources agree that an explanation is a piece of information exchanged in a communication process. In the context of recommender systems, these pieces of information supplement a recommendation with different aims. Following a pragmatic view, the goals for providing explanations in a recommendation process can be identified as follows (Tintarev 2007, Tintarev and Masthoff 2007):

Transparency. Explanations supporting the transparency of recommendations aim to provide information so the user can comprehend the reasoning used to generate a specific recommendation. In particular, the explanation may provide information as to why one item was preferred over another. For example, consider the case in which you wonder why a film recommender assumes you like Westerns, when in fact you do not. Transparency explanations may indicate, for example, that you purchased country songs and that this information is being exploited for recommending Western films, giving you the chance to change false assumptions.

Validity. Explanations can be generated to allow a user to check the validity of a recommendation. For example, "I recommend this type of car because you have four children and the Jumbo-Family-Van of Rising-Sun has seven seats. Because of the number of children, I cannot recommend the Dinki-coupe of SpeedyGECars, as it has only four seats." The ability to check validity is not necessarily related to transparency. For instance, a neural network may have decided that a product is an almost perfect match to a set of customer requirements. Transparency in the computation process, disclosing how the neural network computed the recommendation, will not help a customer validate the recommendation. However, showing a comparison of the required and offered product features allows the customer to validate the quality of the product recommendation.

Trustworthiness. Following Grabner-Kräuter and Kaluscha (2003), trust building can be viewed as a mechanism for reducing the complexity of human decision making in uncertain situations. Explanations aiming to build trust in recommendations reduce the uncertainty about the quality of a recommendation – for example, "The drug Kural cured thousands of people with your disease; therefore, this drug will also help you."[1]

Persuasiveness. Computing technology (Fogg 1999) is regarded as persuasive if the system is intentionally designed to change a person's attitude or behavior in a predetermined way. In this sense, persuasive explanations for recommendations aim to change the user's buying behavior. For instance, a recommender may intentionally dwell on a product's positive aspects and keep quiet about various negative aspects.

Effectiveness. In the context of recommender systems, the term *effectiveness* refers to the support a user receives for making high-quality decisions. Explanations for improving effectiveness typically help the customer discover his or her preferences and make decisions that maximize satisfaction with respect to the selected product. Effective recommenders help users make better decisions.

Efficiency. In the context of recommender systems, the term *efficiency* refers to a system's ability to support users to reduce the decision-making effort. Thus explanations aiming to increase efficiency typically try to reduce the time needed for decision making. However, a measure for efficiency might also be the perceived cognitive effort, which could be different than efficiency based on the time taken to make the recommendation and select a product.

Satisfaction. Explanations can attempt to improve the overall satisfaction stemming from the use of a recommender system. This aim may not be linked to any other explanation goals, such as persuasiveness. The motivation behind this goal may be manifold – for instance, to increase the customer return rate.

Relevance. Additional information may be required in conversational recommenders. Explanations can be provided to justify why additional information is needed from the user.

---

[1] The recommender literature usually does not distinguish between trust and confidence.

Comprehensibility. Recommenders can never be sure about the knowledge of their users. Explanations targeting comprehension support the user by relating the user's known concepts to the concepts employed by the recommender.

Education. Explanations can aim to educate users to help them better understand the product domain. Deep knowledge about the domain helps customers rethink their preferences and evaluate the pros and cons of different solutions. Eventually, as customers become more informed, they are able to make wiser purchasing decisions.

The aforementioned aims for generating explanations can be interrelated. For example, an explanation generated for improving the transparency of a recommendation can have a positive effect on trust. Conversely, explanations aimed at persuasiveness may result in a loss of trust. Consequently, the first step in designing explanation generators is to define the goals of explanations. To assess the utility of explanations, all effects on the various communication aims of the recommendation process must be crosschecked.

As noted earlier, explanations are used in a communication process. Therefore, the suitability of an explanation depends on the goals of both the explanation sender and the receiver. As a consequence, the quality of explanations can be improved by modeling the receiving agent. For instance, to make explanations comprehensible, it is necessary to have information about the knowledge level of the receiver. Generally, the better the model of a receiving agent is, the more effective the arguments generated by a persuasive sending agent will be.

Furthermore, what is regarded as a valid (or good) explanation depends on the communication process itself, a fact that was neglected in early explanation generation attempts and that may lead to spurious explanations, as will be shown later. To summarize, the following factors influence the generation of explanations by a recommender agent communicating with an agent receiving recommendations:

- The piece of information to be explained.
- The goals of the agent in providing (receiving) an explanation.
- The model of the receiving agent, including its behavior and knowledge.
- The state of communication – the exchanged information, including provided recommendations.

How explanation goals can be achieved depends on the recommendation method employed. In particular, if knowledge is available as to how a conclusion was derived, then this knowledge can be exploited for various explanation goals – for example, to increase trust by providing a logically correct argument. However, if such knowledge is not available, then trust building must be supported by other means, such as by referring the receiver to previous high-quality recommendations.

In the next section, we explore the standard explanation approaches for various recommendation methods, such as constraint-based recommenders, case-based recommenders, and recommenders based on collaborative filtering.

6.2 Explanations in constraint-based recommenders

The generation of explanations has a long history in expert systems (Shortliffe 1974). Constraint-based recommenders can be seen as a descendant of expert systems in general, and therefore can draw on established methods. However, we show how these methods have to be extended to work properly for constraint-based recommenders.

For example, in the area of sales support systems (Jannach 2004), constraint-based software applications have been developed to help customers find the right product (configuration) for their needs in domains as diverse as financial products and digital cameras. Constraint-based methods have become a key technology for the implementation of knowledge-based recommender systems because they offer enough expressive power to represent the relevant knowledge. In addition, extensive research has provided a huge library of concepts and algorithms for efficiently solving the various reasoning tasks.

In such applications a solution represents a product or a service a customer can purchase. Explanations are generated to give feedback about the process used to derive the conclusions – for example, "Because you, as a customer, told us that simple handling of a car is important to you, we included a special sensor system in our offer that will help you to park your car easily." Such explanations are exploited by customers in various ways – for instance, to increase confidence in a solution or to facilitate trade-off decisions (Felfernig et al. 2004).

In this section we focus on answering two typical questions a customer is likely to pose. In the case that the recommendation process requires input from the customer, the customer could ask why this information is needed. This corresponds to the classical *why-explanations* of expert systems (Buchanan and Shortliffe 1984). Conversely, when a recommender proposes a set of solutions (e.g., products) then a customer might ask for an explanation why a proposed solution would be advantageous for him or her. Traditionally, this type of explanation is called a *how-explanation* (Buchanan and Shortliffe 1984) because classical expert systems exploited information as to how a conclusion was deduced – for example, the sequence of rules activated in the decision-making process.

We now examine methods for answering these two types of questions by exploiting the reasoning methods of constraint-based systems. We start with an introductory example and subsequently elaborate on the basic principles.[2]

6.2.1 Example

Consider an example from the car domain. Assume two different packages are available for a car; a business package and a recreation package. The customer can decide if he or she wants one, both, or neither of these packages.

The recreation package includes a coupling device for towing trailers and a video camera on the rear of the car, which allows the driver to ascertain the distance to an obstacle behind the car. This camera supports the customer-oriented product function *easy parking*. Customers may not be interested in the technical details a priori but may request functions for describing the wanted product features – for example, customers may not be interested in how the function *easy parking* is implemented, but they are interested that the car gives assistance for parking. Furthermore, functions are exploited to characterize products in an abstract way to justify consumer confidence in a solution – for instance, a sales rep can argue that a car supports easy parking because it includes a video camera at the rear. Functions of products are used to describe them in a more abstract way, which allows the hiding of technical details. Communicating with nontechnical customers requires this abstraction.

The business package includes a sensor system in the back bumper, which also supports easy parking. However, the sensor system is incompatible with the recreation package for technical reasons (the coupling device of the recreation package prevents the sensor system from being mounted). From the customer's point of view, the video camera and the sensor system provide the same functionality. Therefore, if the customer orders the business package and the recreation package, the car includes the video camera, which implements the *easy parking* function. In this configuration, the sensors are not only forbidden (because they are incompatible with the coupling device), but also dispensable. In addition, the business package includes a radio with a GSM telephone (GSM radio), which supports hands-free mobile communication.

---

[2] Parts are reprinted from Proceedings of the 16th European Conference on Artificial Intelligence (ECAI 2004), G. Friedrich, Elimination of Spurious Explanations, pp. 813–817. Copyright (2004), with permission from IOS Press.

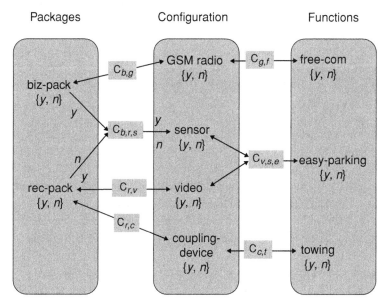

Figure 6.1. Constraint network of car example.

This domain can be modeled as a constraint satisfaction problem using the following variables and constraints: The example constraint network is depicted in Figure 6.1. The set of variables $\mathcal{V}$ is {*biz-pack, rec-pack, GSM-radio, sensor, video, coupling-device, free-com, easy-parking, towing*}. Unary constraints define the domains of the variables. For simplicity, it is assumed that for each variable the domain is {y, n}.

Further constraints are specified by the following tables:

$c_{r,v}$[3]: If *rec-pack* is chosen, then *video* must also be included (and vice versa).

$c_{b,r,s}$: If *biz-pack* is chosen and *rec-pack* is not chosen, then *sensor* is included. *rec-pack* and *sensor* are incompatible.

$c_{r,v}$:		$c_{b,r,s}$:		
rec-pack	*video*	*biz-pack*	*rec-pack*	*sensor*
y	y	y	y	n
n	n	y	n	y
		n	y	n
		n	n	n
		n	n	y

---

[3] Variables are abbreviated by their first letter(s).

$c_{v,s,e}$: If *video* or *sensor* is included, then *easy-parking* is supported (and vice versa).

$c_{v,s,e}$:

video	sensor	easy-parking
n	n	n
y	n	y
n	y	y
y	y	y

The constraint connecting the variables *biz-pack* and *GSM-radio* is called $c_{b,g}$. $c_{g,f}$ connects *GSM-radio* and *free-com*. The constraint connecting the variables *rec-pack* and *coupling-device* is called $c_{r,c} \cdot c_{c,t}$ connects *coupling-device* and *towing*. The tables for these four constraints are identical to the table of $c_{r,v}$.

To find out which packages are appropriate for a given customer, assume that the recommender system asks the customer if he or she wants the car to be able to tow trailers. In return, the customer may ask for the motivation behind this question. In our example, the answer would be: "If you want the car to be able to tow trailers, then the recreation package is appropriate for you." Likewise, the customer is asked if he or she wants hands-free mobile communication.

Assume that the customer wants both towing and hands-free communication. Consequently, the car would come with the business package and the recreation package, including both a video camera and a GSM radio. Functions supported by such a car would include easy parking, hands-free mobile communication, and towing. More formally, if the customer sets {*free-com* = y, *towing* = y}, then the solution to the constraints $C = \{c_{r,v},$ $c_{b,r,s},$ $c_{v,s,e},$ $c_{b,g},$ $c_{g,f},$ $c_{r,c},$ $c_{c,t}\}$ representing the configured car would be to assign {*video* = y, *sensor* = n, *GSM-radio* = y, *coupling-device* = y, *easy-parking* = y, *free-com* = y, *towing* = y, *biz-pack* = y, *rec-pack* = y}.

Assume that this solution is presented to the customer. If the customer asks which choices led to the parking capabilities of the specific configured car, clearly the following answer must be provided: easy parking is supported because the car comes with a video camera. This video camera is included because it is included in the recreation package. The recreation package was chosen because the car should support towing. The business package is not suitable because the sensor system cannot be included.

In the following sections we present concepts and algorithms that are able to compute such explanations automatically. In particular, those parts of the user input and knowledge base must be identified that can be used to explain the features of a given solution. We also review the standard approach for generating explanations and introduce the concept of well-founded explanations.

6.2.2 Generating explanations by abduction

Abduction is the widely accepted concept for generating explanations (Console et al. 1991, Junker 2004). The basic idea of these proposals is to use entailment ($\models$) to explain the outputs of a problem-solving process. Following Junker (2004), the approach of Friedrich (2004) is based on the concept of constraint satisfaction.

More formally, a constraint satisfaction problem (CSP) $(\mathcal{C}, \mathcal{V}, \mathcal{D})$ (Junker 2004) is defined by a set of variables $\mathcal{V}$, a set of constraints $\mathcal{C}$, and a global domain $\mathcal{D}$. Each constraint has the form $c(x_i, \ldots, x_j)$, where $x_i, \ldots, x_j$ are n variables in $\mathcal{V}$ and c is an n-ary constraint predicate. Each n-ary constraint predicate has an associated n-ary relation $R(c) \subseteq \mathcal{D}^n$. A mapping $v : \mathcal{V} \to \mathcal{D}$ of variables to values represented by a set of values associated to variables $\{(x_k = v_{x_k}) | x_k \in \mathcal{V} \wedge v_{x_k} = v(x_k)\}$ satisfies a constraint $c(x_i, \ldots, x_j)$ if and only if (iff) $(v_{x_i}, \ldots, v_{x_j}) \in R(c)$. Such a mapping v is a solution of the CSP iff it satisfies all constraints in $\mathcal{C}$.

A set of constraints C is satisfiable iff the CSP with variables $V(C)$ and constraints C has a solution. A CSP $(\mathcal{C}, \mathcal{V}, \mathcal{D})$ is trivially satisfied if $\mathcal{V}$ or $\mathcal{C}$ is empty. A mapping $v : \mathcal{V} \to \mathcal{D}$ is a solution of $(C, \mathcal{V}, \mathcal{D})$ iff $C \cup \{(x_k = v_{x_k}) | x_k \in V(C) \wedge v_{x_k} = v(x_k)\}$ is satisfied. Consequently, finding a solution to a CSP can be mapped to the problem of checking the consistency of a set of constraints.

Entailment is defined as usual:

Definition 1. (Junker) *A constraint ϕ is a (logical) consequence of a set of constraints C with variables $V(C)$ iff all solutions of the CSP with variables $V(C \cup \phi)$ and constraints C also satisfy the constraint ϕ. We write $C \models \phi$ in this case.*

The following standard definition of an abductive explanation is based on entailment:

Definition 2. *Let $(\mathcal{C}, \mathcal{V}, \mathcal{D})$ be a consistent CSP, and ϕ a constraint.*
A subset C of $\mathcal{C}$ is called an explanation for ϕ in $(\mathcal{C}, \mathcal{V}, \mathcal{D})$ iff $C \models \phi$. C is a minimal explanation for ϕ in $(\mathcal{C}, \mathcal{V}, \mathcal{D})$ iff no proper subset of C is an explanation for ϕ in $(\mathcal{C}, \mathcal{V}, \mathcal{D})$.

For the computation of minimal explanations, QUICKXPLAIN Junker (2004) can be applied, as introduced in Chapter 4. The required reasoning services, such as checking whether a set of constraints is satisfiable, generating a solution, and proving entailment, are supported by standard constraint satisfaction problem solvers. Given a set of constraints C, which are a minimal explanation for ϕ, then usually these constraints may be translated to natural language by exploiting the natural language descriptions of constraints C and ϕ and the values of their variables. This method is termed the *canned text* approach and was originally applied in MYCIN-like expert systems (Buchanan and Shortliffe 1984).

Assume that the recommender asks questions to deduce which packages are appropriate for the customer because cars are manufactured according to selected packages. After a question is posed, the customer may ask for its reason. We consider the case in which the recommender asks whether the feature *towing* is required. In this example, the explanation for *rec-pack* $= y$ is $\{towing = y, c_{c,t}, c_{r,c}\} \models rec\text{-}pack = y$. User inputs are considered to be additional constraints. This explanation serves as a justification for why a customer should provide information about the need to tow trailers. In particular, if towing is required, then by constraint $c_{c,t}$ it can be deduced that a coupling device is needed and by $c_{r,c}$ it may be deduced that the car must come with the recreation package option.

To sum up, minimal explanations can be exploited to give reasons for questions asked by the recommender. If a question is posed to deduce the product characteristics necessary to recommend products, then we can compute minimal explanations for these product characteristics as justifications for the questions posed by the recommender.

Now, assume that a solution (i.e., the recommended car) is presented to the customer and advertised to support not only towing and hands-free communication, but also easy parking (remember, the customer requirements were towing and hands-free communication). Suppose the customer wants to know why the car supports easy parking and which input leads to the inclusion of this function. If we follow strictly the definition of abductive explanations (Definition 2), then there are two minimal explanations (arguments) for *easy-parking* $= y$:

EXP1: $\{towing = y, c_{c,t}, c_{r,c}, c_{r,v}, c_{v,s,e}\} \models easy\text{-}parking = y$, which is intended. If the customer wants the feature *towing*, then the car needs a coupling device. The need for a coupling device implies that the car will be fitted with the recreation package. The recreation package includes a video camera at the rear, which supports easy parking. Hence requiring towing includes a video camera, which supports easy parking as a side effect. However, there is a second minimal explanation:

EXP2: $\{free\text{-}com = y, c_{g,f}, c_{b,g}, c_{b,r,s}, c_{r,v}, c_{v,s,e}\} \models easy\text{-}parking = y$. This explanation states that hands-free communication implies the inclusion of a

GSM telephone, which in turn implies the selection of the business package. The business package then causes either the sensor or the video camera to be included, depending on the choice regarding the recreation package. In both variants of such a car, easy parking is supported.

The original solution (for which we are generating explanations) includes a video camera. Clearly, the second *abductive* explanation is not correct with respect to the original solution, as easy parking is provided by a video camera and the video camera is included in the recreation package and not in the business package. An explanation of the consequences of user inputs must be based on the solution implied by these user inputs. The question is how such spurious explanations can be avoided.

6.2.3 Analysis and outline of well-founded explanations

In this section we elaborate on the reasoning of Friedrich (2004) and outline the basic ideas for eliminating spurious explanations. More specifically, the concept of projection as defined in database relational algebra is applied. Let constraint c have variables $x_i, \ldots, x_j$ (and possibly others). The projection of constraint c on $x_i, \ldots, x_j$ (written as $c\{x_i, \ldots, x_j\}$) is a constraint with variables derived from the variables of c by removing all variables not mentioned in $x_i, \ldots, x_j$, and the allowed-tuples of c are defined by a relation $R(c\{x_i, \ldots, x_j\})$ consisting of all tuples $(v_{x_i}, \ldots, v_{x_j})$ such that a tuple appears in $R(c)$ with $x_i = v_{x_i}, \ldots, x_j = v_{x_j}$. A constraint with no variables is trivially satisfied.

Subsequently, the solutions of the original problem and those of the two minimal explanations EXP1 and EXP2 are compared. The only solution of the original CSP based on the set of all constraints C and all user inputs *towing* $= y$ and *free-com* $= y$ is:

tow	fre	eas	rec	biz	vid	sen	gsm	cou
y	y	y	y	y	y	n	y	y

For the explanation of *easy-parking* $= y$ where EXP1 is used (i.e., the user input *towing* $= y$ and $c_{c,t}$, $c_{r,c}$, $c_{r,v}$, $c_{v,s,e}$), the solutions implied for the original variables $\mathcal{V}$ are

solution	tow	fre	eas	rec	biz	vid	sen	gsm	cou
1	y	y	y	y	y	y	n	y	y
2	y	n	y	y	n	y	n	n	y

In both solutions *easy-parking* is y. Solution 1 is identical to the solution of the original CSP. However, solution 2 differs in the variables {*biz-pack*,

GSM-radio, free-com} from the original CSP. One might argue that an explanation that possibly exploits variable values that are out of the scope of the original solution might lead to a spurious explanation. However, to derive *easy-parking* $= y$, we need only the constraints $c_{c,t}$, $c_{r,c}$, $c_{r,v}$, $c_{v,s,e}$. Consequently, variables {*biz-pack, GSM-radio, free-com*} are superfluous in the derivation of *easy-parking* $= y$. In addition, not all variables in $c_{r,v}$, $c_{v,s,e}$ are necessary for the derivation. If we analyze $c_{v,s,e}$ then we recognize that setting *video* to *y* implies *easy-parking* $= y$, regardless of the value of *sensor*. Consequently, the relevant variables in our case are *towing, coupling-device, rec-pack, video*, and *easy-parking*. The solutions of {*towing* $= y, c_{c,t}, c_{r,c}, c_{r,v}, c_{v,s,e}$} and the solutions of the original CSP projected on these relevant variables are identical.

For the explanation of *easy-parking* $= y$ where EXP2 is used (i.e., the user input *free-com* $= y$ and the constraints $c_{g,f}$, $c_{b,g}$, $c_{b,r,s}$, $c_{r,v}$, $c_{v,s,e}$), the solutions implied for the original variables $\mathcal{V}$ are

solution	tow	fre	eas	rec	biz	vid	sen	gsm	cou
1	y	y	y	y	y	y	n	y	y
2	n	y	y	n	y	n	y	y	n

Because only $c_{g,f}$, $c_{b,g}$, $c_{b,r,s}$, $c_{r,v}$, and $c_{v,s,e}$ are needed for the explanation, the variables not included in these constraints are irrelevant for this explanation. All other variables in these five constraints (i.e., *free-com, GSM-radio, biz-pack, rec-pack, video, sensor, easy-parking*) are needed. For example, if the variable *video* is deleted, then {*free-com* $= y, c_{g,f}, c_{b,g}, c_{b,r,s}, c_{r,v}\{r\}, c_{v,s,e}\{s, e\}$} $\not\models$ *easy-parking* $= y$ because there are solutions in which *easy-parking* $= n$.

The solutions with respect to these relevant variables are

solution	fre	eas	rec	biz	vid	sen	gsm
1	y	y	y	y	y	n	y
2	y	y	n	y	n	y	y

The explanation for *easy-parking* $= y$ must show that *easy-parking* $= y$ is contained in both solutions (in both logical models). In particular, the user input *free-com* $= y$ implies *biz-pack* $= y$. The constraints $c_{b,r,s}$, and *biz-pack* $= y$ imply either *sensor* $= y$ or *rec-pack* $= y$. *rec-pack* $= y$ implies *video* $= y$. In both cases, *easy-parking* $= y$ is implied by $c_{v,s,e}$. This explanation uses variable assignments that are not contained in the original solution, such as the car being supplied with video and not sensor equipment. Such an explanation is considered to be spurious because the reasoning includes a scenario (also called a possible world) that is apparently not possible given the current settings.

The principal idea of well-founded explanations is that an explanation C for a constraint ϕ and for a specific solution f must imply ϕ (as required for an abductive explanation); additionally, possible solutions of the explanation must be consistent with the specific solution f (with respect to the relevant variables).

6.2.4 Well-founded explanations

The definitions of *explanation* presented in this section provide more concise explanations compared with previous approaches (Junker 2004). Friedrich (2004) not only considers the relevance of constraints but also investigates the relevance of variables. The goal is to compute a minimal explanation consisting of the constraints and variables needed to deduce a certain property; these variables are exploited to construct an *understandable* chain of argument for the user (Ardissono et al. 2003). To introduce the following concepts, the projection operation on sets of constraints needs to be specified. $C\{V\}$ is defined by applying the projection on $V \subseteq \mathcal{V}$ to all $c \in C$ – that is, $C\{V\} = \{c\{V \cap V(c)\}|c \in C\}$. $V(c)$ are the variables of c.

Definition 3. *Let* $(\mathcal{C}, \mathcal{V}, \mathcal{D})$ *be a satisfiable CSP,* ϕ *a constraint.*

A-tuple (C, V) *where* $C \subseteq \mathcal{C}$ *and* $V \subseteq \mathcal{V}$ *is an explanation for* ϕ *in* $(\mathcal{C}, \mathcal{V}, \mathcal{D})$ *iff* $C\{V\} \models \phi$.

(C, V) *is a minimal explanation for* ϕ *in* $(\mathcal{C}, \mathcal{V}, \mathcal{D})$ *iff for all* $C' \subset C$ *and all* $V' \subset V$ *it holds that neither* (C', V) *nor* (C, V') *nor* (C', V') *is an explanation for* ϕ *in* $(\mathcal{C}, \mathcal{V}, \mathcal{D})$.

$(\{towing = y,\ c_{c,t},\ c_{r,c},\ c_{r,v},\ c_{v,s,e}\},\ \{towing,\ coupling\text{-}device,\ rec\text{-}pack,\ video,\ easy\text{-}parking\})$ is a minimal explanation for $easy\text{-}parking = y$.

For the computation of minimal explanations, the following monotonicity property is employed.

Remark 1. *If* $C\{V\} \not\models \phi$ *then for all* $V' \subseteq V$ *it holds that* $C\{V'\} \not\models \phi$. *The same applies for deleting constraints. However, it could be the case that* (C', V) *and* (C, V') *are minimal explanations for* ϕ *in* $(\mathcal{C}, \mathcal{V}, \mathcal{D})$ *and* $C' \subset C$ *and* $V' \subset V$.

A CSP solver to find a solution for the user is employed. Such a solution is described by a set of *solution-relevant* variables S which consists of all or a subset of variables of the CSP. Friedrich (2004) makes the reasonable assumption that sufficient information has been provided by the user (or about the user) such that the CSP unambiguously defines the values of variables S. More formally, $f = \{(x_k = v_{x_k})|x_k \in S \wedge (\mathcal{C} \models x_k = v_{x_k})\}$. For example, in the presented car configuration case, the user has to provide enough information so a car is well defined. Information gathering is the task of an elicitation process (Pu et al. 2003, Ardissono et al. 2003). The approach of Friedrich (2004) deals with the generation of explanations for the properties of a (possible) solution. Consequently, a user is free to explore various solutions and can ask for an explanation of the relation between user decisions and properties of a specific solution.

Subsequently, the projection $\int\{V\}$ of a solution $\int$ on variables V is defined as $\{(x_k = v_{x_k}) | x_k \in V \wedge (x_k = v_{x_k}) \in \int\}$.

The definition of well-founded explanations for a property ϕ with respect to a solution $\int$ is based on the following idea. First, an explanation (C, V) for ϕ – that is, $C\{V\} \models \phi$ – must show that every solution (also known as a *logical model* or sometimes as a *possible world model*) of the set of constraints $C\{V\}$ is a solution (a model) of ϕ. This follows the standard definition of abductive explanations. Second, if the explanation $C\{V\}$ permits some possible world models (i.e., solutions of $C\{V\}$) with value assignments of solution-relevant variables S *other* than those assigned in $\int$ (i.e., the solution of the unreduced set of constraints C), then it follows that the explanation of ϕ is based on possible world models that are in conflict with the original solution $\int$ (which was presented to the user). Therefore it must be ensured that every possible world (solution) of $C\{V\}$ is consistent with the variable assignment of $\int$.

Definition 4. *Let $(\mathcal{C}, \mathcal{V}, \mathcal{D})$ be a satisfiable CSP, $\int$ the solution of $(\mathcal{C}, \mathcal{V}, \mathcal{D})$ for the solution relevant variables S, (C, V) an explanation for ϕ.*

A-tuple (C, V) is a WF (WF) explanation for ϕ with respect to $\int$ iff every solution $s\{S\}$ of $(C\{V\}, \mathcal{V}, \mathcal{D})$ is a part of $\int$ (i.e., $s\{S\} \subseteq \int$).

(C, V) is a minimal WF (MWF) explanation for ϕ with respect to $\int$ iff for all $C' \subset C$ and for all $V' \subset V$ it holds that neither (C', V) nor (C, V') nor (C', V') is a WF explanation for ϕ in $(\mathcal{C}, \mathcal{V}, \mathcal{D})$ with respect to $\int$.

Remark 2. *Let $(\mathcal{C}, \mathcal{V}, \mathcal{D})$ be a satisfiable CSP, (C, V) an explanation for ϕ and $\int$ the solution of $(\mathcal{C}, \mathcal{V}, \mathcal{D})$ for the solution relevant variables S.*

(a) *An explanation (C, V) is a WF explanation for $(\mathcal{C}, \mathcal{V}, \mathcal{D})$ with respect to $\int$ iff $C\{V\} \models \int\{V\}$.*

(b) *If $(\mathcal{C}, \mathcal{V}, \mathcal{D})$ is satisfiable and $C \models \phi$ then there exists a WF explanation for ϕ.*

(c) *It could be the case that, for a satisfiable $(\mathcal{C}, \mathcal{V}, \mathcal{D})$ and a ϕ such that $C \models \phi$ and $\int$ the solution of $(\mathcal{C}, \mathcal{V}, \mathcal{D})$ for S, no minimal explanation of ϕ exists that is also WF.*

By applying Definitions 3 and 4 and Remark 2, the subsequent corollary follows immediately, which characterizes well-founded explanations based on logical entailment:

Corollary 1. *Let* $(\mathcal{C}, \mathcal{V}, \mathcal{D})$ *be a satisfiable CSP and* $\mathfrak{f}$ *the solution of* $(\mathcal{C}, \mathcal{V}, \mathcal{D})$ *for the solution relevant variables S.*
A-tuple (C, V) *where* $C \subseteq \mathcal{C}$ *and* $V \subseteq \mathcal{V}$ *is a WF explanation for* ϕ *with respect to* $\mathfrak{f}$ *iff* $C\{V\} \models \phi \wedge \mathfrak{f}\{V\}$.

Let a car be characterized by the solution-relevant variables *coupling-device*, *video*, *sensor*, and *GSM-radio*, which describe the configuration requested by the customer. ({*towing* = *y*, $c_{c,t}$, $c_{r,c}$, $c_{r,v}$, $c_{v,s,e}$}, {*towing*, *coupling-device*, *rec-pack*, *video*, *easy-parking*}) is a MWF explanation for *easy-parking* = *y* with respect to the solution (car configuration) *coupling-device* = *y*, *video* = *y*, *sensor* = *n*, *GSM-radio* = *y*. It entails *easy-parking* = *y* as well as *coupling-device* = *y* and *video* = *y*.

({*free-com* = *y*, $c_{g,f}$, $c_{b,g}$, $c_{b,r,s}$, $c_{r,v}$, $c_{v,s,e}$}, {*free-com*, *GSM-radio*, *biz-pack*, *rec-pack*, *video*, *sensor*, *easy-parking*}) is a minimal explanation for *easy-parking* = *y*, but it is not WF because it does not entail *video* = *y*, *sensor* = *n*.

The computation of MWF explanations is described by Friedrich (2004). The basic idea follows Corollary 1. First the variables and then the constraints are minimized so each further deletion of a constraint or variable will result in the loss of entailment of ϕ or $\mathfrak{f}\{V\}$.

In this section we have shown how an improved version of abductive reasoning can be used to compute explanations in constraint-based recommenders. In the next section, we review state-of-the-art explanation approaches for case-based recommenders, which are typically classified as knowledge-based recommenders. However, because this type of knowledge relies on similarity functions instead of logical descriptions, abduction cannot be exploited for explanation generation.

6.3 Explanations in case-based recommenders

The generation of solutions in case-based recommenders is realized by identifying the products that best fit a customer's query. Each item of a product database corresponds to a case. Typically a customer query puts constraints on the attributes of products – for example, a customer is interested only in digital cameras that cost less than a certain amount of money. In particular, given a query Q about a subset A_Q of attributes A of a case (product) description, the

Table 6.1. *Example product assortment: digital cameras.*

id	price	mpix	opt-zoom	LCD-size	movies	sound	waterproof
p_1	148	8.0	4×	2.5	no	no	yes
p_2	182	8.0	5×	2.7	yes	yes	no
p_3	189	8.0	10×	2.5	yes	yes	no
p_4	196	10.0	12×	2.7	yes	no	yes
p_5	151	7.1	3×	3.0	yes	yes	no
p_6	199	9.0	3×	3.0	yes	yes	no
p_7	259	10.0	3×	3.0	yes	yes	no
p_8	278	9.1	10×	3.0	yes	yes	yes

similarity of a case C to Q is typically defined (see McSherry 2005) as

$$sim(C, Q) = \sum_{a \in A_Q} w_a \, sim_a(C, Q) \tag{6.1}$$

The function $sim_a(C, Q)$ describes the similarity of the attribute values of the query Q and the case C for the attribute a. This similarity is weighted by w_a, expressing the importance of the attribute to the customer.

A recommendation set is composed of all cases C that have a maximal similarity to the query Q. Usually this recommendation set is presented directly to the customer, who may subsequently request an explanation as to why a product is recommended or why the requirements elicitation conversation must be continued. The typical approach used to answer a why-question in case-based recommenders is to compare the presented case with the customer requirements and to highlight which constraints are fulfilled and which are not (McSherry 2003b).

For example, if a customer is interested in digital cameras with a price less than €150, then p_1 is recommended out of the products depicted in Table 6.1. Asking *why* results in the explanation that the price of camera p_1 satisfies the customer constraint. The definition of similarities between attributes and requirements depends on the utility they provide for a customer. McSherry (2003b) distinguishes between more-is-better, less-is-better, and nominal attributes. Depending on the type of attribute, the answer to a why-question could be refined by stating how suitable the attribute value of an item is compared with the required one. Furthermore, the weights of the attributes can be incorporated into the answers – for instance, if the customer requires a price less than €160 and LCD size of more than 2.4 inches, where LCD size is weighted much more than price, then p_5 is recommended. The answer to a why-question can reflect this by stating, "p_5 is recommended because the price is €9 less and the LCD size is 0.6 inches greater than requested. Furthermore, emphasis was placed on LCD size." Of course, a graphical representation of the similarity between values and weights makes such recommendations easier to comprehend.

Basically, the similarity function can be viewed as a utility function over the set of possible cases. Consequently, more elaborated utility functions can be applied. Carenini and Moore (2001) expressed the utility function using an additive multiattribute value function based on the multiattribute utility theory. This function is exploited to generate customer-tailored arguments (explanations) as to why a specific product is advantageous for a customer. A statistical evaluation showed that arguments that are tailored to the preferences of customers are more effective than nontailored arguments or no arguments at all. Effectiveness in this context means that the user accepts the recommended product as a good choice.

As discussed previously, the requirements of a customer might be too specific to be fulfilled by any product. In this circumstance, why-explanations provide information about the violated constraints (McSherry 2003b). For example, if the customer requires a price less than €150 and a movie function, then no product depicted in Table 6.1 fulfills these requirements. However, p_1 and p_5 can be considered as most similar products for a given similarity function, although one of the user requirements is not satisfied. A why-explanation for p_1 would be, "p_1 is within your price range but does not include your movie requirement." The techniques presented in Chapter 4 can be used to generate minimal sets of customer requirements that explain why no products fit and, moreover, to propose minimal changes to the set of requirements such that matching products exist.

Conversational recommender systems help the customer to find the right product by supporting a communication process. Within this process various explanations can be provided. In ExpertClerk (Shimazu 2002), three sample products are shown to a shopper while their positive and negative characteristics are explained. Alternatives are then compared to the best product for a customer query – for example, "Camera p_4 is more expensive than p_1, but p_4 has a sensor with greater resolution (*mpix*)."

The concept of conversational recommenders is further elaborated on by McSherry (2005), while discussing the Top Case recommender, where a customer is queried about his or her requirements until a single best-matching product is identified, regardless of any additional requirements that may be posed by the customer. For example, if a customer requires a camera to be waterproof and to support movies and sound, but price and sensor resolution are of little importance, then p_8 is recommended regardless of any additional requirements about other attributes such as *LCD-size*, *opt-zoom*, *mpix*, and *price*. The systems described by McSherry (2005) take this into account by providing explanations of the form, "Case X differs from your query only in *attribute-set-1* and is the best case no matter what *attribute-set-2* you prefer." In this template, Case X is the recommended case, *attribute-set-1* contains the attributes that differ from the user's query, and *attribute-set-2* is the set of attributes for which preferred values have not been elicited because they are of no consequence to the recommendation.

Furthermore, Top Case can explain why a customer should answer a question by considering two situations. First, questions are asked to eliminate inappropriate products. Second, a question is asked such that a single product is confirmed as the final recommendation. A target case is always shown to the customer along with a number of competing (but similar) cases during requirement elicitation. The target case is selected randomly from the cases that are maximally similar to the customer requirements.

The explanation template used in the first situation (McSherry 2005) is:

If $a = v$ this will increase the similarity of Case X from S_1 to S_2 {and eliminate N cases [including Cases $X_1, X_2, \ldots X_r$]}

where:

- a is the attribute whose preferred value the user is asked to specify,
- v is the value of a in the target case,
- Case X is the target case,
- S_1 is the similarity of the target case to the current query,
- S_2 is the similarity of the target case that will result if the preferred value of a is v,
- N is the number of cases that will be eliminated if the preferred value of a is v, and
- Cases $X_1, X_2, \ldots, X_r$ are the cases that the user was shown in the previous recommendation cycle that will be eliminated if the preferred value of a is v.

The part of the template enclosed in curly brackets is shown only if cases are eliminated. The part enclosed in square brackets is used only if shown cases of the previous recommendations are eliminated.

If a question can confirm the target case then following template is used:

If $a = v$ this will confirm Case X as the recommended case.

where a, v, and Case X are defined as previously.

These approaches assume that a customer can formulate requirements about the attributes of products. However, customers may have only a very rough idea about the product they want to purchase. In particular, they may not have fixed preferences if compromises must be made. For example, if the preferred product is too expensive, then it is not clear which features should be removed. Consequently, help is required while exploring the product space. More generally, the transparency goal applies not only to the reasoning process but also to the case space itself (Sørmo et al. 2005). Transparency of suitable cases and the associated tradeoffs help the customer understand the options available and to reduce uncertainty about the quality of the decision. Tweaking critiquing, as presented in Chapter 4, provides a means exploring the case space effectively (Reilly et al. 2005b). This work was further improved by Pu and Chen (2007), who suggest a new method for ranking alternative products. The idea is to compare potential gains and losses (the "exchange rate") to the target candidate in order to rank alternatives. Moreover, Pu and Chen (2007) show that interfaces based on compound critiques help improve trust in recommender systems. Compared with explanations that merely explain the differences between alternatives and the target case, compound critiques improve the perceived competence and cognitive effort, as well as the intention to return to the interface.

The generation of explanations for case-based recommenders assumes that knowledge about products can be related to customer requirements. However, collaborative filtering recommenders must follow different principles because they generate suggestions by exploiting the product ratings of other customers.

6.4 Explanations in collaborative filtering recommenders

In contrast to the case with knowledge-based recommenders, explicit recommendation knowledge is not available if the collaborative filtering (CF) approach is applied. Consequently, recommendations based on CF cannot provide arguments as to why a product is appropriate for a customer or why a product does not meet a customer's requirements. The basic idea of CF is to mimic the human word-of-mouth recommendation process. Therefore, one approach for implementing explanations in CF recommenders is to give a comprehensible account of how this word-of-mouth approach works. Clearly, this approach aims to increase the transparency of recommendations but also has side effects regarding persuasion, which we subsequently discuss.

In the following we highlight key data and steps in the generation of CF outcomes and discuss their relevance to the construction of explanations. On a highly abstract level there are three basic steps that characterize the operation of CF, as presented in Chapter 2:

- Customers rate products.
- The CF locates customers with similar ratings (i.e., tastes), called neighbors.
- Products that are not rated by a customer are rated by combining the ratings of the customer's neighbors.

In concrete CF systems, these steps are usually implemented by sophisticated algorithms. In theory, the execution of these algorithms can serve as a profound explanation of the outcome. However, such an explanation tends to contribute to the confusion of customers and has negative impact on accepting the results of a recommender, as we will report later on. The art of designing effective explanation methods for CF is to provide the right abstraction level to customers.

Customers rate products. Transparency for this step of the collaborating filtering process is supported by presenting the customer information about the source of the ratings. Some recommenders are clear about the exploited ratings; for example, in the film domain, a system could consider ratings only when the customer explicitly rated a film, neglecting ratings based on other observations. However, some recommenders explore additional information by observing the customer and drawing conclusions based on these observations. For example, if a customer is observed to order mostly science fiction literature, then a movie recommender can exploit this information to hypothesize that the user would rate science fiction films highly. Although explaining this functionality might be an asset, in most cases this could lead to a severe drawback. For example, someone who is not interested in science fiction buys some science fiction books as presents. CF might conclude that the buyer likes this type of literature and starts recommending it. To change this behavior, it is important for the customer to understand the basis of recommendations.

Furthermore, as customers we may be interested in knowing how influential some of our ratings are. For example, if we were to wonder about the frequency of recommendations about science fiction films, we might discover that we had unjustifiably good ratings for some of these films. In addition, the diversity of rated products is essential. For example, if the majority of films rated positively by a customer are science fiction movies, then CF could be misled because the nearest neighbors would tend to be science fiction fans. Informing the customer about the diversity of the rated products gives important feedback about the quality of the customer's ratings.

Locate customers (neighbors) with similar tastes. In the second step, CF exploits the most similar customers to predict customer ratings. Clearly, the quality of this step is tightly linked to the similarity function used to locate customer ratings. From the customer's point of view, the question is whether this similarity function is acceptable – that is whether the selected neighbors have similar tastes. For example, a Beatles fan might not accept pop music ratings derived from other customers who do not like the Beatles themselves, although their ratings on other types of music are almost identical. Consequently, providing information about potential neighbors helps the customer to assess the quality of a prediction and to specify which neighbors are acceptable for predicting ratings.

Compute ratings by combining ratings of neighbors. In the third step, CF combines the ratings of a customer's neighbors. This is usually implemented by a weighted average. The reliability of this calculation is usually ascertained by considering the number and the variance of individual ratings. A large number of ratings of neighbors with a low variance can be regarded as more reliable than a few ratings that strongly disagree. Consequently, providing this information to customers may help them assess the quality of the recommendation, improving their confidence in the system.

In line with these considerations, Herlocker et al. (2000) examined various implementations of explanation interfaces in the domain of the "MovieLens" system. Twenty-one variants were evaluated. In particular, potential customers were asked, on a scale of 1 to 7, how likely they would be to go to see a recommended movie after a recommendation for this movie was presented and explained by one of the twenty-one different explanation approaches. In this comparison, Herlocker et al. (2000) also included the base case in which no additional explanation data were presented. In addition to the base case, an explanation interface was designed that just output the past performance of the recommendation system – for instance, "MovieLens has provided accurate preductions for you 80% of the time in the past".

The results of the study by Herlocker et al. (2000) were

- The best-performing explanation interfaces are based on the ratings of neighbors, as shown in Figures 6.2 and 6.3. In these cases similar neighbors liked the recommended film, and this was comprehensibly presented. The histogram performed better than the table.
- Recommenders using the simple statement about the past performance of MovieLens were the second best performer.
- Content-related arguments mentioning the similarity to other highly rated films or a favorite actor or actress were among the best performers.
- Poorly designed explanation interfaces decreased the willingness of customers to follow the recommendation, even compared with the base case.
- Too much information has negative effects; poor performance was achieved by enriching the data presented in histograms (Figure 6.2) with information about the proximity of neighbors.
- Interestingly, supporting recommendations with ratings from domain authorities, such as movie critics, did not increase acceptance.

Figure 6.2. Histogram of neighbors' ratings.

The study by Herlocker et al. (2000) showed that customers appreciate explanations. In addition, the study analyzed the correctness of decisions by asking the participants if the movie was worth seeing and how they would rate it after seeing the recommended movie. Interestingly, there was no observable difference between the base case (no explanation given) and the cases of the customers who received explanations. Summing up, explanations in the described setting helped persuade customers to watch certain movies but they did not improve the effectiveness of decisions.

A further line of research (O'Donovan and Smyth 2005) deals with trust and competence in the area of CF recommenders. The basic idea is to distinguish between producers and consumers of recommendations and to assess the quality of the information provided by producers. In such scenarios, the consumer of a recommendation is the customer; conversely, the neighbors of a customer are the producers. The quality of a producer's ratings is measured by

Movie: XYZ

Personalized Prediction: ****

Your Neighbors' Ratings for This Movie

Rating	Number of Neighbors
★	2
★★	4
★★★	8
★★★★	20
★★★★★	9

Figure 6.3. Table of neighbors' ratings (Herlocker et al. 2000).

the difference between the producer's ratings and those of consumers. These quality ratings offer additional explanation capabilities that help the customer assess recommendations. Possible explanation forms include "Product Jumbo-Family-Van was recommended by 20 users with a success rate of over 90% in more than 100 cases".

6.5 Summary

As outlined at the beginning of this chapter, there are many types of explanations and various goals that an explanation can achieve. Which type of explanation can be generated depends greatly on the recommender approach applied. A single implementation may, however, contribute to different explanation goals. For example, providing explanations could aim to improve both transparency and persuasiveness of the recommendation process. Great care must be taken to design explanation interfaces to achieve the planned effects.

Various forms of explanations for different approaches were described in this chapter. Because the development of recommender techniques is a highly active research area, many new proposals for supplementary explanation methods can be expected. However, explanations and their effects are an important field in their own right. From psychology it is known that customers do not have a stable utility function. Explanations may be used to shape the wishes and desires of customers but are a double-edged sword. On one hand, explanations can help the customer to make wise buying decisions; on the other hand, explanations can be abused to push a customer in a direction that is advantageous solely for the seller. As a result, a deep understanding of explanations and their effects on customers is of great interest.

7

Evaluating recommender systems

In previous chapters we introduced a variety of different recommendation techniques and systems developed by researchers or already in use on commercial platforms. In the future, many new techniques will claim to improve prediction accuracy in specific settings or offer new ways for users to interact with each other, as in social networks and Web 2.0 platforms.

Therefore, methods for choosing the best technique based on the specifics of the application domain, identifying influential success factors behind different techniques, or comparing several techniques based on an optimality criterion are all required for effective evaluation research. Recommender systems have traditionally been evaluated using offline experiments that try to estimate the prediction error of recommendations using an existing dataset of transactions. Some point out the limitations of such methods, whereas others argue that the quality of a recommender system can never be directly measured because there are too many different objective functions. Nevertheless, the widespread use of recommender systems makes it crucial to develop methods to realistically and accurately assess their true performance and effect on the users. This chapter is therefore devoted to discussing existing evaluation approaches in the light of empirical research methods from both the natural and social sciences, as well as presenting different evaluation designs and measures that are well accepted in the research community.

7.1 Introduction

Recommender systems require that users interact with computer systems as well as with other users. Therefore, many methods used in social behavioral research are applicable when answering research questions such as *Do users find interactions with a recommender system useful?*, *Are they satisfied with the*

Table 7.1. *Basic characteristics of evaluation designs.*

Subject	Online customers, students, historical user sessions, simulated users, computers
Research method	Experimental, quasi-experimental, or nonexperimental
Setting	Real-world scenarios, lab

quality of the recommendations they receive?, What drives people to contribute knowledge such as ratings and comments that boost the quality of a system's predictions? or What is it exactly that users like about receiving recommendations? Is it the degree of serendipity and novelty, or is it just the fact that they are spared from having to search for them? Many more questions like these could be formulated and researched to evaluate whether a technical system is efficient with respect to a specified goal, such as increasing customer satisfaction or ensuring the economic success of an e-commerce platform. In addition, more technical aspects are relevant when evaluating recommendation systems, related, for instance, to a system's technical performance such as responsiveness to user requests, scalability, and peak load or reliability. Furthermore, goals related to the system's life cycle, such as ramp-up efforts, maintainability, and extensibility, as well as lowering the cost of ownership, can be thought of and are of interest for evaluation research.

Because of the diverse nature of possible evaluation exercises in the domain of recommendation systems, we start with very basic properties of research methodologies, as depicted in Table 7.1. The table differentiates empirical research based on the units that are subjected to research methods, such as people or computer hardware. Furthermore, it denotes the top-level taxonomy of empirical research methods, namely experimental and nonexperimental research, as well as the distinction between real-world and lab scenarios where evaluations can be conducted. Each of these meta-level concepts will be explained in more detail in the remainder of this chapter.

7.2 General properties of evaluation research

Empirical research itself has been subject to intense scrutiny from areas as diverse as philosophy and statistics (Pedhazur and Schmelkin 1991). Rather than repeating these principles, guidelines, and procedures here, we focus on some particular aspects and discuss them in the context of evaluating recommender systems. We begin with some general thoughts on rigor and validity of empirical evaluations. Finally, we briefly discuss some selected general criteria that must be kept in mind when evaluating recommendation applications with scientific rigor.

7.2.1 General remarks

Thoroughly describing the methodology, following a systematic procedure, and documenting the decisions made during the course of the evaluation exercise ensure that the research can be repeated and results verified. This answers the question of *how* research has been done. Furthermore, criteria such as the *validity, reliability*, and *sensibility* of the constructs used and measured relate to the subject matter of the research itself, questioning *what* is done. Notably, asking whether the right concepts are measured or whether the applied research design is valid is necessary.

Internal validity refers to the extent to which the effects observed are due to the controlled test conditions (e.g., the varying of a recommendation algorithm's parameters) instead of differences in the set of participants (predispositions) or uncontrolled/unknown external effects. In contrast, *External validity* refers to the extent to which results are generalizable to other user groups or situations (Pedhazur and Schmelkin 1991). When using these criteria to evaluate recommender systems, questions arise such as *Is it valid to exploit users' clicks on pages displaying details of an item as an indicator of their opinion about an item?* External validity examines, for instance, whether the evaluated recommendation scenario is representative of real-world situations in which the same mechanism and user interface of the technique would be used, and whether the findings of the evaluation exercise are transferrable to them. For example, will an increase in users' purchase rate of recommended items because of a new hybrid computation mechanism also be observable when the system is put to the field? *Reliability* is another postulate of rigorous empirical work, requiring the absence of inconsistencies and errors in the data and measurements. Finally, *sensibility* necessitates that different evaluations of observed aspects are also reflected in a difference in measured numbers.

Furthermore, issues surrounding research findings include not only their statistical significance but also information about the size of their effect and thus their significance with respect to the potential impact on real-world scenarios. For instance, what is the impact of a 10 percent increase in the accuracy of predicted ratings? Will this lead to a measurable increase in customer loyalty and lower churn rates of an e-commerce platform? Unfortunately, based on the current state of practice, not all these fundamental questions can be answered, but some guidance for designing evaluations is available, and researchers are urged to critically reflect on their own work and on the work of others.

7.2.2 Subjects of evaluation design

People are typically the subjects of sociobehavioral research studies – that is, the focus of observers. Obviously, in recommender systems research, the populations of interest are primarily specific subgroups such as online customers, web users, or students who receive adaptive and personalized item suggestions.

An experimental setup that is widespread in computer science and particularly, for instance, in subfields such as machine learning (ML) or information retrieval (IR) is datasets with synthetic or historical user interaction data. The basic idea is to have a collection of user profiles containing preference information such as ratings, purchase transactions, or click-through data that can be split into training and testing partitions. Algorithms then exploit the training data to make predictions with the hidden testing partition. The obvious advantage of this approach is that it allows the performance of different algorithms to be compared against each other. Simulating a dataset comes with the advantage that parameters such as distribution of user properties, overall size, or rating sparsity can be defined in advance and the test bed perfectly matches these initial requirements. However, there is significant risk that synthetic datasets are biased toward the design of a specific algorithm and that they therefore treat other algorithms unfairly. For this reason synthetic datasets are advisable only to test recommendation methods for obvious flaws or to measure technical performance criteria such as average computation times – that is, the computer itself becomes subject of the evaluation rather than users.

Natural datasets include historical interaction records of real users. They can be categorized based on the type of user actions recorded. For example, the most prominent datasets from the movie domain contain explicit user ratings on a multipoint Likert scale. On the other hand, datasets that are extracted from web server logs consist of implicit user feedback, such as purchases or add-to-basket actions. The sparsity of a dataset is derived from the ratio of empty and total entries in the user–item matrix and is computed as follows:

$$sparsity = 1 - \frac{|R|}{|I| \cdot |U|} \qquad (7.1)$$

where

$$R = ratings$$

$$I = items$$

$$U = users$$

In Table 7.2 an incomplete list of popular datasets, along with their size characteristics, is given. The well-known MovieLens dataset was derived from

Table 7.2. *Popular data sets.*

Name	Domain	Users	Items	Ratings	Sparsity
BX	Books	278,858	271,379	1,149,780	0.9999
EachMovie	Movies	72,916	1,628	2,811,983	0.9763
Entree	Restaurants	50,672	4,160	N/A	N/A
Jester	Jokes	73,421	101	4.1M	0.4471
MovieLens 100K	Movies	967	4,700	100K	0.978
MovieLens 1M	Movies	6,040	3,900	1M	0.9575
MovieLens 10M	Movies	71,567	10,681	10M	0.9869
Netflix	Movies	480K	18K	100M	0.9999
Ta-Feng	Retail	32,266	N/A	800K	N/A

a movie recommendation platform developed and maintained by one of the pioneers in the field, the GroupLens research group[1] at the University of Minnesota. The EachMovie dataset was published by HP/Compaq and, despite not being publicly available for download since 2004, has still been used by researchers since. One additional movie dataset that has recently been made public is Netflix.[2] Published in conjunction with the Netflix Prize,[3] the company promised $1 million for the first team to provide a 10 percent improvement in prediction accuracy compared with its in-house recommender system. This competition stimulated much research in this direction. Finally, this threshold was reached by the team BellKor's Pragmatic Chaos in 2009. None of the aforementioned movie datasets contain item descriptions such as the movies' plots, actors, or directors. Instead, if algorithms require this additional content information, it is usually extracted from online databases such as the Internet Movie Database – IMDB.[4]

The BX dataset was gathered from a community platform for book lovers and contains explicit and implicit ratings for a large number of books (Ziegler et al. 2005). In contrast, the rating data from the joke recommender Jester represents a very dense dataset with only a few different items (Goldberg et al. 2001). The Entree data collection contains historical sessions from a critique-based recommender, as discussed in Chapter 4. Finally, the Ta-Feng dataset provides a representative set of purchase transactions from the retail domain with a very

---

[1] See http://www.grouplens.org/.
[2] See http://archive.ics.uci.edu/ml/datasets/Netflix+Prize.
[3] See http://www.netflixprize.com/.
[4] See http://www.imdb.com/.

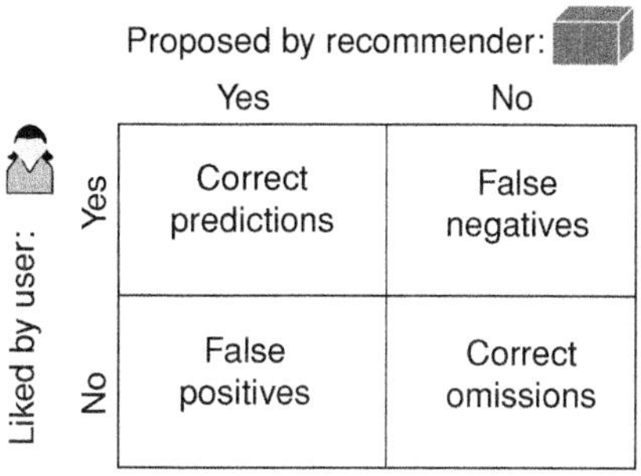

Figure 7.1. Types of errors.

low number of ratings per user. The dataset was exploited to evaluate the hybrid Poisson aspect modeling technique presented by Hsu et al. (2004).

Additional stimuli in the field come from social web platforms that either make their interaction data available to the public or allow researchers to extract this information. From CiteULike[5] and Bibsonomy,[6] tagging annotations on research papers are collected and public datasets can be downloaded from there. The social bookmarking platform del.icio.us[7] is another example for data from the social web that is used for evaluation research.

Nevertheless, the results of evaluating recommender systems using historical datasets cannot be compared directly to studies with real users and vice versa. Consider the classification scheme depicted in Figure 7.1. If an item that was proposed by the recommender is actually liked by a user, it is classified as a *correct prediction*. If a recommender is evaluated using historical user data, preference information is only known for those items that have been actually rated by the users. No assumptions can be made for all unrated items because users might not have been aware of the existence of these items. By default, these unknown item preferences are interpreted as disliked items and can therefore lead to false positives in evaluations – that is, the recommender is punished for recommending items that are not in the list of positively rated items of the historical user session, but that might have been welcomed by the actual user if they were recommended in a live interaction.

In contrast, when recommending items to real users, they can be asked to decide instantly if they like a proposed item. Therefore, both correct predictions and false positives can be determined in this setting. However, one cannot assess whether users would have liked items that were not proposed to them – that is,

---

[5] See http://www.citeulike.org/.
[6] See http://www.bibsonomy.org/.
[7] See http://www.delicious.com/.

the false negatives. Thus, one needs to be aware that evaluating recommender systems using either online users or historical data has some shortcomings. These shortcomings can be overcome only by providing a marketplace (i.e., the set of all recommendable items) that is completely transparent to users who, therefore, rate all items. However, in markets with several hundreds or even thousands of items, dense rating sets are both impossible and of questionable value, as no one would need recommendations if all items are already known by users beforehand.

7.2.3 Research methods

Defining the goals of research and identifying which aspects of the users or subjects of the scientific inquiry are relevant in the context of recommendation systems lie at the starting point of any evaluation. These observed or measured aspects are termed *variables* in empirical research; they can be assumed to be either independent or dependent. A few variables are always independent because of their nature – for example, gender, income, education, or personality traits – as they are, in principle, static throughout the course of the scientific inquiry. Further variables are independent if they are controlled by the evaluation design, such as the type of recommendation algorithm that is applied to users or the items that are recommended to them. Dependent variables are those that are assumed to be influenced by the independent variables – for instance, user satisfaction, perceived utility, or click-through rate can be measured.

In an *experimental research design*, one or more of the independent variables are manipulated to ascertain their impact on the dependent variables:

> An *experiment* is a study in which at least one variable is manipulated and units are randomly assigned to the different levels or categories of the manipulated variables (Pedhazur and Schmelkin 1991, page 251).

Figure 7.2 illustrates such an experiment design, in which subjects (i.e., units) are randomly assigned to different treatments – for instance, different recommendation algorithms. Thus, the type of algorithm would constitute the manipulated variable. The dependent variables (e.g., v_1 and v_2 in Figure 7.2) are measured before and after the treatment – for instance, with the help of a questionnaire or by implicitly observing user behavior. Environmental effects from outside the experiment design, such as a user's previous experience with recommendation systems or the product domain, also need to be controlled – for instance, by ensuring that only users that are sophisticated or novices in the

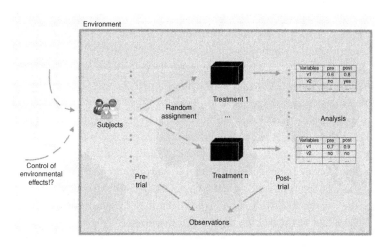

Figure 7.2. Example of experiment design.

product domain participate in the experiment (i.e., by elimination or inclusion) or by factorization (i.e., ensuring that sophisticated and novice users have an equal chance of being assigned to a treatment $1 \leq i \leq n$). For a deeper discussion on conducting live-user experiments and alternate experiment designs, the reader is referred to a textbook specifically focusing on empirical research (Pedhazur and Schmelkin 1991).

When experimenting offline with datasets, units (i.e., historical user sessions) do not need to be randomly assigned to different treatments. Instead, all algorithms can be evaluated on all users in the dataset. Although sequentially assigning real users to several treatments would lead to strongly biased results from repeated measurements (e.g., users might remember their initial answers), in offline experiments the historical user behavior will obviously remain static.

A *quasi-experimental design* distinguishes itself from a real experiment by its lacking random assignments of subjects to different treatments – in other words, subjects decide on their own about their treatment. This might introduce uncontrollable bias because subjects may make the decision based on unknown reasons. For instance, when comparing mortality rates between populations being treated in hospitals and those staying at home, it is obvious that higher mortality rates in hospitals do not allow us to conclude that these medical treatments are a threat to people's lives. However, when comparing purchase rates of e-commerce users who used a recommender system with the purchase rates of those who did not, a methodological flaw is less obvious. On one hand, there could be unknown reasons (i.e., uncontrolled variables) that let users who have a strong tendency to buy also use the recommender system, whereas on the other hand, a higher purchase rate of recommender users could really be an indicator of the system's effectiveness. Therefore, the effectiveness of quasiexperimental designs is not undisputed and, as a consequence, their results must be interpreted with utmost circumspection, and conclusions need to be drawn very carefully (Pedhazur and Schmelkin 1991).

Nonexperimental designs include all other forms of quantitative research, as well as qualitative research. Quantitative research relies on numerical measurements of different aspects of objects, such as asking users different questions about the perceived utility of a recommendation application with answers on a seven-point Likert scale, requiring them to rate a recommended item or measuring the viewing time of different web pages. In contrast, qualitative research approaches would conduct interviews with open-ended questions, record think-aloud protocols when users interact with a web site, or employ focus group discussions to find out about users' motives for using a recommender system. For a more elaborate discussion of qualitative research designs, the reader is referred to Miles and Huberman (1994) and Creswell (2009).

One nonexperimental research design that is quite interesting in the context of evaluating recommender systems is *longitudinal research*, in which the entity under investigation is observed repeatedly as it evolves over time. Such a design allows criteria such as the impact of recommendations on the customer's lifetime value to be measured. Such research endeavors are very complex and costly to carry out, however, as they involve observing subjects over a long period of time. Zanker et al. (2006) conducted longitudinal research in which the sales records of an online store for the periods before and after introducing a recommendation system were analyzed and compared with each other. One of the most interesting results was that the top-seller list (i.e., the items that were most frequently sold) changed considerably and some items that were rarely purchased in the period before the introduction of the recommendation system became top-selling items afterward. Further analysis indicated that the increase in the number of pieces sold for these items correlated positively with the occurrence of these items in actual recommendations.

Cross-sectional research designs can also be very promising in the recommender systems domain, analyzing relations among variables that are simultaneously measured in different groups, allowing generalizable findings from different application domains to be identified.

Case studies (Stake 1995, Yin 2002) represent an additional way of collecting and analyzing empirical evidence that can be applied to recommendation systems research when researchers are interested in more principled questions. They focus on answering research questions about how and why and combine whichever types of quantitative and qualitative methods necessary to investigate contemporary phenomena in their real-life contexts. Therefore, to answer the question of how recommendation technology contributed to Amazon.com's becoming the world's largest book retailer would require a case study research design.

7.2.4 Evaluation settings

The evaluation setting is another basic characteristic of evaluation research. In principle, we can differentiate between *lab studies* and *field studies*. A lab situation is created expressly for the purpose of the study, whereas a field study is conducted in an preexisting real-world environment.

Lab situations come with the major advantage that extraneous variables can be controlled more easy by selecting study participants. However, doubts may exist about study participants who are motivated to participate primarily by money or prizes. Therefore, a study needs to be carefully designed to ensure that participants behave as they would in a real-world environment. In contrast, research that is conducted in the field comes with the advantage that users are intrinsically motivated to use a system or even spend their own money when trusting a recommendation system and purchasing the item that was proposed to them. Nevertheless, researchers tend to have little control over the system, as the commercial interests of the platform operator usually prevail. Typically, one has little choice over the different settings, as other factors, such as the availability of data or real-world platforms, will influence the decision.

7.3 Popular evaluation designs

Up to now, experiment designs that evaluate different algorithm variants on historical user ratings derived from the movie domain form by far the most popular evaluation design and state of practice. To substantiate this claim, we conducted a survey of all research articles that appeared on the topic of recommender systems in the reputed publication *ACM Transactions on Information Systems* (*ACM TOIS*) over a period of five years (2004–2008). Twelve articles appeared, as listed in Table 7.3 in chronological order. The first of them has been the most influential with respect to evaluating recommender systems and, in particular, collaborative filtering systems, as it focuses on comparing different accuracy measures for collaborative filtering algorithm variants. As can be seen from Table 7.3, in three-quarters of these articles, offline experiments on historical user sessions were conducted, and more than half of authors chose movie recommendations as their application domain. Adomavicius et al. (2005)

Table 7.3. *Evaluation designs in ACM TOIS 2004–2008.*

Reference	Approach	Goal (Measures)	Domain
Herlocker et al. (2004)	Offline experiments	Accuracy (MAE,[a] ROC[b] curve)	ML[c]
Middleton et al. (2004)	Experimental user study	Accuracy (hit rate)	Web pages, e-mails
Hofmann (2004)	Offline experiments	Accuracy (MAE, RMSE[d])	EM[e]
Huang et al. (2004)	Offline experiments	Accuracy (Precision, Recall, F1)	Bookstore
Deshpande and Karypis (2004)	Offline experiments	Accuracy (hit rate, rank metric)	EM, ML, mail order purchases
Miller et al. (2004)	Offline experiments	Accuracy (MAE, Recall), catalog coverage	ML
Adomavicius et al. (2005)	Offline experiments	Accuracy (Precision, Recall, F1)	Movie ratings
Wei et al. (2005)	Offline experiments with simulated users	Marketplace efficiency	Synthetic datasets
Lee et al. (2006)	Qualitative user study	Usage analysis and wish list for improved features	Broadcast news
Ma et al. (2007)	Experimental user study	Search efficiency (mean log search time, questionnaire)	Web pages
Im and Hars (2007)	Offline experiments	Accuracy (MAE– NMAE[f])	Movie ratings, research papers, BX-Books, EM
Wang et al. (2008)	Offline experiments	Accuracy (MAE)	EM, ML

[a] MAE: mean absolute error.
[b] ROC: receiver operating characteristic.
[c] ML: MovieLens dataset.
[d] RMSE: root mean square error.
[e] EM: EachMovie dataset.
[f] NMAE: normalized mean absolute error.

and Im and Hars (2007), collected these ratings from specifically designed platforms that also collected situational parameters such as the occasion in which the movie was consumed. The others worked on the then publicly available datasets MovieLens and EachMovie (see Subsection 7.2.2). Experimental studies involving live users (under lab conditions) were done by Middleton et al. (2004) and Ma et al. (2007), who measured the share of clickthroughs from overall recommended items and search efficiency with respect to search time. A qualitative research design was employed only by Lee et al. (2006), who evaluated an automated content-based TV news delivery service and explored the usage habits of a group of sixteen users. The study consisted of pre- and post-trial questionnaires, diaries from each user during the one-month trial, and interaction data. The outcome of the study was a wish list for feature improvements and more insights into the usage patterns of the tool – for example, that users mainly accessed the section on latest news and used the system's search functionality only very rarely.

7.4 Evaluation on historical datasets

Because of the paramount importance of experimental evaluations on historical datasets for recommender systems research, we focus in this section on how they are carried out. Based on a small example, we discuss popular methodologies and metrics, as well as the interpretation of results.

7.4.1 Methodology

For illustrative purposes, we assume that an arbitrary historical user profile contains ten fictitious movie ratings, as depicted in Table 7.4. When evaluating a recommendation method, a group of user profiles is normally used as input to train the algorithm and build a model that allows the system to compute recommendations efficiently at run time. A second group of user profiles, different from the first, is required for measuring or testing the algorithm's performance. To ensure that the measurements are reliable and not biased by some user profiles, the random split, model building, and evaluation steps are repeated several times to determine average results. *N-fold cross-validation* is a stratified random selection technique in which one of N disjunct fractions of the user profiles of size $\frac{1}{N}$ is repeatedly selected and used for evaluation, leaving the remaining $\frac{N-1}{N}$ user profiles to be exploited for building the algorithm's model. Consequently, each user profile is used exactly once to evaluate the algorithm and $N - 1$ times to contribute to the algorithm's model building step. In the

Table 7.4. *Example user ratings.*

Row	UserID	MovieID	Rating
1	234	110	5
2	234	151	5
3	234	260	3
4	234	376	5
5	234	539	4[a]
6	234	590	5
7	234	649	1
8	234	719	5[a]
9	234	734	3
10	234	736	2

[a] Randomly selected ratings for testing.

extreme case, in which N is equal to the total number of user profiles, the splitting method is termed *leave one out*. From the computational point of view this method is the most costly, as the model has to be rebuilt for each user. At the same time, however, it allows the algorithm to exploit the maximum amount of community data for learning. Therefore, in situations in which the user base is only very small – a few hundred different profiles – a leave-one-out strategy can make sense to use as much data as possible for learning a model.

In addition, during the testing step, the user profile must be split into two groups, namely, user ratings to train and/or input the algorithm (i.e., determining similar peers in case of collaborative filtering) and to evaluate the predictions. In our example, we assume that the fifth and the eighth rows (see footnote in Table 7.4) of user number 234 have been randomly selected for testing – that is, they constitute the *testing set* and the other eight rows are part of the *training* or *learning set*.

One of two popular variants may be applied to split the rating base of the currently evaluated user into training and testing partitions. The *all but N* method assigns a fixed number N to the testing set of each evaluated user, whereas the *given N* method sets the size of the training partition to N elements. Both methods have their strengths, especially when one varies N to evaluate the sensitivity of an algorithm with respect to different testing or training set sizes. A fixed training set size has the advantage that the algorithm has the same amount of information from each tested user, which is advantageous when measuring the predictive accuracy. In contrast, fixed testing set sizes establish equal conditions for each user when applying classification metrics.

When evaluating algorithms, such as a simple nonpersonalized recommendation mechanism, that suggest the same set of popular items to every user and therefore do not need to identify similar peers or do not require a set of liked items to query the product catalog for similar instances, the evaluation method is effectively *Given 0* – that is, the training set of past ratings of evaluated users is empty and all ratings can be used for testing the algorithm's predictions. Such an evaluation approach also applies to the constraint-based recommendation paradigm (see Chapter 4).

Based on the scale of historical ratings available – that is, unary (purchase) transactions or ratings on Likert scales – an evaluation can examine the prediction or the classification capability of a recommender system. A *prediction task* is to compute a missing rating in the user/item matrix. The prediction task requires Likert scale ratings that have been explicitly acquired from users, such as the ones specified in Table 7.4. The *classification task* selects a ranked list of *n* items (i.e., the recommendation set) that are deemed to be relevant for the user. The recommendation set typically contains between three and ten items, as users typically tend not to want to scroll through longer lists. To evaluate the accuracy of an algorithm's classifications, Likert scale ratings need to be transformed into relevant and not-relevant items – for instance, classifying only items rated 4 and above as relevant. This leads us back to the discussion in Section 7.2.2 on how to treat items with unknown rating values. The current state of practice assumes that these items are nonrelevant, and therefore evaluation measures reward algorithms only for recommending relevant items from the testing set, as is explained in the next subsection.

7.4.2 Metrics

Herlocker et al. (2004) provide a comprehensive discussion of accuracy metrics together with alternate evaluation criteria, which is highly recommended for reading. We therefore focus only on the most common measures for evaluations based on historical datasets.

Accuracy of predictions. When evaluating the ability of a system to correctly predict a user's preference for a specific item, mean absolute error (MAE) is undisputedly the most popular measure, as confirmed by the outcome of the small survey in Section 7.3. The MAE metric was already discussed in the context of collaborative filtering (see Chapter 2) and when dynamizing a weighted hybridization strategy (Chapter 5). Nevertheless, we restate its computation scheme for reasons of completeness.

$$MAE = \frac{\sum_{u \in U} \sum_{i \in testset_u} |rec(u, i) - r_{u,i}|}{\sum_{u \in U} |testset_u|} \qquad (7.2)$$

MAE computes the average deviation between computed recommendation scores ($rec(u, i)$) and actual rating values ($r_{u,i}$) for all evaluated users $u \in U$ and all items in their testing sets ($testset_u$). Alternatively, some authors, such as Sarwar et al. (2001), compute the root mean square error (RMSE) to put more emphasis on larger deviations or, similar to Goldberg et al. (2001), create a normalized MAE (NMAE) with respect to the range of rating values.

$$NMAE = \frac{MAE}{r_{max} - r_{min}} \tag{7.3}$$

r_{max} and r_{min} stand for the highest and lowest rating values to normalize $NMAE$ to the interval $0 \ldots 1$. Consequently, the normalized deviations should be comparable across different application scenarios using different rating scales. Im and Hars (2007), for example, used NMAE to compare the effectiveness of collaborative filtering across different domains.

Accuracy of classifications. The purpose of a classification task in the context of product recommendation is to identify the n most relevant items for a given user. *Precision* and *Recall* are the two best-known classification metrics; they are also used for measuring the quality of information retrieval tasks in general. Both are computed as fractions of $hits_u$, the number of correctly recommended relevant items for user u. The Precision metric (P) relates the number of hits to the total number of recommended items ($|recset_u|$).

$$P_u = \frac{|hits_u|}{|recset_u|} \tag{7.4}$$

In contrast, the Recall (R) computes the ratio of hits to the theoretical maximum number of hits owing to the testing set size ($|testset_u|$).

$$R_u = \frac{|hits_u|}{|testset_u|} \tag{7.5}$$

According to McLaughlin and Herlocker (2004), measuring an algorithm's performance based on Precision and Recall reflects the real user experience better than MAE does because, in most cases, users actually receive ranked lists from a recommender instead of predictions for ratings of specific items. They determined that algorithms that were quite successful in predicting MAEs for rated items produced unsatisfactory results when analyzing their top-ranked items. Carenini and Sharma (2004a) also argue that MAE is not a good indicator from a theoretical perspective, as all deviations are equally weighted. From the user's perspective, however, the only fact that counts is whether an item is recommended.

Assume that a recommender computes the following item/rating -tuples for user 234, whose rating profile is presented in Table 7.4:

$$recset_{234} = \{(912, 4.8), (47, 4.5), (263, 4.4), \mathbf{(539, 4.1)}, (348, 4), \ldots, \mathbf{(719, 3.8)}\}$$

Although only a single item from the user's test set is recommended among the top five, an MAE-based evaluation would give favorable results, as the absolute error is on average, 0.65. If the evaluation considered only the top three ranked items, however, Precision and Recall would be 0, and if the recommendation set is changed to contain only the five highest ranked items, $P_{234} = \frac{1}{5}$ and $R_{234} = \frac{1}{2}$.

By increasing the size of a recommendation set, the tradeoff between Precision and Recall metrics can be observed. Recall will typically improve as the chance of hitting more elements from the test set increases with recommendation set size, at the expense of lower Precision. For instance, if item 719 was recommended only on the twentieth and last position of the recommendation list to user 234, Recall would jump to 100 percent, but Precision would drop to 10 percent.

Consequently, the F1 metric is used to produce evaluation results that are more universally comparable:

$$F1 = \frac{2 \cdot P \cdot R}{P + R} \qquad (7.6)$$

The $F1$ metric effectively averages Precision and Recall with bias toward the weaker value. Comparative studies on commercial datasets using P, F, and $F1$ have, for example, been conducted by Sarwar et al. (2000b) and Zanker et al. (2007).

Some argue, however, that a classification measure should reflect the proportion of users for which at least one item from the user's test profile is recommended. In other words, the hit rate should be defined as

$$hitrate_u = \begin{cases} 1 & : \text{if } hits_u > 0 \\ 0 & : \text{else} \end{cases} \qquad (7.7)$$

Deshpande and Karypis (2004) used this measure to compare their item-based collaborative filtering variant with a user-based one, whereas O'Sullivan et al. (2004) employed it for measuring the quality of TV program guides. Nguyen and Ricci (2007b) assessed different algorithm variants for a mobile critique-based recommender, also based on hit rate. Interestingly, they presented a simulation model that allows one to evaluate historical critiquing sessions by replaying the query input. The logs were derived from user studies on the mobile recommendation application presented by Nguyen and Ricci (2007a).

Accuracy of ranks. Rank scores extend the results of classification metrics with a finer level of granularity. They differentiate between successful hits by also taking their relative position in recommendation lists into account. Breese et al. (1998) propose a metric that assumes decreasing utilities based on items' rank. The parameter α sets the half-life of utilities, which means that a successful hit at the first position of the recommendation list has twice as much utility to the user than a hit at the $\alpha + 1$ rank. The rationale behind this weighting is that later positions have a higher chance of being overlooked by the user, even though they might be useful recommendations.

$$rankscore_u = \sum_{i \in hits_u} \frac{1}{2^{\frac{rank(i)-1}{\alpha}}} \tag{7.8}$$

$$rankscore_u^{max} = \sum_{i \in testset_u} \frac{1}{2^{\frac{idx(i)-1}{\alpha}}} \tag{7.9}$$

$$rankscore'_u = \frac{rankscore_u}{rankscore_u^{max}} \tag{7.10}$$

The function $rank(i)$ returns the position of item i in the user's recommendation list. $Rankscore_u^{max}$ is required for normalization and returns the maximum achievable score if all the items in the user's test set were assigned to the lowest possible ranks, i.e. ranked according to a bijective index function $idx()$ assigning values $1, \ldots, |testset_u|$ to the test set items. Thus, for our example user 234, with twenty recommendations and hits on the fourth and twentieth positions, the half-life utility rank score would be computed as:

$$rankscore_{234} = \frac{1}{2^{\frac{4-1}{10}}} + \frac{1}{2^{\frac{20-1}{10}}} = 1.08$$

$$rankscore_{234}^{max} = \frac{1}{2^{\frac{1-1}{10}}} + \frac{1}{2^{\frac{2-1}{10}}} = 1.93$$

$$rankscore'_{234} = \frac{1.08}{1.93} = 0.56$$

Another very simple rank accuracy measure is the *lift index*, first proposed by Ling and Li (1998). It assumes that the ranked list is divided into ten equal deciles and counts the number of hits in each decile as $S_{1,u}, S_{2,u}, \ldots, S_{10,u}$, where $\sum_{i=1}^{10} S_i = hits_u$.

$$liftindex_u = \begin{cases} \frac{1 \cdot S_{1,u} + 0.9 \cdot S_{2,u} + \cdots + 0.1 \cdot S_{10,u}}{\sum_{i=1}^{10} S_{i,u}} & : \text{if } hits_u > 0 \\ 0 & : \text{else} \end{cases} \tag{7.11}$$

Compared with the rank score of Breese et al. (1998), the lift index attributes even less weight to successful hits in higher ranks. Consequently, for the example user 234, the lift index is calculated as follows:

$$liftindex_{234} = \frac{0.9 \cdot 1 + 0.1 \cdot 1}{2} = 0.5 \qquad (7.12)$$

Finally, an example of using the lift index on recommendation results is presented by Hsu et al. (2004). For a discussion on additional rank accuracy metrics, we refer readers to Herlocker et al. (2004).

Additional metrics. One metric that allows evaluators to compare different techniques based on their capability to compute recommendations for a large share of the population is user coverage (*Ucov*). It is of particular interest when one wants to analyze an algorithm's behavior with respect to new users with few known ratings.

$$Ucov = \frac{\sum_{u \in U} \rho_u}{|U|} \qquad (7.13)$$

$$\rho_u = \begin{cases} 1 & : \text{if } |recset_u| > 0 \\ 0 & : \text{else} \end{cases} \qquad (7.14)$$

It measures the share of users to whom nonempty recommendation lists can be provided. Obviously, it is sensible to measure user coverage only in conjunction with an accuracy metric, as otherwise recommending arbitrary items to all users would be considered as an acceptable strategy.

A similar coverage metric can be computed on the item universe.

$$Ccov = \frac{|\bigcup_{u \in U} recset_u|}{|I|} \qquad (7.15)$$

Catalog coverage (*Ccov*) reflects the total share of items that are recommended to a user in all sessions (Herlocker et al. 2004) and can be used as an initial indication for the diversity of an algorithm's recommendations.

However, Ziegler et al. (2005) propose a more elaborate measure of the diversity of recommendation lists, termed *intra-list similarity* (ILS).

$$ILS_u = \frac{\sum_{i \in recset_u} \sum_{j \in recset_u, i \neq j} sim(i, j)}{2} \qquad (7.16)$$

For a given similarity function *sim*(*i*, *j*) that computes the similarity between two recommended items, ILS aggregates the pairwise proximity between any two items in the recommendation list. ILS is defined to be invariant for all permutations of the recommendation list, and lower scores signify a higher diversity. Ziegler et al. (2005) employed this metric to compare a topic diversification algorithm on the BX books dataset.

7.4.3 Analysis of results

Having applied different metrics as part of an experimental study, one must question whether the differences are statistically meaningful or solely due to chance. A standard procedure for checking the significance of two deviating mean metrics is the application of a pairwise analysis of variance (ANOVA). The different algorithm variants constitute the independent categorical variable that was manipulated as part of the experiment. However, the null hypothesis H_0 states that the observed differences have been due to chance. If the outcome of the test statistics rejects H_0 with some probability of error – typically $p \leq .05$ – significance of findings can be reported. For a more detailed discussion of the application of test statistics readers are referred to Pedhazur and Schmelkin (1991); textbooks on statistics, as well as articles discussing the application of statistical procedures in empirical evaluation research, such as Demšar (2006) or Garcìa and Herrera (2008).

In a second step, the question as to whether the observed difference is of practical importance must be asked. When contemplating the substantive significance of a fictitious finding, like a 5 percent increase in recommendation list diversity caused by an algorithm modification, statistics cannot help. Instead, additional – and more complex – research is required to find out whether users are able to notice this increase in diversity and whether they appreciate it. The effect of higher recommendation list diversity on customer satisfaction or actual purchase rates must be evaluated, a task that can be performed not by experimenting with historical datasets but rather by conducting real user studies. The next section will provide some examples of these.

7.5 Alternate evaluation designs

As outlined in the previous section, recommender systems are traditionally evaluated using offline experiments to try to estimate the prediction error of the recommendations based on historical user records. Although the availability of well-known datasets such as MovieLens, EachMovie, or Netflix has stimulated the interest of researchers in the field, it has also narrowed their creativity, as newly developed techniques tend to be biased toward what can be readily evaluated with available resources. In this section we therefore refer to selected examples of evaluation exercises on recommender systems that adopt alternate evaluation designs and do not experiment on historical datasets. Furthermore, we structure our discussion according to the taxonomy of research designs presented in Section 7.2.3.

7.5.1 Experimental research designs

User studies use live user interaction sessions to examine the acceptance or rejection of different hypotheses. Felfernig and Gula (2006) conducted an experimental user study to evaluate the impact of different conversational recommender system functions, such as explanations, proposed repair actions, or product comparisons. The study, involving 116 participants, randomly assigned users to different variants of the recommender system and applied pre- and post-trial surveys to identify the effect of user characteristics such as the level of domain knowledge, the user's trust in the system, or the perceived competence of the recommender. The results show that study participants appreciate particular functionality, such as explanations or the opportunity to compare products, as it tends to increase their perceived level of knowledge in the domain and their trust in the system's recommendations. A similar study design was applied by Teppan and Felfernig (2009b), who reported on a line of research investigating the effectiveness of psychological theories in explaining users' behavior in online choice situations; this will be examined in more detail in Chapter 10.

An experimental user study was also conducted by Celma and Herrera (2008), who were interested in comparing different recommendation variants with respect to their novelty as perceived by users in the music domain. One interesting aspect of this work is that it combines an item-centric network analysis of track history with a user-centric study to explore novelty criteria to provide recommendations from several perspectives. An intriguing finding of this study is that both collaborative filtering and a content-based music recommender did well in recommending familiar items to the users. However, the content-based recommender was more successful in identifying music from the long tail of an item catalog ranked by popularity (i.e., the less frequently accessed items) that would be considered novel by the participants. As collaborative filtering focuses on identifying items from similar peers, the recommended items from the long tail are already familiar to the music enthusiasts, whereas content-based music recommendation promises a higher chance to hit interesting similar items in different portions of the long tail, according to this study.

Pu et al. (2008) compared the task completion times of users interacting with two different critiquing-based search interfaces. They employed a within-subjects experiment procedure, in which all twenty-two participants were required to interact with both interfaces. This is opposed to a between-subjects test, in which users are randomly assigned to one interface variant. However, to counterbalance bias from carryover effects from evaluating the first interface prior to the second, the order of interfaces was alternated every two consecutive users. Because of the small number of subjects, only a few differences in measurements were statistically significant; nevertheless, the goal of this study, namely, exploring the support for tradeoff decisions of different critiquing-based recommendation interfaces, is of great interest.

In Chapter 8, an online evaluation exercise with real users is described as a practical reference. It employs a between-subjects experiment design in which users are randomly assigned to a specific personalized or impersonalized recommendation algorithm variant and online conversion is measured. This type of online experiment is also known as *A/B testing*.

7.5.2 Quasi-experimental research designs

A quasi-experimental evaluation of a knowledge-based recommender in the tourism domain was conducted to examine conversion rates – that is, the share of users who subsequently booked products (Zanker et al. 2008 and Jannach et al. 2009). The study strongly confirmed that users who interacted with the interactive travel advisor were more than twice as likely to issue a booking request than those who did not. Furthermore, an interesting cultural difference between Italian- and German-speaking users was detected, namely that Italian users were twice as likely to use interactive search tools such as the travel recommender.

7.5.3 Nonexperimental research designs

Swearingen and Sinha (2001) investigated the human-computer interaction (HCI) perspective when evaluating recommender systems, adopting a mixed approach that included quantitative and qualitative research methods. The subjects were observed while they interacted with several commercial recommendation systems, such as Amazon.com. Afterward they completed a satisfaction and usability questionnaire and were interviewed with the aim of identifying factors that can be used to predict the perceived usefulness of a recommendation system to derive design suggestions for good practice from an HCI perspective. Results of that study included that receiving very novel and unexpected items is welcomed by users and that information on how recommendations are derived by the system should be given.

Experiences from fielded applications are described by Felfernig et al. (2006–07). The authors used a nonexperimental quantitative research design in which they surveyed actual users from two commercial recommender systems in the domains of financial services and electronic consumer goods. In the latter domain, a conversational recommender for digital cameras was fielded. Based on users' replies to an online questionnaire, the hypothesis that interactive sales recommenders help users to better orient themselves when being confronted with large sets of choices was also confirmed. In the financial services domain, the installation of constraint-based recommenders was shown to support sales agents during their interaction with prospective clients. Empirical surveys determined that the time savings achieved by the sales representatives while interacting with clients are a big advantage, which, in turn, allows sales staff to identify additional sales opportunities (Felfernig et al. 2006–07).

Another interesting evaluation exercise with a nonexperimental quantitative design is to compare predictions made by a recommendation system with those made by traditional human advisors. Krishnan et al. (2008) conducted such a study and compared the MovieLens recommender system with human subjects. The results of their user study, involving fifty research subjects, indicated that the MovieLens recommender typically produced more precise predictions (based on MAE) than the group of humans, despite the fact that only experienced MovieLens users with long rating records were invited to participate in the survey. However, a subgroup of the human recommenders (i.e., research subjects) produced consistently better results than the employed system, which could, in turn, be used to further improve the algorithm's ability to mimic the human subjects' problem-solving behavior. An additional aspect of this specific evaluation design is that it supports the credibility of the system in the eyes of its users, as it demonstrates its ability to provide better predictions than human experts.

7.6 Summary

After reflecting on the general principles of empirical research, this chapter presented the current state of practice in evaluating recommendation techniques. We discussed the meta-level characteristics of different research designs – namely, subjects, research method, and setting – and consulted authoritative literature for best research practices.

Furthermore, a small survey of highly reputed publications on recommendation systems in the *ACM TOIS* was presented, which gave an overview of research designs commonly used in practice. As a result, we focused in particular on how to perform empirical evaluations on historical datasets and discussed different methodologies and metrics for measuring the accuracy or coverage of recommendations.

From a technical point of view, measuring the accuracy of predictions is a well-accepted evaluation goal, but other aspects that may potentially affect the overall effectiveness of a recommendation system remain largely underdeveloped. Therefore, Section 7.5 presented several examples of evaluation studies that were based not on historical datasets but rather on real user studies. They were grouped according the classification scheme presented in Section 7.2.3, namely, into experimental, quasi-experimental, and nonexperimental research methods. Although the works discussed in Section 7.5 do not cover the complete range of study designs that have been explored so far, this selection can undoubtedly serve as a helpful reference when designing new evaluation exercises.

7.7 Bibliographical notes

Herlocker et al.'s (2004) article on evaluating collaborative filtering recommender systems is the authority in the field and is therefore one of the most frequently cited articles on recommendation systems. Since then, few works have appeared on the topic of evaluating recommender systems in general. One exception is the work of del Olmo and Gaudioso (2008), who criticize existing accuracy and ranking metrics for being overparticularized and propose a new category of metrics designed to measure the capacity of a recommender to make successful decisions. For this reason they present a new general framework for recommender systems that formalizes their recommendation process into several temporal stages. The essence of their approach is that a recommender system must be able to not only choose *which* items should be recommended, but also decide *when* and *how* recommendations should be shown to ensure that users are provided with useful and interesting recommendations in a timely manner. One interesting aspect of this article is its consideration of the interactivity of a recommender system, a property that has not been evaluated in existing approaches.

Furthermore, literature on empirical research in general, such as Pedhazur and Schmelkin (1991), on the interleaved quantitative processes of measurement, design, and analysis, or Creswell (2009), on mixed research designs focusing on qualitative methods, are also relevant when assessing alternate strategies for evaluating the quality and value of recommender systems.

8

Case study: Personalized game recommendations on the mobile Internet

Although the interest in recommender systems technology has been increasing in recent years in both industry and research, and although recommender applications can nowadays be found on many web sites of online retailers, almost no studies about the actual *business value* of such systems have been published that are based on real-world transaction data.

As described in Chapter 7, the performance of a recommender system is measured mainly based on its accuracy with respect to predicting whether a user will like a certain item. The implicit assumption is that the online user – after establishing trust in the system's recommendations or because of curiosity – will more often buy these recommended items from the shop.

However, a shop owner's key performance indicators related to a personalized web application such as a recommender system are different ones. Establishing a trustful customer relationship, providing extra service to customers by proposing interesting items, maintaining good recommendation accuracy, and so on are only a means to an end. Although these aspects are undoubtedly important for the long-term success of a business, for an online retailer, the important performance indicators are related to (a) the increase of the conversion rate – that is, how web site visitors can be turned into buyers, and (b) questions of how to influence the visitors in a way that they buy more or more profitable items.

Unfortunately, only few real-world studies in that context are available because large online retailers do not publish their evaluations of the business value of recommender systems. Only a few exceptions exist. Dias et al. (2008), for instance, present the results of a twenty-one-month evaluation of an probabilistic item-based recommender system running on a large Swiss e-grocer web portal. Their measures include "shopper penetration", "direct extra revenue", and "indirect extra revenue". Their analysis showed several interesting points. First, a relatively small (when compared with overall sales) extra revenue can be

189

generated directly by the recommender. The fact that direct revenues measurably increased when the probabilistic model went through a periodic update suggests that good recommendation accuracy is still important, despite some legitimate criticism of simple accuracy measures (McNee et al. 2006). The more important business value, however, comes from *indirect* revenues caused by the recommender systems. Indirect revenues include the money spent on repeated purchases of items initially recommended by the system and on items sold from categories to which the customer was newly introduced to through a recommended item. This, in turn, also supports the theory that diversity in recommendation lists is a valuable property, as "unexpected" items in these lists may help to direct users to other, possibly interesting, categories.

An earlier evaluation based on real-world data was presented by Shani et al. (2002), in which the authors performed different experiments on an online bookstore. During their experiment, visitors to the web shop received buying proposals either from a "predictive" or a new Markov decision process recommender. Thus, they were able to compare the respective profits that were generated by different techniques during the observation period. In addition, at least for a period of seven days, the recommendation functionality was fully removed from the web shop. Although this sample is statistically too small to be significant, the comparison of sales numbers of two consecutive weeks (one with and one without the recommender) showed a 17 percent drop in the recommender-free week.

Another initial study on how recommender systems influence the buying behavior of web shop visitors is presented by Zanker et al. (2006). In this work, it was shown that the recommendations of a virtual advisor for premium cigars can stimulate visitors to buy cigars other than the well-known Cohibas and thus increase sales diversity, which is interesting from up-selling and cross-selling perspectives and could also create indirect revenue as described by Dias et al. (2008); see also Fleder and Hosanagar (2007), for a discussion of the role of sales diversity in recommender systems.

In Zanker et al. (2008) and Jannach et al. (2009), a different study using the same recommendation technology was made in the tourism industry, in which it could be observed that the number of accommodation availability enquiries is measurably higher when web site visitors are guided by the virtual advisor. Another evaluation of how different information types and recommendation sources influence consumers can be found in Senecal and Nantel (2004).

Similar to these works, the case study presented in this chapter[1] focuses on evaluating the business value of recommender systems in a commercial context.

---

[1] The work was also presented at the 7th Workshop on Intelligent Techniques for Web Personalization and Recommender Systems at IJCAI'09 (Hegelich and Jannach 2009); a summary of the results of the study can also be found in Jannach and Hegelich (2009).

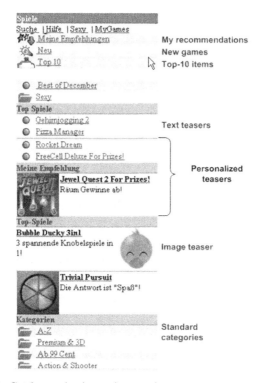

Figure 8.1. Catalog navigation and categories.

In addition, it aims to answer the question whether certain algorithms perform better than others in a certain environment and application domain in the line of the work of, for example, Breese et al. (1998) or Zanker et al. (2007).

8.1 Application and personalization overview

The study presented in this chapter was conducted in the context of a mobile Internet portal of a large telecommunications provider in Germany. Customers access this portal through their mobile devices and are offered a wide range of applications and games, which they can purchase directly and download to their cell phones.

Figure 8.1 shows the entry screen of the games area of the portal. Customers explore the item catalog in the following ways:

- Through manually edited *or* nonpersonalized lists such as "New items" or "Top 10 items" (top area of screen).
- Through direct text or image links (teasers) to certain items that are shown on the middle area of the start screen.

- Through predefined standard categories (lower area) such as "A–Z", "From 99 Cents", or "Action & Shooter".
- In addition, after a purchase, when the payment confirmation is displayed, customers are presented with a list of other, possibly interesting items (postsales recommendation).

Accordingly, the portal was extended with personalized content as follows:

(a) A new top-level link, "My Recommendations", was introduced, which leads to a personalized recommendation list ("Meine Empfehlungen" in German).
(b) The games presented in the lower two of the four text teasers and the first image teaser on the start page were personalized. Because of existing contracts, the first two text links and the two lower image links were manually predefined. The manually edited links remained the same during the whole experiment, which made it possible to analyze the effects of personalizing the other links independently.
(c) The lists in the standard categories such as "99 Cents" were personalized except for categories such as "A–Z", which have a "natural" ordering.
(d) The games presented on the postsales page were also personalized.

During the experiments, different algorithms were used to calculate the personalized recommendations. To measure the effect of personalization, members of the control group were shown nonpersonalized or manually edited lists that were based on the release date of the game.

Customers can immediately purchase and download games through the portal by choosing items from the presented lists. The relation between their navigation and buying behavior can therefore be easily determined, as all portal visitors are always logged in. Several thousand games (across all categories) are downloaded each day through the platform. The prices for the games range from free evaluation versions (demos) to "99 Cent Games" to a few euros for premium games; the amounts are directly charged to the customer's monthly invoice. In contrast to the study by Dias et al. (2008), in which users purchased the same goods repeatedly, customers in this domain purchase the same item only once – in other words, the domain is similar to popular recommender systems application areas such as books and movies.

From the perspective of the application domain, the presented game portal stands in the line of previous works in the area of recommender systems for mobile users. Recent works in the field of mobile recommenders include, for instance, Miller et al. (2003), Cho et al. (2004), van der Heijden et al. (2005), Ricci and Nguyen (2007), Li et al. (2008), and Nguyen and Ricci (2008).

Content personalization approaches for the mobile Internet are presented also by Pazzani (2002), Billsus and Pazzani (2007), and Smyth et al. (2007). In Smyth and Cotter (2002), finally, the effects of personalizing the navigational structure on a commercial Wireless Access Protocol (WAP) portal are reported.

Overall, it can be expected that this area will attract even more attention in the future because of the rapid developments in the hardware sector and the increasing availability of cheap and fast mobile Internet connections. In contrast to some other approaches, the recommender system on this platform does not exploit additionally available information such as the current geographical position or demographic and other customer information known to the service provider. Standard limitations of mobile Internet applications, such as relatively small network capacity and limited display sizes, apply, however.

8.2 Algorithms and ratings

During the four-week evaluation period, customers were assigned to one of seven different groups when they entered the games section of the portal. For each group, the item lists were generated in a different way. For the first four groups, the following recommendation algorithms were used:

- Item-based collaborative filtering (CF) (Sarwar et al. 2001) as also used by Amazon.com (Linden et al. 2003).
- The recent and comparably simple Slope One algorithm (Lemire and Maclachlan 2005).
- A content-based method using a TF-IDF representation of the item descriptions and the cosine similarity measure.
- A "switching" hybrid algorithm (Burke 2002b) that uses the content-based method when fewer than eight item ratings are available, and item-based collaborative filtering otherwise.

Two groups received nonpersonalized item lists, one based on the average item rating ("Top Rating") and one based on the sales numbers (top sellers). For the final group, the control group, the recommendation lists were manually edited as they were before the personalization features were introduced. Within most categories, the ordering was based on the release date of the game or chosen based on existing contracts. The top-level link "My Recommendations" was not available for the control group. During the entire evaluation period, customers remained in their originally assigned groups.

From all customers who visited the games portal during the evaluation, a representative sample of more than 155,000 was included in the experiment, so each group consisted of around 22,300 customers. Only customers for which all algorithms were able to produce a recommendation were chosen – that is, users for whom a minimum number of ratings already existed. The catalog of recommendable items consisted of about 1,000 games.

A five-point rating scale from -2 to $+2$ was used in the experiments. Because the number of explicit item ratings was very low and only about 2 percent of the customers issued at least one rating, implicit ratings were also taken into account: both clicks on item details as well as actual purchases were interpreted as implicit ratings. When no explicit rating was given, a view on item details was interpreted as a rating of 0 (medium); several clicks on the same item were not counted. An actual purchase was interpreted as a rating of 1 (good) for the item. Explicit ratings overrode these implicit ratings.

To achieve the best possible recommendation accuracy, the item similarities and the average differences for the collaborative filtering and the Slope One techniques were computed using the full customer base and not only the 155,000-customer subsample.

8.3 Evaluation

The following hypotheses are in the center of the evaluation:

- H1: Personalized recommendations attract more customers to detailed product information pages (item view conversion rate).
- H2: Personalized recommendations help turn more visitors into buyers (sales conversion rate).
- H3: Personalized recommendations stimulate individual customers to view more items.
- H4: Personalized recommendations stimulate individual customers to buy more items.

The detailed evaluation will show that depending on the navigational situation of the portal visitor, different phenomena with respect to the effectiveness of recommendation algorithms can be observed. Before considering the overall effect of the use of recommendation technology on the portal, the individual results obtained for these different situations will be discussed.

8.3.1 Measurement 1: "My Recommendations"

The following results are related to the personalized recommendation list that is presented when the customer clicks on the "My Recommendations" link, as

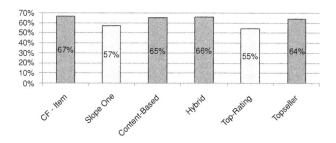

Figure 8.2. Conversion rate: item views to "My Recommendations" visits.

shown in the top area of Figure 8.1. Throughout the evaluation, different levels of gray will be used to highlight data rows in the charts that are significantly different ($p < 0.01$) from each other.

The conversion rate measurements (hypotheses H1 and H2) are given in Figure 8.2, which depicts the item view conversion rate for visitors to the "My Recommendations" list, and Figure 8.3, which shows how many of the users who visited the "My Recommendations" section actually purchased an item[2].

In Figure 8.2 it can be seen that the different algorithms fall into two groups: one in which about two-thirds of the customers actually click on at least one of the presented items and one in which only 55 percent are interested in the recommended items. Considering the actual numbers, the differences between the two groups are significant ($p < 0.01$).

From the personalized methods, only the Slope One algorithm did not attract significantly more visitors than the nonpersonalized list of top-rated items. Interestingly, the nonpersonalized top-seller list also has a good item view conversion rate – in other words, placing generally liked, top-selling items in a recommendation list seems to work quite well in the domain.

When the sales conversion rate is considered, it can be observed from Figure 8.3 that only the CF method helps to turn more visitors into buyers (Hypothesis H2).

The evidence for our hypotheses H3 (more item views per customer) and H4 (more purchases per customer) in the context of the "My Recommenda-tions" section can be seen in Figures 8.4 and 8.5. Figure 8.4 shows that all recommendation algorithms (except for Slope One) stimulate users to click on more items. Compared with the findings with respect to the conversion rates,

---

[2] In Figures 8.2 to 8.5, the control group is not depicted, because the "My Recommendations" section, which was newly introduced for measuring the impact of personalization, was not available for them.

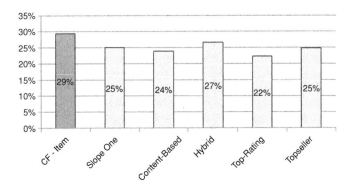

Figure 8.3. Conversion rate: buyers to "My Recommendations" visits.

this can be interpreted that personalized lists seem to contain more items that are interesting to a customer.

When it comes to actual purchases (game downloads), Figure 8.5 shows that most personalized methods, and even the simple Slope One algorithm, outperform the nonpersonalized approaches.

For some of the games provided on the mobile portal, free evaluation versions (demos) are available. If not mentioned otherwise, all numbers given with respect to conversion rates and sales figures are related to all item downloads – free demos plus actual game purchases. Figure 8.6 repeats the numbers of Figure 8.5, but also shows the fraction of demo downloads and purchased games. Because of the nature of the algorithms and the particularities of the application (see more details in Measurement 4), the recommendation lists produced by the TopRating and Slope One methods contain a relatively high portion of demo games. Given the high number of actual downloads, these demo recommendations seem to be well accepted, but unfortunately, these two techniques perform

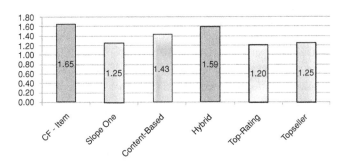

Figure 8.4. Item views per "My Recommendations" visits.

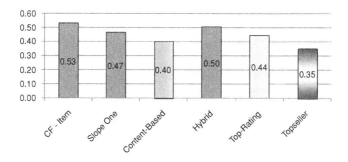

Figure 8.5. Item purchases per "My Recommendations" visits.

particularly poorly when the games are not free. The item-based, content-based, and hybrid techniques, on the other hand, not only help to sell as many items as a simple top-seller promotion but also make users curious about demo games. The TopRating method raises interest only in demo versions. The list of top-selling items is generally dominated by non-free, mainstream games, which explains the fact that nearly no demo games are chosen by the users.

8.3.2 Measurement 2: Post-sales recommendations

The next navigational situation in which product recommendations are made is when a customer has purchased an item and the payment confirmation has just finalized the transaction. About 90,000 customers who actually bought at least one item during the evaluation period were involved in the experiment. Overall, the evaluation sample contains more than 230,000 views of the post-sales

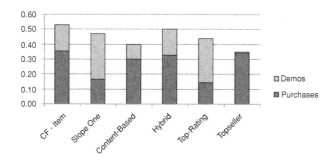

Figure 8.6. Game purchases and demo downloads in "My Recommendations" visits.

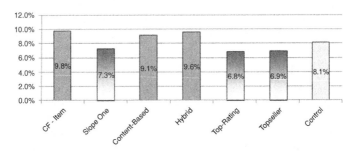

Figure 8.7. Conversion rate: item views to post-sales list views.

five-item recommendation lists, meaning that, on average, customers bought more than one item.

The experimental setup is nearly identical with that for Measurement 1; customers received their recommendations based on different recommendation algorithms. The recommendation list of the control group was manually edited and ordered by game release date. Items that the current customer had already purchased before were removed from these lists.

The same hypotheses were tested in this experiment – that is, to what extent recommender systems stimulate customers to view and buy more items. The results are shown in Figures 8.7 through 8.10.

With respect to the conversion rates, the following observations can be made. First, the manually edited list of recent items (viewed by the control group) worked quite well and raised more customer interest than the nonpersonalized techniques and even the Slope One algorithm (Figure 8.7). When it comes to actual purchases (Figure 8.8), however, the manually edited list did not help turn more visitors into buyers. Interestingly, the relative improvement caused by personalized recommendations with respect to this conversion rate is higher on the post-sales recommendation page than in the "My Recommendations"

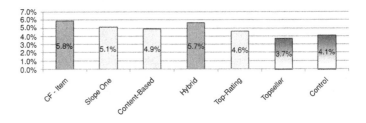

Figure 8.8. Conversion rate: Buyers to post-sales list views.

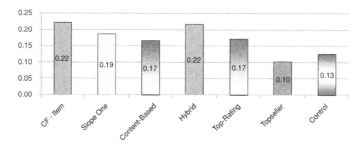

Figure 8.9. Item visits per post-sales list views.

sections. Again, the CF algorithm worked best; in absolute numbers, the differences between the various techniques are significant ($p < 0.01$).

With respect to the number of item visits and purchases per customer (Figures 8.9 and 8.10), it can again be observed that the different recommendation techniques not only stimulated visitors to view more items but actually also helped to increase sales. It can also be seen that displaying a list of top-selling items after a purchase leads to a particularly poor effect with respect to the overall number of downloads.

Another observation is that the items that are recommended by the Slope One technique and the TopRating method are also downloaded very often (see Figure 8.10), presumably because the recommendation lists again contain many free demos. Figure 8.11 therefore shows the ratio of demo downloads to game purchases, which is quite similar to the one from the "My Recommendations" section – that is, recommending top-selling or newly released items does not stimulate additional interest in free evaluation versions (demo games). The trend toward interest in demo versions seems to be a bit more amplified than

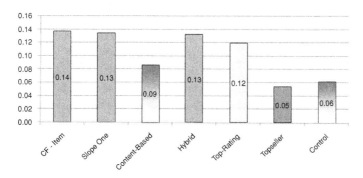

Figure 8.10. Item purchases to post-sales list visits.

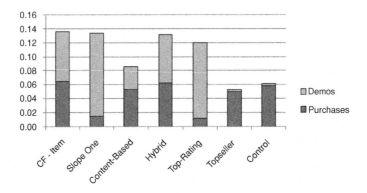

Figure 8.11. Game purchases and demo downloads on post-sales page.

in the "My Recommendations" section, which indicates that after a purchase transaction, customers first have a look at another, but free, game.

Finally, in this navigational context, the content-based method could raise some initial customer interest (Figure 8.9), perhaps because games are recommended that are quite similar to previously downloaded ones. However, although customers viewed some of the items, they had no strong tendency to purchase them, probably because the games were – according to the general tendency of content-based methods – too similar to games they already knew. The list of top-selling items again contained mostly non-free games, which explains the small fraction of demo games here; the same holds for the control group.

8.3.3 Measurement 3: Start page recommendations

This measurement analyzes the effect of the personalized recommendations on the start page, as shown in Figure 8.1. Remember that some elements in these lists are edited manually but were static during the experiment. Thus, item visits or purchases from these links (that could have been other banner advertisements as well) were not included in the evaluation.

During the experiment, the personalized elements of the list – the last two text teasers and the first image teaser – were determined based on the top-three list of the individual recommendation algorithms or based on the nonpersonalized lists of top-selling and top-rated items. Customers assigned to the control group received manually determined recommendations that were ranked by release date.

For this experiment, only the conversion rate figures for the different teaser elements on the start page will be shown.

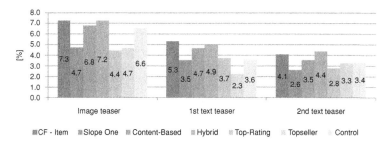

Figure 8.12. Conversion rate: item views to start page visits.

Figure 8.12 shows the percentage of portal visitors who followed one of the personalized product links on the start page. On average, the image teaser was clicked on by around 6 percent of the users. Although the image represents only the third-ranked item of the recommendation algorithms and is also positioned after the text links, its conversion rate is significantly higher than that for the text links. As this also holds for the nonpersonalized methods, the attractiveness of the third link can be attributed to its visual representation. Interestingly, however, the image teaser leads to a good conversion rate with respect to actual sales (Figure 8.13). With respect to these conversion rates, both the CF method and the content-based method lead to a significant increase of item detail clicks and purchases. It can also be observed that the conversion rates of the first text teaser can even be better than the image teaser when the text links are personalized. Thus, personalization can partially even outweigh the disadvantages of the unflashy representation.

Another particularity of this measurement on the start page is that the manually selected items used for the control group lead to comparably good conversion rates, especially with respect to item visits. A possible explanation could be that customers have no special expectations with respect to the offers on the

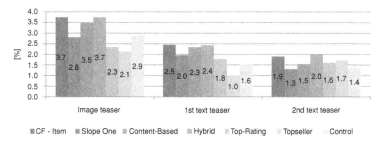

Figure 8.13. Conversion rate: purchases from start page visits.

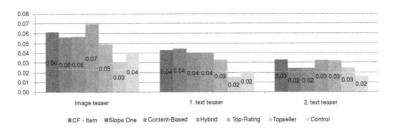

Figure 8.14. Purchases per start page visits.

start page. The fact that the manually selected items are newly released ones might further contribute to the good acceptance.

Although recommending items based on their average customer rating (as done by the Slope One and the TopRating techniques) worked well in the first two experiments, this approach does not work particularly well on the start page – customers seem to prefer either new items or items that are somehow related to their previous buying history.

Finally, when it comes to the number of purchases induced by the recommendation lists, the personalized techniques clearly outperformed the manually defined lists, at least for the first two teaser elements (see Figure 8.14). The item click and sales numbers of the other four, and statically defined image and text teasers with the personalized ones, were also compared. It could be seen that although the personalized items are partially placed lower on the screen and are thus harder to select, they received significantly more clicks and led to more sales than the nonpersonalized links.

8.3.4 Measurement 4: Overall effect on demo downloads

In Measurements 1 and 2, it could be seen that Slope One and the nonpersonalized technique based on item ratings led to significantly more views and downloads of demo games. In this measurement, the goal was to analyze whether this trend also exists when the entire platform is considered, including, for instance, all other personalized and nonpersonalized navigation possibilities.

No explicit category in the navigation tree for "free demos" exists. Games for which free evaluation versions exist can, however, appear in all other personalized and nonpersonalized item listings in the portal. In addition, customers are pointed to demos in two additional ways: (a) through direct-access links that are sent to them in sales promotions and (b) through pointers to other demo games that are displayed after a demo has been downloaded.

The distribution of views and downloads of demo games during the four-week evaluation period for the different recommendation groups is shown in

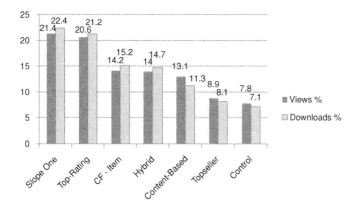

Figure 8.15. Distribution of demo game item views and downloads.

Figure 8.15. Overall, about 38,000 downloads were observed for the selected subsets of customers. When considering the actual downloads, it can be seen that the ranking of the algorithms remains the same; the differences are even amplified.

As already briefly mentioned in previous sections, this result can be explained by different facts that are related to the particular application setting and the nature of Slope One and the top-rating algorithm, which both tend to rank demo games highly in the different categories described previously, for the following reasons. First, as demo games can be downloaded at no cost and user ratings are possible on the platform only after a download, more explicit ratings are available for these games. Next, explicit ratings tend to be above average also in this domain. A similar phenomenon can also be observed in other datasets such as the MovieLens rating database. Finally, as customers receive a nonpersonalized pointer to another demo after downloading a free game, a reinforcement of the effect occurs.

An in-depth analysis of whether the downloads that were stimulated by the different algorithms led to significantly different demo-download/purchase conversion rates was not done in this case study. What could, however, be observed in a first analysis is that the demo/purchase conversion rate was significantly higher when the demo was promoted by a recommendation list (as opposed to a banner advertisement).

8.3.5 Measurement 5: Overall effects

In the final measurement reported in this study, the overall effect of the personalized recommendations (as an add-on to the other navigational options) was evaluated. Again, the interesting figures are related to item view and sales conversion rates (H1 and H2) as well as to the question of whether more items were viewed and purchased by individual customers (H3 and H4).

With respect to the conversion rates (hypotheses H1 and H2), no significant differences between the personalized and nonpersonalized variants could be observed on the platform as a whole. On average, about 80 percent of all observed customers viewed at least one item, and around 57 percent bought at least one game, independent of the recommendation algorithm group they were assigned to. These figures are nearly identical for all seven test groups. For the item view conversion rate, for instance, the numbers only range from 79.6 percent to 80.3 percent. Thus, although slight improvements could be observed in individual (personalized) situations, as described earlier, the influence on the overall conversion rate is too small, and thus the percentage of portal visitors who view or purchase items could not be significantly increased by the additional use of personalized recommendation lists.

There could be different reasons for this non-effect. First, besides the described personalized lists, there are various other ways in which customers can access the item catalogs. Many customers, for instance, use the built-in search functionality of the portal; the ranking of search results is not personalized. The list of new items (see Figure 8.1) is also one of the most popular ways of browsing the catalog and is used by significantly more people than, for instance, the new "My Recommendations" section. An analysis showed that personalizing this particular list does not improve the conversion rates, as customers always prefer to see the latest releases at the top of such a list. Second, in this evaluation, only customers have been considered for whom a minimum number of ratings already existed – that is, users who are in generally interested in games. An evaluation of whether more *new* users can be tempted to purchase items was not in the focus of the evaluation.

With respect to hypotheses H3 and H4 (increased number of item views and sales per customer), the following observations can be made. Regarding the average number of item views per customer (H3), it could be seen that all personalized algorithms outperform the nonpersonalized top-seller list and the control group. Similar to the effect of Measurement 4, Slope One and the simple ranking based on average customer rating raised the most attention. Thus, H3 could be only partially validated at the global scale as the nonpersonalized top-rating technique was also successful.

The observations made with respect to the number of purchased/downloaded items per customer (H4) are shown in Figure 8.16.

The figure shows that the additional attention raised by Slope One and the TopRating algorithm also leads to a measurably increased number of items

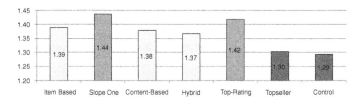

Figure 8.16. Average number of purchases, including free downloads, per customer on entire platform.

purchased and downloaded per customer. Figure 8.17 shows the number of downloaded items (including the demos) for the different algorithms. Finally, if we look at the actual sales numbers for non-free games only (Figure 8.18), it can be seen that although the Top-Rating list raised attention for free demos, it did not lead to increased sales for non-free items. Overall, all personalized techniques were more successful than the nonpersonalized one. On the global scale, however, the difference was – a bit surprisingly – significant only for the content-based method, which indicates that customers tend to spend money on items that are similar to those they liked in the past. In fact, a closer look on the performance of the algorithms in specific subcategories shows that the content-based method often slightly outperforms other methods with respect to non-free games. Although the differences were not significant in the individual situations, these slightly higher sales numbers add up to a significant difference on the global scale. Examples of categories in which the content-based method worked slightly better with respect to non-free games are the "new games", "half-price", or "erotic games" sections of the download portal.

Overall, the increase in actual sales that are directly stimulated by the recommender system is between 3.2 percent when compared to the Top-Rating

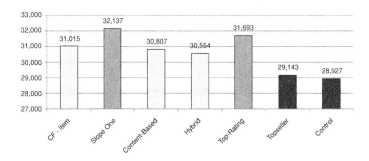

Figure 8.17. Total number of purchases and downloads.

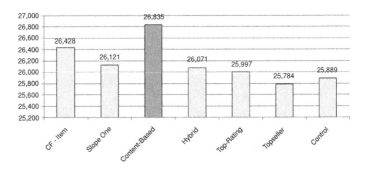

Figure 8.18. Total number of purchases (without demos).

technique, and around 3.6 percent when no personalized recommendation is available.

In general, these last observations suggest that in situations in which the user has no strong expectations on a certain genre (such as the "My Recommendations" section), collaborative methods – which also recommend items in categories that the user has not seen before – work particularly well. In many other situations, however, users tend to prefer recommendations of game subcategories that they already know. One exception is the post-sales situation, in which users are, not surprisingly, not interested in purchasing games that are very similar to the ones they have just bought.

8.4 Summary and conclusions

In this study, the effects of personalized item recommendation in various navigational contexts on a mobile Internet game portal were analyzed. Different standard recommendation techniques were implemented on the portal and deployed in parallel in a real-world setting for a period of four weeks. In addition, nonpersonalized techniques based on top-selling or top-rated items were used for comparison purposes.

The findings can be summarized as follows:

Ratings in the mobile Internet. The number of explicit item ratings was very low on the considered mobile Internet portal, and only about 2 percent of the users issued explicit ratings. Although no studies are available that compare the willingness of customers to rate items in different settings, it can be suspected that the relatively high effort for submitting an item vote using a mobile device compared with a web browser discourages users from participating in this community process.

Recommending in a navigational context. The effects of personalized recommendations have been measured in different navigational situations, such as the start page of the portal or the post-sales situation. In addition, a differentiation was made between the interest that was raised by the recommendations and the actual effect on the buying behavior of the customers.

With respect to the navigational context, customers seem to react slightly differently to recommendations, probably because of different expectations. In the dedicated "My Recommendations" section of the portal, classical CF and the hybrid technique are particularly good at raising customer interest, as customers view many of the recommended items. Although customers are also easily stimulated to download free games by the comparably simple Slope One and TopRating methods, these techniques do not lead to a significant increase in non-free games. A similar effect can be observed in the post-sales situation; the trend toward free demo downloads is even amplified in this situation. Thus, the item-based, content-based, and hybrid techniques that lead to a good number of purchases but also raise additional interest in demos seem to be a good choice here.

On the portal entry page, the recommendation of top-rated (or top-selling) items has a particularly poor effect, and the personalized methods lead to significantly better results. A listing of newly released items on the start page, however, also worked quite well.

In certain navigational situations, it could be observed that personalization worsens the conversion rates and sales numbers. In the section on new items, which contains games of the last three weeks, the strict chronological order, with the newest items on top, works best. Most probably, the visitors to the "New" category enter this section regularly and check only the first few lines for new arrivals.

Finally, when measuring the number of game downloads, including the demos, on the entire platform, it can be seen that naive approaches such as TopRating and the comparably simple Slope One technique work sufficiently well to raise the users' interest in individual games. The important result, however, is that with respect to actual sales, the content-based and the item-based methods were clearly better than all others. Overall, it could be demonstrated that recommender systems are capable of stimulating a measurable increase in overall sales by more than 3 percent on the entire platform.

PART II

Recent developments

9

Attacks on collaborative recommender systems

When we discussed collaborative filtering techniques in previous chapters of this book, we made the implicit assumption that everyone in the user community behaves honestly and is fair and benevolent. If this assumption holds, all the participants profit: customers are receiving good buying proposals, well-appreciated items get some extra promotion, and the recommendation service itself will be appreciated by web site visitors if its proposals are of high quality and correctly reflect the opinions of the user community.

In the real world, however, the assumption of having only honest and fair users may not hold, in particular when we consider that the proposals of a recommender system can influence the buying behavior of users and real money comes into play. A malevolent user might, for instance, try to influence the behavior of the recommender system in a such way that it includes a certain item very often (or very seldom) in its recommendation list. We shall call this an attack on the recommender system. When a person expresses his or her genuine negative opinion – which can be based on any reasons – this is not seen as an attack. An attack occurs when an agent tries to influence the functioning of the system intentionally.

In general, attackers might have different goals, such as to increase the sales of an item, to cause damage to a competitor, or to sabotage the system as a whole so it is no longer able to produce helpful recommendations. In this chapter we focus on situations in which the goal of the attacker is to promote a certain item or bias the system not to recommend items of a competitor. Note that the manipulation of the "Internet opinion" – for instance, in product review forums – is not a new problem and is therefore not limited to automated recommender systems as discussed in this book. Recent research in marketing supplies evidence not only that consumers' buying behavior can be influenced by community reviews (Chevalier and Mayzlin 2006) but also that the strategic manipulation of opinion forums has a measurable impact on customers and firms (Dellarocas 2006, Mayzlin 2006). Lam and Riedl (2004), report examples of real-world manipulation attempts on such systems, including attacks on one of Amazon.com's buying tip features[1] or eBay's trust and reputation system[2].

Technically, an "attacker" on a community-based recommender system could try to bias the system's behavior by creating several fake user accounts (profiles) and give the target item of the attack a particularly good or bad rating value. However, such a simple strategy will not work very well in most real systems, which are based on nearest-neighbors methods. Profiles that consist only of one particularly good or bad rating will not be similar to any other user's profile, meaning that their rating values for the target item will never have a significant impact. An attacker therefore needs to find better ways to shill the system (Lam and Riedl 2004) and to make sure that the opinions carried in the fake profiles will take effect.

In the following sections, we discuss the different ways in which collaborative filtering systems can be attacked by malevolent users and summarize recent research that analyzes the vulnerability of different recommendation approaches. Afterward, possible countermeasures are discussed before we briefly review the privacy aspects of recommender systems. Our discussion will naturally be limited to community-based approaches, because since only their behavior can be influenced by a manipulated set of user ratings. Content- and knowledge-based systems can be manipulated only by those involved in the setup of the system unless their knowledge sources are mined from public sources.

9.1 A first example

In the following simple example, we sketch the general idea of a profile injection attack; see Mobasher et al. (2007) for a more detailed example. In this simplified scenario, we assume that a simplified version of a memory-based collaborative filtering method is used, which uses Pearson correlation as a similarity measure and a neighborhood size of 1 – that is, only the opinion of the most similar user will be used to make a prediction. Without the fake profile in the last row of the ratings matrix, *User2* is the most similar user, and this user's rating value 2 (dislike) for the target item will be taken as a prediction for *Alice*. However, in the situation of a successful attack, as shown in Table 9.1, the fake profile becomes the most similar one, which means that the particularly high rating for the target item will be taken as a prediction for Alice.

---

[1] news.com.com/2100-1023-976435.html.
[2] http://www.auctionbytes.com/cab/abn/y03/m09/i17/s01.

Table 9.1. *A profile injection attack.*

	Item1	Item2	Item3	Item4	...	...	Target	*Pearson*
Alice	5	3	4	1	...	...	?	
User1	3	1	2	5	...	...	5	−0.54
User2	4	3	3	3	...	...	2	0.68
User3	3	3	1	5	...	...	4	−0.72
User4	1	5	5	2	...	...	1	−0.02
Attack	**5**	**3**	**4**	**3**			**5**	**0.87**

In realistic settings, however, attacking a recommender system by inserting fake profiles to influence its predictions is not that easy. Consider only the following two aspects of that point. First, to be taken into account in the neighborhood formation process, a fake profile must be similar to an existing profile. In general, however, an attacker has no access to the ratings database when trying to determine good values for the selected items (see Table 9.1) that are used to establish the similarity with existing profiles. On the other hand, one single attack profile will not influence the prediction of the system very much when a larger neighborhood size is used. Therefore, several fake profiles must be inserted, which, however, might be also not so easy for two reasons: First, the automatic insertion of profiles is prohibited by many commercial systems (e.g., by using a so-called Captcha, see also Section 9.5 on countermeasures). Second, attack situations, in which many user profiles are created in a relatively short time, can easily be detected by an attack-aware recommender system.

In the following section, we discuss the different possible dimensions of attacks according to Lam and Riedl (2004), as well as more elaborate attack methods, and, finally, show how they influence the predictions of a recommender system as discussed by Mobasher et al. (2007).

9.2 Attack dimensions

A first differentiation between possible attack types can be made with respect to the goal of the attack – that is, whether the goal is to increase the prediction value of a target item (*push attack*) or decrease it (called a *nuke attack*). Although by intuition there seems to be no technical difference between these types of goals, we will see later on that push and nuke attacks are actually not always equally effective. Finally, one further possible intent of an attacker can simply be to make the recommender system unusable as a whole.

Another differentiation factor between attacks is whether they are focused only on particular users and items. Targeting a subset of the items or users might be less suspicious. Such more focused (segmented) attacks may also be more effective, as the attack profiles can be more precisely defined in a way that they will be taken into account for the system's predictions with a higher probability. Existing research on vulnerabilities of recommender systems has therefore focused on attack models in single items.

Whether one is able to attack a recommender system in an effective way also depends on the amount of knowledge the attacker has about the ratings database. Although it is unrealistic to assume that the ratings database is publicly available, a good estimate of the distribution of values or the density of the database can be helpful in designing more effective attacks.

Finally, further classification criteria for recommender system attacks include

- *Cost*: How costly is it to insert new profiles? Can the profile injection task be automated, or do we need manual interaction? How much knowledge about the existing ratings in the database is required to launch an attack of a certain type?
- *Algorithm dependence*: Is the attack designed for a specific algorithm, such as a memory-based filtering technique, or is it independent of the underlying algorithm?
- *Detectability*: How easily can the attack be detected by the system administrator, an automated monitoring tool, or the users of the system itself?

9.3 Attack types

Consider the most common profile injection attack models (for formal definitions, see Mobasher et al. 2007). The general form of an attack profile, consisting of the target item, a (possibly empty) set of selected items that are used in some attack models, a set of so-called filler items, and a set of unrated items is shown in Table 9.2. The attack types discussed on the following pages differ from each other basically in the way the different sections of the attack profiles are filled.

Note that in the subsequent discussion, we follow the attack type classification scheme by Mobasher et al. (2007); other attack types, which partially also require detailed knowledge about the ratings database, are described by O'Mahoney et al. (2005) and Hurley et al. (2007).

Table 9.2. *Structure of an attack profile.*

Item1	...	ItemK	...	ItemL	...	ItemN	Target
r_1	...	r_k	...	r_l	...	r_n	X
selected items			filler items		unrated items		

9.3.1 The random attack

In the *random attack*, as introduced by Lam and Riedl (2004), all item ratings of the injected profile (except the target item rating, of course) are filled with random values drawn from a normal distribution that is determined by the mean rating value and the standard deviation of all ratings in the database.

The intuitive idea of this approach is that the generated profiles should contain "typical" ratings so they are considered as neighbors to many other real profiles. The actual parameters of the normal distribution in the database might not be known: still, these values can be determined empirically relatively easily. In the well-known MovieLens data set, for example, the mean value is 3.6 on a five-point scale (Lam and Riedl 2004). Thus, users tend to rate items rather positively. The standard deviation is 1.1. As these numbers are publicly available, an attacker could base an attack on similar numbers, assuming that the rating behavior of users may be comparable across different domains. Although such an attack is therefore relatively simple and can be launched with limited knowledge about the ratings database, evaluations show that the method is less effective than other, more knowledge-intensive, attack models (Lam and Riedl 2004).

9.3.2 The average attack

A bit more sophisticated than the random attack is the *average attack*. In this method, the average rating per item is used to determine the rating values for the profile to be injected. Intuitively, the profiles that are generated based on this strategy should have more neighbors, as more details about the existing rating datasets are taken into account.

In fact, experimental evaluations and a comparison with the random attack show that this attack type is more effective when applied to memory-based user-to-user collaborative filtering systems. The price for this is the additional knowledge that is required to determine the values – that is, one needs to estimate the average rating value for every item. In some recommender systems, these average rating values per item can be determined quite easily, as they are explicitly provided when an item is displayed. In addition, it has also been shown that such attacks can already cause significant harm to user-based recommenders, even if only a smaller subset of item ratings is provided in the injected profile – that is, when there are many unrated items (Burke et al. 2005).

9.3.3 The bandwagon attack

The *bandwagon attack* exploits additional, external knowledge about a rating database in a domain to increase the chances that the injected profiles have many neighbors. In nearly all domains in which recommenders are applied, there are "blockbusters" – very popular items that are liked by a larger number of users. The idea, therefore, is to inject profiles that – besides the high or low rating for the target items – contain only high rating values for very popular items. The chances of finding many neighbors with comparable mainstream choices are relatively high, not only because the rating values are similar but also because these popular items will also have many ratings. Injecting a profile to a book recommender with high rating values for the *Harry Potter* series (in the year 2007) would be a typical example of a bandwagon attack. Another noteworthy feature of this attack type is that it is a low-cost attack, as the set of top-selling items or current blockbuster movies can be easily determined.

In an attack profile as shown in Table 9.2, we therefore fill the slots for the selected items with high rating values for the blockbuster items and add random values to the filler items to ensure that a sufficient overlap with other users can be reached. As discussed by Mobasher et al. (2007), the bandwagon attack seems to be as similarly harmful as the average attack but does not require the additional knowledge about mean item ratings that is the basis for the average attack.

9.3.4 The segment attack

The rationale for segment attack (Mobasher et al. 2005) is straightforwardly derived from the well-known marketing insight that promotional activities can be more effective when they are tailored to individual market segments. When designing an attack that aims to push item A, the problem is thus to identify a subset of the user community that is generally interested in items that are similar to A. If, for example, item A is the new *Harry Potter* book, the attacker will include positive ratings for other popular fantasy books in the injected profile. This sort of attack will not only increase the chances of finding many neighbors in the database, but it will also raise the chances that a typical fantasy book reader will actually buy the book. If no segmentation is done, the new

Harry Potter book will also be recommended to users who never rated or bought a fantasy novel. Such an uncommon and suspicious promotion will not only be less effective but may also be easier to detect; see Section 9.5 for an overview of automated attack detection techniques.

For this type of attack, again, additional knowledge – for example, about the genre of a book – is required. Once this knowledge is available, the attack profiles can be filled with high ratings for the selected items and low filler ratings. The segment attack was particularly designed to introduce bias toward item-based collaborative filtering approaches, which – as experiments show (Mobasher et al. 2007) – are, in general, less susceptible to attacks than their user-based counterparts. In general, however, this type of attack also works for user-based collaborative filtering.

9.3.5 Special nuke attacks

Although all of the aforementioned attack types are, in principle, suited to both push and nuke individual items, experimental evaluations show that they are more effective at pushing items. Mobasher et al. (2007) therefore also proposed special nuke attack types and showed that these methods are particularly well suited to bias a recommender negatively toward individual items.

- *Love/hate attack*: In the corresponding attack profiles, the target item is given the minimum value, whereas some other randomly chosen items (those in the filler set) are given the highest possible rating value. Interestingly, this simple method has a serious effect on the system's recommendations when the goal is to nuke an item, at least for user-based recommenders. However, it has been shown that if we use the method the other way around – to push an item – it is not effective. A detailed analysis of the reasons for this asymmetry has not been made yet, however.
- *Reverse bandwagon*: The idea of this nuke attack is to associate the target item with other items that are disliked by many people. Therefore, the selected item set in the attack profile is filled with minimum ratings for items that already have very low ratings. Again, the amount of knowledge needed is limited because the required small set of commonly low-rated items (such as recent movie flops) can be identified quite easily.

9.3.6 Clickstream attacks and implicit feedback

The attack types described thus far are based on the assumption that a "standard" recommender system is used that collects explicit ratings from registered users.

Although most of today's recommender systems fall into this category, we also briefly discuss attacks on systems that base their recommendations on implicit feedback, such as the user's click behavior.

In the Amazon.com example, a clickstream-based recommender would probably inform you that "users who viewed this book also viewed these items". In such scenarios, the personalization process is typically based on mining the usage logs of the web site. In the context of attacks on recommender systems, it is therefore interesting to know whether and how easily such systems can be manipulated by a malevolent user. This question is discussed in detail by Bhaumik et al. (2007), who describe two possible attack types and analyze how two different recommendation algorithms react to such attacks.

Let us briefly sketch the basic idea of recommending web pages based on nearest-neighbor collaborative filtering and usage logs. The required steps are the following: first, the raw web log data is preprocessed and the individual user sessions are extracted. A user session consists of a session ID and a set of visited pages; time and ordering information are often neglected. Based on these sessions, typical navigation patterns are identified in the mining step, which is commonly based on algorithms such as clustering or rule mining. At run time, a recommender system then compares the set of viewed pages of the current user with these navigation patterns to predict the pages in which the active user most probably will be interested. To calculate the predictions, a procedure similar to the one commonly used for user-based collaborative filtering with cosine similarity measure can be employed.

Attacks on such systems can be implemented by employing an automated crawler that simulates web browsing sessions with the goal of associating a target item (the page to be "pushed") with other pages in such a way that the target items appear on recommendation lists more often. Bhaumik et al. (2007) devise and evaluate two attack types in their experiments. In the segment attack, the target page is visited by the crawler, together with a specific subset of pages that are of interest to a certain subcommunity of all site users. In the *popular page* attack, which is a sort of bandwagon attack, the target page is visited together with the most popular pages.

A first evaluation of these attack types and two recommendation algorithms (kNN and one based on Markov models) showed that recommenders (or, more generally, personalization techniques) based on usage logs are susceptible to crawling attacks and also that – at least in this evaluation – small attack sizes are sufficient to bias the systems. Until now, no advanced countermeasures have been reported; further research in the broader context of usage-based web personalization is required.

9.4 Evaluation of effectiveness and countermeasures

Mobasher et al. (2007) present the result of an in-depth analysis of the described attack types on different recommender algorithms. To measure the actual influence of an attack on the outcome of a recommender system, the measures "robustness" and "stability" have been proposed by O'Mahony et al. (2004). *Robustness* measures the shift in the overall accuracy before and after an attack; the *stability* measure expresses the attack-induced change of the ratings predicted for the attacked items. Mobasher et al. (2007) introduce two additional measures. For the push attack, they propose to use the "hit ratio", which expresses how often an item appeared in a top-N list; see also the *ExpTopN* metric of Lam and Riedl (2004). For nuke attacks, the change in the predicted rank of an item is used as a measure of the attack's effectiveness.

9.4.1 Push attacks

User-based collaborative recommenders. When applied to user-based collaborative systems, the evaluation of the different attacks on the MovieLens data set shows that both the average and the bandwagon attacks can significantly bias the outcome of the recommender system. In particular, it was shown that in both these approaches, a relatively small number of well-designed item ratings in the attack profiles (selected items or filler items) are required to achieve a change in the bias of the recommender. In fact, having too many items in the filler set in the average attack even decreases the achievable prediction shift.

Besides a good selection of the item ratings in the profile, the size of the attack is a main factor influencing the effectiveness of the attack. With an attack size of 3 percent – that is, 3 percent of the profiles are faked after the attack – a prediction shift of around 1.5 points (on a five-point scale) could be observed for both attack types. The average attack is a bit more effective; however, it requires more knowledge about average item ratings than the bandwagon attack. Still, an average increase of 1.5 points in such a database is significant, keeping in mind that an item with a "real" average rating of 3.6 will receive the maximum rating value after the attack.

Although an attack size of only 3 percent seems small at first glance, this of course means that one must inject 30,000 fake profiles into a one-million-profile rating database to reach the desired effect, which is something that probably will not remain unrecognized in a real-world setting. The same holds for the 3 percent filler size used in the experiments. Determining the average item ratings for 3 percent of the items in a several-million-item database may be a problematic task as well.

A detailed analysis of the effects of different attacks on user-based collaborative filtering systems and parameterized versions, as well as augmented variants thereof (using, e.g., significance weighting or different neighborhood sizes), can be found in O'Mahony et al. (2004).

Model-based collaborative recommenders. When attacking a standard item-based algorithm (such as the one proposed by Sarwar et al. (2001)) with the same sets of manipulated profiles, it can be observed that such algorithms are far more stable than their user-based counterparts. Using the same datasets, a prediction shift of merely 0.15 points can be observed, even if 15 percent of the database entries are faked profiles.

The only exception here is the segment attack, which was designed specifically for attacks on item-based methods. As mentioned previously, an attacker will try to associate the item to be pushed with a smaller set of supposedly very similar items – the target segment. Although the experiments of Mobasher et al. (2007) show that the prediction shift for all users is only slightly affected by such an attack, these types of attacks are very effective when we analyze the prediction shift for the targeted users in the segment. Interestingly, the impacts of a segment attack are even higher than those of an average attack, although the segment attack requires less knowledge about the ratings database. Furthermore, although it is designed for item-based systems, the segment attack is also effective when user-based collaborative filtering is employed.

The results of further experiments with additional model-based collaborative filtering techniques are reported by Sandvig et al. (2007) and Mobasher et al. (2006); a summary of the findings is given by Sandvig et al. (2008). The effects of different attacks have been evaluated for a k-means clustering method, for clustering based on probabilistic latent semantic analysis (pLSA) (Hofmann and Puzicha 1999), for a feature reduction technique using principal component analysis (PCA), as well as for association rule mining based on the Apriori method. The evaluation of different attacks showed that all model-based approaches are more robust against attacks when compared with a standard user-based collaborative filtering approach. Depending on the attack type and the respective parameterizations, slight differences between the various model-based algorithms can be observed.

Finally, Mobasher et al. (2007) also report on an attack experiment of a hybrid recommender system based on item-based filtering and "semantic similarity" (see Mobasher et al. 2004). Not surprisingly, it can be observed that such combined approaches are even more stable against profile injection attacks, because the system's predictions are determined by both the ratings of the user community and some additional domain knowledge that cannot be influenced by fake profiles.

9.4.2 Nuke attacks

Another observation that can be derived from the experiments by Mobasher et al. (2007) is that most attack types are efficient at pushing items but have a smaller impact when they are used to nuke items. The specifically designed nuke methods are, however, quite effective. The very simple love/hate method described previously, for instance, causes a higher negative prediction shift than the knowledge-intensive average method, which was one of the most successful ones for pushing items. Also, the bandwagon attack is more efficient than other methods when the goal is to nuke items, which was not the case when the purpose was to push items. A detailed explanation of the reasons of this asymmetry of attack effectiveness has not yet been found.

Item-based methods are again more stable against attacks, although some prediction shifts can also be observed. Interestingly, only the love/hate attack type was not effective at all for nuking items in an item-based recommender; the reverse bandwagon turns out to be a method with a good nuke effect.

Overall, the questions of possible attacks on recommender systems and the effectiveness of different attack types have been raised only in recent years. A definite list of attack models or a ranking with respect to effectiveness cannot be made yet, as further experiments are required that go beyond the initial studies presented, for example, by Mobasher et al. (2007).

9.5 Countermeasures

Now that we are aware of the vulnerabilities of current recommender system technology, the question arises: how we can protect our systems against such attacks?

Using model-based techniques and additional information. So far, our discussion shows that one line of defense can be to choose a recommendation technique that is more robust against profile injection attacks. Most model-based approaches mentioned in the preceding sections not only provide recommendation accuracy that is at least comparable with the accuracy of memory-based kNN approaches, but they are also less vulnerable.

In addition, it may be advisable to use a recommender that does not rely solely on rating information that can be manipulated with the help of fake

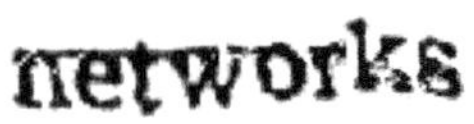

networks

Figure 9.1. Captcha to prevent automated profile creation.

profiles. Additional information, for instance, can be a semantics-based similarity measure, as described above. Alternatively, the recommender could also exploit information about the trust among the different participants in the community, as proposed by Massa and Avesani (2007). Such trust networks among users can not only help to improve the recommendation accuracy, in particular for users who have only a few neighbors, but they could also be used to increase the weights of the ratings of "trusted" friends, thus making it at least harder to attack the recommender system because additional information must be injected into the system's database.

Increasing injection costs. A straightforward defense measure is to simply make it harder to automatically inject profiles. The aforementioned experiments show that a certain attack size must be chosen to achieve the desired push or nuke effects. This typically means that the profile creation task must be automated in realistic scenarios because thousands of profiles cannot easily be entered by hand. Standard mechanisms to prevent the automatic creation of accounts include the usage of a Captcha (Von Ahn et al. 2003), as shown in Figure 9.1. A Captcha (Completely Automated Public Turing test to tell Computers and Humans Apart) is a challenge-response test designed to find out whether the user of a system is a computer or a human. A very common test is to present the user with a distorted image showing a text, as in Figure 9.1, and asking the user to type the letters that are shown in the image. Although such techniques are relatively secure as of today, advances are also being made in the area of automatic analysis of such graphical images (see, e.g., Chellapilla and Simard 2004). In addition, it is of course possible to have the Captchas solved by low-cost outsourced labor. Given today's labor cost in some regions, the cost of resolving one Captcha by hand can be as low as one cent per piece.

As another relatively simple measure, the providers of the recommender service can also increase the costs by simply limiting the number of allowed profile creation actions for a single IP address within a certain time frame. However, with the help of onion routing techniques or other source obfuscation and privacy-enhancing protocols, this protection mechanism can also be defeated.

Automated attack detection. Defense measures of this type aim to automatically detect suspicious profiles in the ratings database. Profiles can raise suspicion for various reasons, such as because the given ratings are "unusual" when compared with other ratings or because they have been entered into the system in a short time, causing the prediction for a certain item to change quickly. In this section, we briefly summarize different methods that have been proposed to detect fake profiles.

Su et al. (2005) propose a method to detect group shilling attacks, in which several existing users of the system cooperate to push or nuke certain items, an attack type we have not discussed so far because it typically involves human interaction and is not automated, as are the other attack types. Their approach works by detecting clusters of users who have not only co-rated many items, but also have given similar (and typically unusual) ratings to these items. After such clusters are identified based on an empirically determined threshold value, their recommendations can be removed from the database. In contrast to other shilling attacks, the particular problem of "group shilling" is that it is not based solely on the injection of fake profiles and ratings. Instead, in this scenario, users with "normal" profiles (that also contain fair and honest ratings) cooperate in a particular case, so that simple fake profile detection methods might miss these ordinary-looking profiles.

Like other attack-prevention methods, the technique proposed by Su et al. (2005) is designed to cope with a certain type of attack (i.e., group shilling). In principle, we could try to develop specific countermeasures for all known attack types. Because the attack methods will constantly improve, however, the goal is to find detection methods that are more or less independent of the specific attack types.

An approach to detect fake profiles, which is independent from the attack type, is proposed by Chirita et al. (2005). Their work is based on the calculation and combination of existing and new rating metrics, such as the degree of agreement with other users, degree of similarity with top neighbors, or rating deviation from mean agreement. Depending on an empirically determined probability function, the ratings of users are classified as normal or faked. An evaluation of the approach for both the random and average attacks on the MovieLens dataset showed that it can detect fake "push" profiles quite accurately, in particular, in situations in which items are pushed that had received only few and relatively low ratings by other users in the past.

Zhang et al. (2006) take a different approach that is based on the idea that every attack type (known or unknown) will influence the rating distribution of some items over time. Therefore, instead of analyzing rating patterns statically, they propose to monitor the ratings for certain items over time to detect anomalies. In particular, time series for the following two properties are constructed and analyzed: *sample average* captures how the likability of an item changes over time; *sample entropy* shows developments in the distribution of the ratings for an item. To detect attacks more precisely, changes in these values are observed within limited time windows. The optimal size of such time windows is determined in a heuristic procedure. An experimental evaluation, based again on the MovieLens dataset and simulated attacks, showed that a relatively good detection rate can be achieved with this method and the number of false alarms is limited. Again, however, certain assumptions must hold. In this case, the assumptions are that (a) attack profiles are inserted in a relatively small time window and (b) the rating distributions for an item do not significantly change over time. If this second assumption does not hold, more advanced methods for time series analysis are required. A knowledgeable attacker or group of attackers will therefore be patient and distribute the insertion of false profiles over time.

Finally, a general option for distinguishing real profiles from fake ones is to use a supervised learning method and train a classifier based on a set of manually labeled profiles. Realistic rating databases are too sparse and high-dimensional, however, so standard learning methods are impractical (Mobasher et al. 2007). Mobasher et al. therefore propose to train a classifier on an aggregated and lower-dimensional data set (Bhaumik et al. 2006). In their approach, the attributes of each profile entry do not contain actual item ratings, but rather describe more general characteristics of the profile, which are derived from different statistics of the data set. Similar to Chirita et al. (2005), one part of this artificial profile contains statistics such as the rating deviation from mean agreement. In addition, to better discriminate between false profiles and real but "eccentric" ones, the training profiles contain attributes that capture statistics that can be used to detect certain attack types (such as the random attack). Finally, an additional intraprofile attribute, which will help to detect the concentration of several profiles on a specific target item, is calculated and included in the profile. The training dataset consists of both correct profiles (in this case, taken from the MovieLens dataset) and fake profiles that are generated according to the different attack types described previously. The evaluation of attacks with varying attack sizes, attack models, and filler sizes shows that in some situations, fake profiles can be detected in a relatively precise manner – that is, with few authentic profiles being excluded from the database and most of the false ones detected. For special attack types, however, such as the segment or love/hate attacks, which in particular do not require large attack sizes, the achievable prediction shifts still seem to be too high, even if automated detection of fake profiles is applied.

9.6 Privacy aspects – distributed collaborative filtering

A different aspect of security and trustworthiness of recommender systems is the question of user privacy. Rating databases of collaborative filtering recommender systems contain detailed information about the individual tastes and preferences of their users. Therefore, collaborative filtering recommenders – as with many personalization systems – face the problem that they must store and manage possibly sensitive customer information. Especially in the recommender systems domain, this personal information is particularly valuable in monetary terms, as detailed customer profiles are the basis for market intelligence, such as for the segmentation of consumers. On the other hand, ensuring customer privacy is extremely important for the success of a recommender system. After a particular system's potential privacy leaks are publicly known, many users will refrain from using the application further or at least from providing additional personal details, which are, however, central to the success of a community-based system.

The main architectural assumption of collaborative filtering recommender systems up to now were that

- there is one central server holding the database, and
- the plain (unobfuscated and nonencrypted) ratings are stored.

Given such a system design, there naturally exists a central target point of an attack; even more, once the attacker has achieved access to that system, all information can be directly used. Privacy-preserving collaborative filtering techniques therefore aim to prevent such privacy breaches by either distributing the information or avoiding the exchange, transfer, or central storage of the raw user ratings.

9.6.1 Centralized methods: Data perturbation

Privacy-preserving variants of CF algorithms that work on centralized, but obfuscated, user data have been proposed for nearest-neighbor methods (Polat and Du 2003), SVD-based recommendation (Polat and Du 2005) and, the Eigentaste method (Yakut and Polat 2007).

The main idea of such approaches can be summarized as follows (Polat and Du 2003). Instead of sending the raw ratings to the central server, a user (client) first obfuscates (disguises) his ratings by applying random data perturbation (RDP), a technique developed in the context of statistical databases to preserve users' privacy. The idea behind this approach is to scramble the original data in such a way that the server – although it does not know the exact values of the customer ratings but only the range of the data – can still do some meaningful computation based on the aggregation of a large number of such obfuscated data sets. Zhang et al. (2006b) describe this process as preserving privacy by "adding random noise while making sure that the random noise preserves enough of the signal from the data so that accurate recommendations can still be made".

Consider the simple example from Polat and Du (2003) in which the server must do some computation based on the sum of a vector of numbers $A = (a_1, \ldots, a_n)$ provided by the clients. Instead of sending A directly, A is first disguised by adding a vector $R = (r_1, \ldots, r_n)$, where the r_is are taken from a uniform distribution in a domain $[-\alpha, \alpha]$. Only these perturbed vectors $A' = (a_1 + r_1, \ldots, a_n + r_n)$ are sent to the server. The server therefore does not know the original ratings but can make a good estimate of the sum of the vectors if the range of the distribution is known and enough data are available, as in the long run

$$\Sigma_{i=1}^{n}(a_i + r_i) = \Sigma_{i=1}^{n}(a_i) + \Sigma_{i=1}^{n}(r_i) \approx \Sigma_{i=1}^{n}(a_i) \qquad (9.1)$$

Similarly, such an estimate can be made for the scalar product, as again, the contribution of the r_is (taken from the uniform distribution $[-\alpha, \alpha]$) will converge to zero. Based on the possibility of approximating the sum and the scalar product of vectors, Polat and Du (2003) devised a nearest-neighbor collaborative filtering scheme that uses z-scores for rating normalization. In the scheme, the server first decides on the range $[-\alpha, \alpha]$, which is communicated to the clients. Clients compute z-scores for the items they have already rated and disguise them by adding random numbers from the distribution to it. The server aggregates this information based on the above observations for the sum and the scalar product and returns – upon a request for the prediction of an unseen item – the relevant parts of the aggregated information back to a client. The client can then calculate the actual prediction based on the privately owned information about the client's past ratings and the aggregated information retrieved from the server.

The main tradeoff of such an approach is naturally between the degree of obfuscation and the accuracy of the generated recommendations. The more "noise" is included in the data, the better users' privacy is preserved. At the same time, however, it becomes harder for the server to generate a precise enough approximation of the real values. In Polat and Du (2003), results of several experiments with varying problem sizes and parameterizations for the random number generation process are reported. The experimental evaluations show that for achieving good accuracy, when compared with a prediction based on the original data, a certain number of users and item ratings are required, as the server needs a certain number of ratings to approximate the original data. Given a sufficient number of ratings and a carefully selected distribution function, highly precise recommendations can be generated with this method, in which the original user's ratings are never revealed to others.

Later research, however, showed that relatively simple randomization schemes, such as the one by Polat and Du, may not be sufficiently robust against privacy-breaching attacks, as much of the original information can be derived by an attacker through advanced reconstruction methods; see Zhang et al. (2006a).

Zhang et al. (2006b) therefore propose an extended privacy-preserving CF scheme, in which the main idea is not to use the same perturbation level for all items, as proposed by Polat and Du (2003). Instead, the client and the server exchange more information than just a simple range for the random numbers. In the proposed approach, the server sends the client a so-called perturbation guidance, on which the client can intelligently compute a perturbation that takes the relative importance of individual item ratings into account. The intuition behind this method is that when using an equal perturbation range for all items, this range might be too large for "important" items that are crucial for detecting item similarities; at the same time, the range will be too small for items that are not critical for generating good recommendations (thus causing unnecessary rating disclosure). Technically, the perturbation guidance matrix that captures the importance level for item similarities is a transformation matrix computed by the server based on SVD.

The experiments by Zhang et al. (2006b) show that, with this method, accuracy comparable to the simple randomization method can be achieved even if much more overall "noise" is added to the data, meaning that less of the private information has to be revealed.

9.6.2 Distributed collaborative filtering

Another way of making it harder for an attacker to gain access to private information is to distribute the knowledge and avoid storing the information in one central place. Agents participating in such a recommendation community may then decide by themselves with whom they share their information and, in addition, whether they provide the raw data or some randomized or obfuscated version thereof.

Peer-to-peer CF. One of the first approaches in that direction was described by Tveit (2001), who proposed to exchange rating information in a scalable peer-to-peer (P2P) network such as Gnutella. In that context, the recommendation

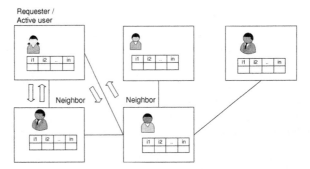

Figure 9.2. Collaborative filtering in P2P environment.

problem was viewed as a search problem, in which the active user broadcasts a query (i.e., a vector of the user's item ratings) to the P2P network. Peers who receive a rating vector calculate the similarity of the received vector with the other known (cached) vectors. If the similarity exceeds a certain threshold, the known ratings are returned to the requester, who can use them to calculate a prediction. Otherwise, the query is forwarded to the neighboring peers and thus spread over the network. Figure 9.2 illustrates a P2P network, in which every user maintains his or her private information and communicates it to his or her neighbors on demand.

Tveit's early approach to distributed CF was proposed in the context of the then-evolving field of mobile commerce but did not include mechanisms for data obfuscation; furthermore, other questions of scalability, cache consistency, and fraudulent users remained open.

Distributed CF with obfuscation. In the work by Berkovsky et al. (2007), therefore, an approach was proposed and evaluated that combines the idea of P2P data exchange and data obfuscation. Instead of broadcasting to the network the "raw" profile and the identification of the target item for which a recommendation is sought, only an obfuscated version is published. Members of the network who receive such a request compute the similarity of the published profile with their own and return a prediction for the target item (if possible) alongside a number expressing the degree of profile similarity. The request-seeking user collects these answers and calculates a prediction using a standard nearest-neighbor method.

In this protocol, disclosure of private profile data occurs in two situations. When the requester sends out his or her profile, he or she inevitably needs to include some personal rating information. On the other hand, the responding agent's privacy is also endangered when he or she returns a rating for the target item. Although this is merely one single rating, an attacker could use a series of requests (probe attack) to incrementally reconstruct the profile of an individual respondent. Obfuscation (e.g., through randomization) will help to preserve the privacy of the participants. It is advisable, however, to perturb only the profiles of the respondent agents to a significant extent, as obfuscation of the critical requester profile quickly deteriorates recommendation accuracy.

Again, a tradeoff between privacy protection and recommendation accuracy exists: the more the profiles are obfuscated, the more imprecise the results are going to be. Berkovsky et al. (2007) analyze different obfuscation variants and their effects on the recommendation accuracy on the MovieLens dataset. The obfuscation schemes can be varied along different dimensions. First, when replacing some of the original values with fake values, one can use different value distributions. The options range from fixed and predefined values over uniform random distributions to distributions that are similar to the original rating (bell-curve) distributions. Another dimension is the question of whether all ratings should be perturbed or whether it is better to obfuscate only the extreme ratings (see also the earlier discussion of important ratings). Finally, the returned predictions may or may not also be obfuscated.

The following effects of the different obfuscation strategies can be observed. First, it does not seem to matter much whether the fake values are taken from a random or bell-curve distribution or we replace original values with the "neutral" value (e.g., 3 on a 1-to-5 scale). The only major influence factor on the accuracy is the percentage of randomized values. With respect to the effects of obfuscation on "moderate" and "extreme" ratings, the prediction accuracy for extreme ratings (e.g., 1 and 5 on the 1-to-5 scale) quickly deteriorates but remains rather stable for moderate ratings. Still, when the percentage of obfuscated values increases too much, accuracy worsens nearly to the level of unpersonalized recommendations. Overall, the experiments also underline the assumption that obfuscating the extreme ratings, which intuitively carry more information and are better suited to finding similar peers, can quickly degrade recommendation accuracy. Perturbing moderate ratings in the data set does not affect accuracy too much. Unfortunately, however, the unobfuscated extreme ratings are those that are probably most interesting to an attacker.

Distributed CF with estimated concordance measures. Lathia et al. (2007) pick up on this tradeoff problem of privacy versus accuracy in distributed collaborative filtering. The main idea of their approach is not to use a standard similarity measure such as Pearson correlation or cosine similarity. Instead, a so-called concordance measure is used, which leads to accuracy results comparable to those of the Pearson measure but that can be calculated without breaching the user's privacy.

Given a set of items that have been rated by user A and user B, the idea is to determine the number of items on which both users have the same opinion (concordant), the number of items on which they disagree (discordant), and the number of items for which their ratings are tied – that is, where they have the same opinion or one of the users has not rated the item. In order to determine the concordance, the deviation from the users' average is used – that is, if two users rate the same item above their average rating, they are concordant, as they both like the item. Based on these numbers, the level of association between A and B can be computed based on Somers' d measure:

$$d_{A,B} = \frac{NbConcordant - NbDiscordant}{NbItemRatingsUsed - NbTied} \qquad (9.2)$$

One would intuitively assume that such an implicit simplification to three rating levels (agree, disagree, no difference) would significantly reduce recommendation accuracy, as no fine-grained comparison of users is done. Experiments on the MovieLens dataset (Lathia et al. 2007), however, indicate that no significant loss in recommendation precision (compared with the Pearson measure) can be observed.

The problem of privacy, of course, is not solved when using this new measure, because its calculation requires knowledge about the user ratings. To preserve privacy, Lathia et al. (2007) therefore propose to exploit the transitivity of concordance to determine the similarity between two users by comparing their ratings to a third set of ratings. The underlying idea is that when user A agrees with a third user C on item i and user B also agrees with C on i, it can be concluded that A and B are concordant on this item. Discordant and tied opinions between A and B can be determined in a similar way.

The proposed protocol for determining the similarity between users is as follows:

- Generate a set of ratings r of size N using random numbers taken uniformly from the rating scale. Ensure that all values are different from the mean of the rating set r.
- Determine the number of concordant, discordant, and tied ratings of user A and user B with respect to r.
- Use these value pairs to determine the upper and lower bounds of the real concordance and discordance numbers between A and B.
- Use the bounds to approximate the value of Somers' d measure.

The calculations of the upper and lower bounds for the concordance and discordance numbers are based on theoretical considerations of possible overlaps in the item ratings. Consider the example for the calculation of the bounds for the *tied* ratings. Let $T_{A,r}$ be the number of ties of A with r and $T_{B,r}$ be the number of ties of B with r. The problem is that we do not know on which items A and B were tied with r. We know, however, that if by chance A and B tied with r on exactly the same items, the lower bound is $max(T_{A,r}, T_{B,r})$; accordingly, the upper bound is $(T_{A,r} + T_{B,r})$ if they are tied on different items. If $(T_{A,r} + T_{B,r})$ is higher than the number of items N, the value of N is, of course, the upper bound. Similar considerations can be made for the bounds for concordant and discordant pairs.

Given these bounds, an estimate of the similarity based on Somers' d measure can be made by using the midpoint of the ranges and, in addition, weighting the concordance measures higher than the discordance measure as follows (midpoint values are denoted with an overline):

$$predicted(d(A, B)) = \frac{\overline{NbConcordant} - 0.5 \times \overline{NbDiscordant}}{\overline{NbItemRatingsUsed} - \overline{NbTied}} \tag{9.3}$$

With respect to recommendation accuracy, first experiments reported by Lathia et al. (2007) show that the privacy-preserving concordance measure yields good accuracy results on both artificial and real-world data sets.

With respect to user privacy, although with the proposed calculation scheme the actual user ratings are never revealed, there exist some theoretical worst-case scenarios, in which an attacker can derive some information about other users. Such situations are very unlikely, however, as they correspond to cases in which a user has not only rated all items, but by chance there is also full agreement of the random sets with the user's ratings. The problem of probing attacks, however, in which an attacker requests ratings for all items from one particular user to learn the user model, cannot be avoided by the proposed similarity measure alone.

Community-building and aggregates. Finally, we mention a relatively early and a slightly different method to distribute the information and recommendation process for privacy purposes, which was proposed by Canny 2002a, 2002b. In contrast to the pure P2P organization of users described earlier, Canny proposed that the participants in the network form knowledge communities that may share their information inside the community or with outsiders. The shared information, from which the active user can derive predictions, is only an aggregated one, however, based, for example, on SVD (Canny 2002a). Thus, the individual user ratings are not visible to a user outside the community who requests information. In addition, Canny proposes the use of cryptographic schemes to secure the communication between the participants in the network, an idea that was also picked up by Miller et al. later on (2004) in their PocketLens system.

Overall, Canny's work represents one of the first general frameworks for secure communication and distributed data storage for CF. The question of how to organize the required community formation process – in particular, in the context of the dynamic web environment – is not yet fully answered and can be a severe limitation in practical settings (Berkovsky et al. 2007).

9.7 Discussion

Recommender systems are software applications that can be publicly accessed over the Internet and are based on private user data. Thus, they are "natural" targets of attacks by malicious users, particularly because in many cases real monetary value can be achieved – for example, by manipulating the system's recommendation or gaining access to valuable customer data.

First, we discussed attacks on the correct functioning of the systems, in which attackers inject fake profiles into the rating database to make a system unusable or bias its recommendations. An analysis of different attack models showed that, in particular, standard, memory-based techniques are very vulnerable. For model-based methods, which are based, for instance, on item-to-item correlation or association rules, no effective attack models have been developed so far. Intuitively, one can assume that such systems are somehow harder to attack, as their suggestions are based on aggregate information models rather than on individual profiles.

Hybrid methods are even more stable, as they rely on additional knowledge that cannot be influenced by injecting false profiles.

Besides the proper choice of the recommendation technique, other countermeasures are possible. Aside from making it harder (or impossible) to inject a sufficient number of profiles automatically, one option is to monitor the evolution of the ratings database and take a closer look when atypical rating patterns with respect to the values or insertion time appear.

Unfortunately, no reports of attacks on real-world systems are yet publicly available, as providers of recommendations services are, of course, not interested in circulating information about attacks or privacy problems because there is money involved. It is very likely, however, that different attacks have been launched on popular recommender systems, given the popularity of such systems and the amounts of money involved. Future research will require cooperation from industry to crosscheck the plausibility of the research efforts and to guide researchers in the right direction.

Gaining access to (individual) private and valuable user profiles is the other possible goal of an attack on a recommender system discussed in this chapter. Different countermeasures have been proposed to secure the privacy of users. The first option is to obfuscate the profiles – for instance, by exchanging parts of the profile with random data or "noise". Although this increases privacy, as the real ratings are never stored, it also reduces recommendation accuracy. The other option is the avoidance of a central place of information storage through the distribution of the information. Different P2P CF protocols have therefore been developed; typically, they also support some sort of obfuscation or additional measures that help to ensure the user's privacy.

What has not been fully addressed in distributed CF systems is the question of recommendation performance. Today's centralized, memory-based, and optimized recommendation services can provide recommendations in "near real time." How such short response times, which are crucial for the broad acceptance of a recommender system, can be achieved in a distributed scenario needs to be explored further in future work.

10

Online consumer decision making

10.1 Introduction

Customers who are searching for adequate products and services in bricks-and-mortar stores are supported by human sales experts throughout the entire process, from preference construction to product selection. In online sales scenarios, such an advisory support is given by different types of recommender systems (Häubl and Murray 2006, Xiao and Benbasat 2007). These systems increasingly take over the role of a profitable marketing instrument, which can help to increase a company's turnover because of intelligent product and service placements. Users of online sales environments have long been identified as a market segment, and the understanding of their purchasing behavior is of high importance for companies (Jarvenpaa and Todd 1996, Thompson and Yeong 2003, Torkzadeh and Dhillon 2002). This purchasing behavior can be explained by different models of human decision making (Gigerenzer 2007, Payne et al. 1993, Simon 1955); we discuss selected models in the following sections.

Traditional models of human decision making are based on the assumption that consumers are making optimal decisions on the basis of rational thinking (Grether and Plott 1979, McFadden 1999). In those models, consumers would make the optimal decision on the basis of a formal evaluation process. One major assumption is that preferences remain consistent and unchangeable. In contradiction to those economic models, research has clearly pointed out that preference stability in decision processes does not exist. For instance, a customer who purchases a digital camera could first define a strict upper limit for the price of the camera, but because of additional technical information about the camera, the customer could change his or her mind and significantly increase the upper limit of the price. This simple example clearly indicates the nonexistence of stable preferences, which led to the development of different alternative decision models (Gigerenzer 2007, Payne et al. 1993, Simon 1955). The most important models are discussed here.

Effort accuracy framework. This model focuses on cost-benefit aspects, in which a decision process is interpreted as a tradeoff between the decision-making effort and the accuracy of the resulting decision. It is based on the idea that human decision behavior is adaptive (Payne et al. 1993) and that consumers dispose of a number of different decision heuristics that they apply in different decision contexts. The selection of a heuristic depends on the decision context, specifically on the tradeoff between decision quality (accuracy) and related cognitive efforts. The effort accuracy framework clearly contradicts the afore-mentioned economic models of decision making, in which optimality aspects are predominant and cognitive efforts in decision processes are neglected. The quality of consumer decision support in terms of perceived usefulness and ease of use has an important impact on a consumer's behavioral intentions – for example, in terms of reusing the recommender system in the future. Explanations regarding the interdependencies between usefulness and usability factors and behavioral intentions are included in the so-called technology acceptance model (TAM); for a related discussion see, for example, Xiao and Benbasat (2007).

Preference construction. The idea of interpreting consumer choice processes in the light of preference construction has been developed by Bettman et al. (1998). Their work takes into account the fact that consumers are not able to clearly identify and declare their preferences before starting a decision process – decision making is more characterized by a process of preference construction than a process of preference elicitation, which is still the predominant interpretation of many recommender applications. As a consequence of these findings, the way in which a recommender application presents itself to the user has a major impact on the outcome of a decision process.

To make recommenders even more successful, we must integrate technical designs for recommender applications with the deep knowledge about human decision-making processes. In this chapter, we analyze existing theories of decision, cognitive, personal, and social psychology with respect to their impacts on preference construction processes. An overview of those psychological theories and their role in recommender systems is given in Tables 10.1 and 10.2. Table 10.1 enumerates cognitive and decision psychological phenomena that have a major impact on the outcome of decision processes but are not explicitly taken into account in existing recommender systems. Table 10.2 enumerates

Table 10.1. *Theories from cognition and decision psychology.*

Theory	Description
Context effects	Additional irrelevant (inferior) items in an item set significantly influence the selection behavior.
Primacy/recency effects	Items at the beginning and the end of a list are analyzed significantly more often than items in the middle of a list.
Framing effects	The way in which different decision alternatives are presented influences the final decision taken.
Priming	If specific decision properties are made more available in memory, this influences a consumer's item evaluations.
Defaults	Preset options bias the decision process.

relevant phenomena from personality and social psychology that also play a role in the construction of recommender applications. All these theories will be discussed and analyzed in the following subsections.

10.2 Context effects

The way in which we present different item sets to a consumer can have an enormous impact on the outcome of the overall decision process. A decision

Table 10.2. *Theories from personality and social psychology.*

Theory	Description
Internal vs. external LOC	Externally influenced users need more guidance; internally controlled users want to actively and selectively search for additional information.
Need for closure	Describes the individual pursuit of making a decision as soon as possible
Maximizer vs. satisficer	Maximizers try to find an optimal solution; satisficers search for solutions that fulfill their basic requirements.
Conformity	A person's behavior, attitudes, and beliefs are influenced by other people.
Trust	A person's behavioral intention is related to factors such as the willingness to buy.
Emotions	Mental states triggered by an event of importance for a person
Persuasion	Changing attitudes or behaviors

Table 10.3. *Asymmetric dominance effect.*

Product	A	B	D
price per month	30	20	35
download limit	10GB	6GB	9GB

is always made depending on the context in which item alternatives are presented. Such context effects have been intensively investigated, for example, by Huber et al. (1982), Simonson and Tversky (1992), and Yoon and Simonson (2008). In the following sections we present different types of context effects and then show how these context effects can influence decision behavior in recommendation sessions. The important thing to note about context effects is that additions of completely inferior item alternatives can trigger significant changes in choice behaviors; this result provides strong evidence against traditional economic choice models that focus on optimal decisions. Superiority and inferiority of items are measured by comparing the underlying item properties. For example, in Table 10.3, item A dominates item D in both aspects (price per month and download limit).

Compromise effect. Table 10.4 depicts an example of the *compromise effect*. In this scenario, the addition of alternative D (the decoy alternative) increases the attractiveness of alternative A because, compared with product D, A has only a slightly lower download limit but a significantly lower price. Thus A appears to be a compromise between the product alternatives B and D. If we assume that the selection probability for A out of the set $\{A, B\}$ is equal to the selection probability of B out of $\{A, B\}$, – that is, $P(A, \{A, B\}) = P(B, \{A, B\})$ – then the inclusion of an additional product D causes a preference shift toward A: $P(A, \{A, B, D\}) > P(B, \{A, B, D\})$. In this context, product D is a so-called decoy product, which represents a solution alternative with the lowest attractiveness.

Table 10.4. *Compromise effect.*

Product	A	B	D
price per month	30	25	50
download limit	10GB	3GB	12GB

Table 10.5. *Attraction effect.*

Product	A	B	D
price per month	30	250	28
download limit	10GB	36GB	7GB

Asymmetric dominance effect. Another type of context effect is *asymmetric dominance* (see Table 10.4): in this case, product *A* dominates *D* in both dimensions (price and download limit), whereas product *B* dominates alternative *D* in only one dimension (price). In this case, the additional inclusion of *D* into the choice set could trigger an increase of the selection probability of *A*.

Attraction effect. Finally, the *attraction effect* occurs in situations in which product *A* is a little bit more expensive but of significantly higher quality than *D* (see Table 10.5). In this situation as well, the introduction of product *D* would induce an increased selection probability for *A*.

Table 10.6 summarizes the major properties of the context effects we have discussed so far. These effects can be exploited for different purposes within the scope of recommendation sessions:

- *Increased selection share of a target product.* As already mentioned, the selection probabilities change in the case that additional (inferior) items are added to a result set. This effect has been shown in empirical studies in a number of application domains such as *financial services*, *e-tourism*, or *consumer electronics*.
- *Increased confidence in a decision*: context effects in recommendation result sets can not only increase the selection probability for target items (also with more than two items in result set) but also increase a consumer's confidence in her own decision.
- *Increased willingness to buy*: decision confidence is strongly correlated with the willingness to buy (Chen and Pu 2005). Consequently, on the basis of an increased decision confidence, context effects can be exploited for increasing the purchase probability.

From a theoretical point of view these results are important, but the question remains how to really exploit those effects in recommendation scenarios. To predict selection behavior on a recommender result page, we must calculate dominance relationships among different item alternatives. Models for the prediction of item dominances have been introduced, for example, by Teppan

Table 10.6. *Summary of context effects: A is the target item, B represents the competitor, and D is the decoy item.*

Effect	Description
Compromise effect	Product A is of slightly lower quality but has a significantly lower price.
Asymmetric dominance effect	Product A dominates D in both dimensions (product B does not).
Attraction effect	Product A is a little more expensive but has a significantly higher quality.

and Felfernig (2009a) and Roe et al. (2001). The major outcome of those models are dominance relationships between the items of a consideration set (*CSet*).

Formula 10.1[1] allows the calculation of a dominance value for an item x in the item set *CSet*: $d(x, CSet)$. This value is calculated by a pairwise comparison of the item property values of x with each y in *CSet*. When we apply Formula 10.1 to the item set depicted in Table 10.4, we receive the dominance values that are depicted in Table 10.7. For example, item B is better than item A regarding the property *price*; if we interpret $x = A$, then the corresponding dominance value for the property *price* is -0.67 (the factor is negative if the value of x is worse than the value of y). These values provide an estimation of how dominant an item x appears in *CSet*. The values in Table 10.7 clearly show the dominance of item A over the items B and D; furthermore, D is dominated by the other alternatives.

$$d(x, CSet) = \sum_{y \in CSet-x} \sum_{a \in properties} \frac{x_a - y_a}{a_{max} - a_{min}} \tag{10.1}$$

Such dominance relationships can be exploited for configuring result sets (Felfernig et al. 2008c). If a recommendation session results, for example, in $n = 10$ possible items (the *candidate items*, or *consideration set*) and a company wants to increase the sales of specific items (the *target items*) in this set, this can be achieved by an optimal result set configuration that is a subset of the items (e.g., five items) retrieved by the recommender system. The optimization criterion in this context is to maximize the dominance values for the target items.

A further potential application of the aforementioned dominance model is the automated detection of context effects in result sets, with the goal to avoid unintended biases in decision processes. If such item constellations are detected,

---

[1] Simplified version of the dominance metric presented by Felfernig et al. (2008a).

Table 10.7. *Calculation of dominance values for x ∈ CSet (items from Table 10.3).*

	x	y_1	y_2	sum	$d(x, CSet)$
	A	**B**	**D**		
price per month		−0.67	0.33	−0.33	
download limit		1.0	0.25	1.25	
					0.92
	B	**A**	**D**		
price per month		0.67	1.0	1.67	
download limit		−1.0	−0.75	−1.75	
					−0.08
	D	**A**	**B**		
		−0.33	−1.0	−1.33	
		−0.25	0.75	0.5	
					−0.83

additional items (neutralizing items) must be added or identified decoy items must be deleted – this can be interpreted as a type of result set configuration problem (Teppan and Felfernig 2009b). Exploiting models of decision biases for neutralizing purposes is extremely important for more customer-centered recommendation environments that are aware of decision biases and try to avoid those when otherwise suboptimal decisions would lead to unsatisfied customers and sales churn. Currently, research focuses on understanding different effects of decision biases but not on how to effectively avoid them. Initial results of empirical studies regarding avoidance aspects are reported by Teppan and Felfernig (2009a).

10.3 Primacy/recency effects

Primacy/recency effects as a cognitive phenomenon describe situations in which information units at the beginning and at the end of a list of items are more likely remembered than information units in the middle of the list[2].

---

[2] These effects are also called *serial position effects* (Gershberg and Shimamura 1994, Maylor 2002).

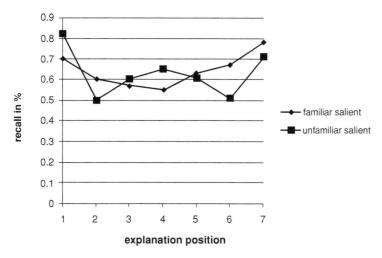

Figure 10.1. Primacy/recency effects in item explanations.

Thus, primacy/recency shows two recall patterns: on one hand, elements at
the beginning of a list (primacy) and, on the other hand, elements at the end
of a list (recency) are recalled more often than those positioned in the middle
of a list. Primacy/recency effects in recommendation dialogs must be taken
into account because different question sequences can potentially change the
selection behavior of consumers (Häubl and Murray 2003). The existence
of different types of serial position effects in knowledge-based recommender
applications is also analyzed by Felfernig et al. (2007a). This study indicates
that product explanations positioned at the beginning and at the end of a list of
explanations are remembered more often than those explanations positioned in
the middle of a list (see Figure 10.1). The two curves in Figure 10.1 represent
two different explanation lists. The first one is denoted as *familiar salient*, in
which explanations related to well-known item properties are positioned at the
beginning and the end of the list. The second one is denoted as *unfamiliar
salient*, in which explanations related to less familiar properties are positioned
at the beginning and the end of the list.

Primacy/recency effects as a decision phenomenon describe situations in
which items presented at the beginning and at the end of a list are evaluated
significantly more often compared with items in the middle of a list. Significant
shifts in selection behavior that are triggered by different element orderings on
a recommender result page are reported by Felfernig et al. (2007a). The same

Table 10.8. *Utility function for primacy/recency effects.*

item position(i)	1	2	3	4	5
posutility(i)	5	3	1	3	5

phenomenon exists as well in the context of web search scenarios (Murphy et al. 2006): web links at the beginning and the end of a list are activated significantly more often than those in the middle of the list. Typically, users are not interested in evaluating large lists of items to identify those that best fit their wishes and needs. Consequently, a recommender application must calculate rankings that reduce the cognitive overheads of a user as much as possible.

An approach to take into account primacy/recency effects in the presentation of items on product result pages was introduced by Felfernig et al. (2008c). Assuming that we have *n* items in a result set, we have *n*! permutations of orders in which items can be presented. Typically, in knowledge-based recommender applications, the utility of items is determined on the basis of multiattribute utility theory (MAUT; Winterfeldt and Edwards 1986), which calculates a ranking *r* for each item contained in the result set. MAUT-based ranking does not take primacy/recency effects because items are presented in the order of decreasing utility values. A way to take serial position effects in MAUT-based rankings into account is presented in Formula 10.2 (Felfernig et al. 2008c).

$$orderutility([p_1, p_2, \ldots, p_n]) = \sum_{i=1}^{n} utility(p_i) * posutility(i) \qquad (10.2)$$

In Formula 10.2, *orderutility*($[p_1, p_2, \ldots, p_n]$) specifies the overall utility of the sequence $[p_1, p_2, \ldots, p_n]$, utility(p_i) specifies the MAUT-based utility of a specific item contained in the result set, and *posutility*(i) specifies the utility of a specific position *i*. To take primacy/recency effects into account, the function *posutility*(i) could be specified as shown in Table 10.8: prominent positions at the beginning and the end of the product list have a higher value determined by *posutility*. Note that in this simple example we assume that every recommendation result consists of exactly five items.

This principle of ranking items in the evaluation of result sets is also applied in the context of automated explanation generation in the domain of buildings (Carenini and Moore 2006). Such explanations include a set of arguments that help a user understand why a certain item has been recommended.

10.4 Further effects

Framing denotes the effect that the way a decision alternative is presented influences the decision behavior of the user (see, e.g., Tversky and Kahneman 1986).

The way pricing information is presented to a user significantly influences the way in which other attributes of a certain decision alternative are evaluated (Levin et al. 1998). This specific phenomenon is denoted as *price framing*: if price information is provided for subcomponents of a product (e.g., the memory unit), then users put more focus on the evaluation of those subcomponents because price information is provided on a more detailed level. Conversely, if an all-inclusive price is presented (e.g., the price for the camera including the memory unit), then users focus their evaluation on important item properties (e.g., the resolution or zoom).

Attribute framing denotes the phenomenon that different but equivalent descriptions of a decision task lead to different final decisions. For example, a financial service described with 0.98 probability of no loss is evaluated better than an equivalent service described with 0.02 probability of loss (valence consistency shift [Levin et al. 1998]). As another example, consumers prefer to buy meat that is 80 percent lean compared with meat that is 20 percent fat. Consumers who are highly familiar with a specific type of product are less amenable to framing effects, as they have clear preferences that should be fulfilled by the recommended product (Xiao and Benbasat 2007).

Priming denotes the idea of making some properties of a decision alternative more accessible in memory, with the consequence that this setting will directly influence the evaluations of a consumer (McNamara 1994; Yi 1990). *Background priming* (Mandel and Johnson 1999) exploits the fact that different page backgrounds can directly influence the decision-making process. An example of background priming is provided by Mandel and Johnson (1999), in which one version of an online furniture selling environment had a background with coins and the second version had a cloudy background (cirrocumulus), which triggered feelings such as *comfort* or *silence*. Users who interacted with the first version chose significantly less expensive products compared with those who interacted with the cloudy-background version.

Priming effects related to the inclusion or exclusion of certain product attributes are discussed by Häubl and Murray (2003). Participants in a user study had the task of selecting a backpacking tent in an online store. The participants were supported by a recommendation agent that collected preference information regarding different properties of tents. One group of participants was asked to specify importance values regarding the properties *durability* and *flynet*; the other group had to specify importance values regarding the properties *warranty* and *weight*. The recommender application then ordered the set of attributes to conform to the preferences specified by the participant (all available tents were visible for both groups). Finally, the participants had to select their favorite item. The result of this study was that participants chose items that outperformed other items in exactly those properties asked by the recommender application. Consequently, human decision-making processes can be influenced systematically by selective presentations of properties in the dialog.

An aspect related to priming is the reduction of questions in a recommendation dialog with the goal to reduce factors such as the dissatisfaction with a choice made or even the aversion to make a decision (Fasolo et al. 2007). Furthermore, the systematic reduction of choice alternatives in choice sets can lead to increased purchase rates (Hutchinson 2005).

Defaults play an important role in decision-making processes because people often tend to favor the status quo compared with other potentially equally attractive decision alternatives (Ritov and Baron 1992, Samuelson and Zeckhauser 1988). This tendency to maintain decisions and being reluctant to change the current state is also called *status quo bias* (Samuelson and Zeckhauser 1988). Potential changes to the current state are always related to some kind of losses or expected gains – and people are typically loss-averse (Tversky and Kahneman 1984, Ritov and Baron 1992). If default options are used in the presentation of decision alternatives, users are reluctant to change this setting (the current state). This phenomenon is able to trigger biases in decision processes (Herrmann et al. 2007, Ritov and Baron 1992). Consumers tend to associate a certain risk with changing a default, as defaults concerns are interpreted to be a central part of a company's product design. Thus a typical application of defaults concerns properties with an associated risk if not selected (e.g., safety equipment in cars, investment protection, or warranties with electronic equipment).

Besides triggering biasing effects, defaults can also reduce the overall interaction effort with the recommender application and actively support consumers in the product selection process – especially in situations in which consumers do not have a well-established knowledge about the underlying product assortment. For example, when buying a digital camera, a default value regarding the needed storage medium could be helpful. Furthermore, defaults can increase the subjectively felt granularity of recommender knowledge bases, as consumers will tend to think that companies really tried to do their best to explain and present the product assortment. Finally, defaults could be used to manipulate the customer in the sense that options are recommended that are of very low or no value for the customer but of value to the seller.

10.5 Personality and social psychology

Besides the cognitive and decision psychological phenomena discussed in the previous sections, different personality properties pose specific requirements on the design of recommender user interfaces. A detailed evaluation of personality properties is possible but typically goes along with considerably efforts related to the answering of large questionnaires. A widespread questionnaire used for identification of personality properties is the NEO Five-Factor Inventory (NEO-FFI; McCrae and Costa 1991), which requires the answering of sixty different questions. There are few application scenarios in which users will accept such an overhead. In this section, we focus on scenarios in which such a detailed personality analysis is not needed.

Locus of control (LOC) can be defined as the amount a human being is able to control occurring events (Duttweiler 1984; Kaplan et al. 2001). The relevance of LOC for the design of recommender user interfaces is explained simply by the fact that users should be able to decide on their own with which type of interface they prefer to interact. Predefined and static dialogs better support users without a special interest in controlling the recommendation process (*external LOC*), whereas more flexible dialogs better support users with a strong interest in controlling the recommendation process (*internal LOC*). More flexible recommender user interfaces not only let the user select the parameters they want to specify but also actively propose interesting parameters and feature settings (Mahmood and Ricci 2007, Tiihonen and Felfernig 2008). Recent research starts to differentiate among influence factors on LOC. For example, the application domain has a major impact on the orientation of LOC – a user could be an expert in the domain of digital cameras but be a nonexpert in the domain of financial services. Such factors are systematically analyzed in the *attribution theory* developed by Weiner (2000).

Need for closure (NFC) denotes the individual's need to arrive at a decision as soon as possible and to get feedback on how much effort is still needed to successfully complete a decision task (Kruglanski et al. 1993). It also refers to a tendency of people to prefer predictability and to narrow down the efforts of an information search as much as possible. Recommender applications can take into account the NFC, for example, by the inclusion of progress bars that inform about the current status of the overall process and the still open number of questions. An example for such a *progress indication* is shown in Figure 4.9, in which the user gets informed about the current status of the recommendation process in terms of the currently active phase. Another concept that helps to take into account the NFC is an immediate display of temporary recommendation results such that the user has the flexibility to select an item for detailed inspection whenever he or she wants. Finally, automated repair actions (see the chapter on knowledge-based recommendation) also help to take into account the NFC (immediate help to get out from the dead end).

Maximizer and satisficer (MaxSat) are two further basic behavioral patterns (Kruglanski et al. 1993, Schwartz et al. 2002). *Maximizers* interacting with a recommender application typically need a longer time span for completing a session because they prefer to know many technical details about the product and, in general, tend to identify an optimal solution that requires an exhaustive search over the available decision alternatives. In contrast, *satisficers* are searching for "good enough" solutions until one solution is found that is within an acceptability threshold. A simple example for the behavioral pattern of maximizers and satisficers is the selection of TV channels (Iyengar et al. 2006). Satisficers focus on the identification of a channel that offers the first acceptable program, whereas maximizers spend most of the time on selection activities such that, compared to satisficers, significantly less viewing time is available for them.

Maximizer and satisficer personality properties can be derived directly by analyzing the interaction behavior of the current user. For example, if a user continually focuses on a detailed analysis of technical product properties, the user can be categorized as a maximizer. Personality properties can then be used, on one hand, when presenting recommendations on a result page by giving more application-oriented or more technical explanations, and on the other hand, throughout the dialog phase by giving more technical or nontechnical explanations, hints, information about already given answers, or information about open questions. Interestingly, Botti and Iyengar (2004) and Iyengar et al. (2006) report results of studies in which maximizers have a tendency to more negative subjective evaluations of decision outcomes ("post-decision regret"), which makes the outcome of a decision process harder to enjoy. This could be explained by the fact that maximizers tend to underestimate the affective costs of evaluating as many options as possible, which contradicts with the assumption of psychologists and economists that the provision of additional alternatives always is beneficial for customers (Botti and Iyengar 2004, Iyengar et al. 2006).

Conformity is a process in which a person's behaviors, attitudes, and beliefs are influenced by other people (Aronson et al. 2007). In the line of this definition, recommenders have the potential to affect users' opinions of items (Cosley et al. 2003).

Empirical studies about conformity effects in a user's rating behavior in CF applications are presented by Cosley et al. (2003). The authors investigated whether the display of item predictions affects a user's rating behavior. The outcome of this experiment was that users confronted with a prediction significantly changed (adapted) their rating behavior. The changed rating behavior can be explained by the fact that the display of ratings simply influences people's beliefs. This occurs in situations in which item evaluations are positively or negatively manipulated: in the case of higher ratings (compared with the original ratings), users tend to provide higher ratings as well. The effect also exists for lower ratings compared with the original ratings.

In summary, the recommender user interface can have a strong impact on a user's rating behavior. Currently, with a few exceptions (Beenen et al. 2004), collaborative recommender systems research focuses on well-tuned algorithms, but the question of user-centered interfaces that create the best experiences is still an open research issue (Cosley et al. 2003). Another example for related research is that by Beenen et al. (2004), who investigated the impact of positive user feedback on the preparedness for providing item ratings. The result of the study was significant: users who got positive feedback (ratings are really needed in order to achieve the overall goal of high-quality recommendations) on their ratings rated more frequently.

Trust is an important factor that influences a consumer's decision whether to buy a product. In online sales environments, a direct face-to-face interaction between customer and sales agent is not possible. In this context, trust is very hard to establish but easy to lose, which makes it one of the key issues to deal with in online selling environments. Notions of trust concentrate mainly on improvements in the dimensions of security of transactions, privacy preserving applications, reputation of the online selling platform, and competence of the recommendation agents (Chen and Pu 2005, Grabner-Kräuter and Kaluscha 2003). A customer's willingness to buy or return to a web site are important trust-induced benefits (Chen and Pu 2005, Jarvenpaa et al. 2000).

Trust-building processes in the context of recommender systems depend strongly on the design of the recommender user interface and the underlying recommendation algorithms (Chen and Pu 2005, Felfernig et al. 2006). Major elements of a recommender user interface that support trust building are explanations, product comparisons, and automated repair functionalities (Felfernig et al. 2006a). Explanation interfaces (Pu and Chen 2007) are an important means to support recommender system transparency in terms of arguments as to why a certain item has been recommended or why certain critiques[3] have been proposed. Significant increases in terms of trust in the recommender application have been shown in various user studies – see, for example, Felfernig et al. (2006a) and Pu and Chen (2007). Product comparisons also help a user establish a higher level of trust in the recommender application. This can be explained simply by the fact that comparison functionalities help to decrease the mental workload of the user because differences and commonalities among items are clearly summarized. Different types of product comparison functionalities are available on many e-commerce sites – for example, www.amazon.com or www.shopping.com (Häubl and Trifts 2000). Finally, repair actions are exploited by users with a low level of product domain knowledge; in this context, repair actions help to increase the domain knowledge of the user because they provide explanations why no recommendation could be found for certain combinations of requirements (Felfernig et al. 2006a). Repair actions are typically supported by constraint-based recommender applications – an example for a commercially available application is discussed by Felfernig et al. (2007b). The second major factor that influences the perceived level of trust is the overall quality of recommendations – the higher the conformity with the user's real preferences, the higher is the trust in the underlying recommender algorithm (Herlocker et al. 2004).

Emotions. Although the importance of extending software applications with knowledge about human emotions is agreed on (Picard 1997), most of the existing recommender applications still do not take this aspect into account (Gonzalez et al. 2002). User profiles typically do not include information about human emotions, and as a consequence, recommender applications are, in many cases, unable to adapt to the constantly changing and evolving preferential states. An *emotion* can be defined as "a state usually caused by an event of importance to the subject. It typically includes (a) a conscious mental state with a recognizable quality of feeling and directed towards some object, (b) a bodily perturbation of some kind, (c) recognizable expressions of the face, tone of voice, and gesture [and] (d) a readiness for certain kinds of action" (Oatley and Jenkins 1996). There are different, but not well agreed on, approaches to categorizing emotional states (Parrot 2001). One categorization that is also applied by the commercially available movie recommendation environment

---

[3] Critiquing has been described earlier.

Table 10.9. *Emotion categories used in* MovieProfiler *(originally, this set of emotions was developed by Plutchik and Hope [1997]).*

Emotion	Description
Fear	A feeling of *danger* and/or *risk* independent of the fact of being real or not
Anger	Status of *displeasure* regarding an action and/or an idea of a person or an organization
Sorrow	Status of *unhappiness* and/or *pain* because of an unwanted condition and the corresponding emotion
Joy	Status of being *happy*
Disgust	Associated with things and actions that appear *"unclean"*
Acceptance	Related to *believability*, the degree to which something is accepted as true
Anticipation	Expectation that *something "good" will happen*
Surprise	Emotion triggered by an *unexpected event*

MovieProfiler[4] has been developed by Plutchick (see, e.g., Plutchik and Hope 1997) – the corresponding emotion types are explained in Table 10.9. The search engine of MovieProfiler supports item search on the basis of an emotional profile specified by the user (see Figure 10.2). The search engine follows a case-based approach in which the most similar items are retrieved by the application. The innovative aspect of MovieProfiler is the search criteria that are represented as emotional preferences that define which expectations a user has about a recommended movie. A user indicates on a five-point psychometric scale which specific emotions should be activated by a film. In a similar way, users are able to evaluate movies regarding the emotions *fear, anger, sorrow, joy, disgust, acceptance, anticipation,* and *surprise* (see Figure 10.3).

Persuasion. Behavioral decision theory shows that human decision processes are typically based on adaptive decision behavior (Payne et al. 1993). This type of behavior can be explained by the *effort-accuracy tradeoff*, which states that people typically have limited cognitive resources and prefer to identify optimal choices with as little effort as possible. The aspect of the availability of limited cognitive resources is also relevant in the theory of decision making under bounded rationality (Simon 1955); bounded rationality can act as a door opener for different nonconscious influences on the decision behavior of a consumer. This way of explaining human decision processes has a

---

[4] www.movieprofiler.com.

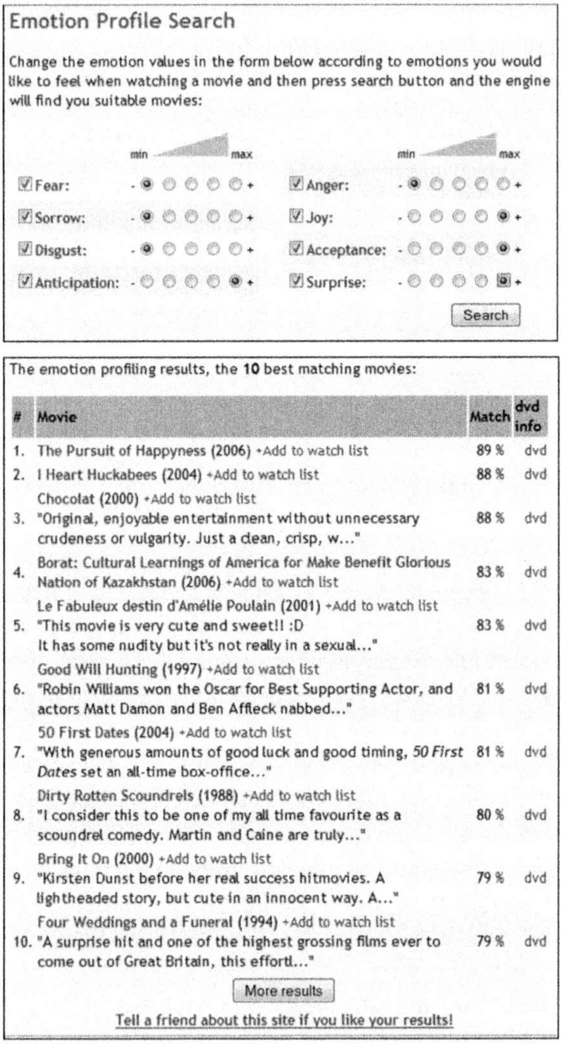

Emotion Profile Search

Change the emotion values in the form below according to emotions you would like to feel when watching a movie and then press search button and the engine will find you suitable movies:

	min ▱ max			min ▱ max
☑ Fear:	· ⦿ ⊙ ⊙ ⊙ ⊙ +	☑ Anger:	· ⦿ ⊙ ⊙ ⊙ ⊙ +	
☑ Sorrow:	· ⦿ ⊙ ⊙ ⊙ ⊙ +	☑ Joy:	· ⊙ ⊙ ⊙ ⊙ ⦿ +	
☑ Disgust:	· ⦿ ⊙ ⊙ ⊙ ⊙ +	☑ Acceptance:	· ⊙ ⊙ ⊙ ⊙ ⦿ +	
☑ Anticipation:	· ⊙ ⊙ ⊙ ⊙ ⦿ +	☑ Surprise:	· ⊙ ⊙ ⊙ ⊙ ▣ +	

[Search]

The emotion profiling results, the **10** best matching movies:

#	Movie	Match	dvd info
1.	The Pursuit of Happyness (2006) +Add to watch list	89 %	dvd
2.	I Heart Huckabees (2004) +Add to watch list	88 %	dvd
3.	Chocolat (2000) +Add to watch list "Original, enjoyable entertainment without unnecessary crudeness or vulgarity. Just a clean, crisp, w..."	88 %	dvd
4.	Borat: Cultural Learnings of America for Make Benefit Glorious Nation of Kazakhstan (2006) +Add to watch list	83 %	dvd
5.	Le Fabuleux destin d'Amélie Poulain (2001) +Add to watch list "This movie is very cute and sweet!! :D It has some nudity but it's not really in a sexual..."	83 %	dvd
6.	Good Will Hunting (1997) +Add to watch list "Robin Williams won the Oscar for Best Supporting Actor, and actors Matt Damon and Ben Affleck nabbed..."	81 %	dvd
7.	50 First Dates (2004) +Add to watch list "With generous amounts of good luck and good timing, *50 First Dates* set an all-time box-office..."	81 %	dvd
8.	Dirty Rotten Scoundrels (1988) +Add to watch list "I consider this to be one of my all time favourite as a scoundrel comedy. Martin and Caine are truly..."	80 %	dvd
9.	Bring It On (2000) +Add to watch list "Kirsten Dunst before her real success hitmovies. A lightheaded story, but cute in an innocent way. A..."	79 %	dvd
10.	Four Weddings and a Funeral (1994) +Add to watch list "A surprise hit and one of the highest grossing films ever to come out of Great Britain, this effort..."	79 %	dvd

[More results]

Tell a friend about this site if you like your results!

Figure 10.2. MovieProfiler recommender – emotion-based search: the user specifies requirements in terms of emotions (*fear, anger, sorrow, joy, disgust, acceptance, anticipation,* and *surprise*) that should be activated when watching the movie.

major impact on the perceived role of recommender applications, which now moves from the traditional interpretation as tools for supporting preference *elicitation* toward an interpretation of tools for preference *construction*. As discussed in the previous sections, the design of a recommender application can have a significant impact on the outcome of a decision process (Gretzel and

The Green Mile (1999)

›Emotion Profile ›Target Segment Profile

› Horror:

› Anger:

› Sorrow:

› Joy:

› Disgust:

› Acceptance:

› Anticipation:

› Surprise:

Number of reviews: 1

ADD TEXTUAL REVIEW!

OR

ADD EMOTION REVIEW!

› Released:
1999

› Runtime: 188
mins

› IMDb rating:
8.10/10

› Movies.go.com
Average
Reader
Rating:
4.62/5

› Add to Watch
list

The Green Mile
Tom Hanks, David M...
Best Price $0.72
or Buy New

Buy from

Privacy Information

Add Movie Evaluation

Low-resolution | High-resolution

Name of the movie: **The Green Mile**
IMDB link: http://www.imdb.com/title/tt0120689/
Production year: 1999
Runtime (in minutes): 188

For textual review, please follow this link! Otherwise for emotion value review, fill
in the following (2 page) form.

› Give your evaluation of the following emotion in this movie: **terror,
horror:**(What's this?)
Least ○ ○ ◉ ○ ○ **Most**
› Give your evaluation of the following emotion in this movie: **Anger,
rage:**(What's this?)
Least ○ ○ ◉ ○ ○ **Most**
› Give your evaluation of the following emotion in this movie: **sorrow,
grief, sadness or depressing:**(What's this?)
Least ○ ○ ◉ ○ ○ **Most**
› Give your evaluation of the following emotion in this movie: **joy and
happiness:**(What's this?)
Least ○ ○ ◉ ○ ○ **Most**
› Give your evaluation of the following emotion in this movie: **disgust,
disgusting:**(What's this?)
Least ○ ○ ◉ ○ ○ **Most**
› Give your evaluation of the following emotion in this movie: **acceptance,
credibility, believability (of the characters and the storyline):**(What's this?)
Least ○ ○ ◉ ○ ○ **Most**
› Give your evaluation of the following emotion in this movie: **anticipation:**
(What's this?)
Least ○ ○ ◉ ○ ○ **Most**
› Give your evaluation of the following emotion in this movie: **surprising:**
(What's this?)

Figure 10.3. MovieProfiler recommender – evaluation of movies: movies
can be evaluated regarding the emotions *fear*, *anger*, *sorrow*, *joy*, *disgust*,
acceptance, *anticipation*, and *surprise*.

Fesenmaier 2006). Consequently, recommender technologies can be interpreted as persuasive technologies in the sense of Fogg: "Persuasive technology is broadly defined as technology that is designed to change attitudes or behaviors of the users through persuasion and social influence, but not through coercion" (Felfernig et al. 2008c, Fogg 2003). This interpretation is admissible primarily if recommendation technologies are applied with the goal of supporting (not manipulating) the customer in finding the product that fits his or her wishes and needs. Obviously, persuasive applications raise ethical considerations, as all of the effects mentioned here could be applied to stimulate the customer to purchase items that are unnecessary or not suitable.

10.6 Bibliographical notes

Consumer buying behavior and decision making have been studied extensively in different research areas, such as cognitive psychology (Gershberg and Shimamura 1994, Maylor 2002), decision psychology (Huber et al. 1982, Yoon and Simonson 2008), personality psychology (Duttweiler 1984, Weiner 2000), social psychology (Beenen et al. 2004, Cosley et al. 2003), and marketing and e-commerce (Simonson and Tversky 1992, Xiao and Benbasat 2007). Predominantly, the reported results stem from experiments related to isolated decision situations. In many cases, those results are not directly applicable to recommendation scenarios without further empirical investigations that take into account integrated recommendation processes. The exception to the rule is Cosley et al. (2003), who analyze the impact of social-psychological effects on user behavior in the interaction with collaborative filtering recommenders. Furthermore, Felfernig et al. (2007, 2008a, 2008c), and Häubl and Murray (2003, 2006) focus on the integration of research results from marketing and decision psychology into the design of knowledge-based recommender applications. The investigation of culture-specific influences on the acceptance of user interfaces has been the focus of a number of studies – see, for example, Chau et al. (2002), Chen and Pu (2008), and Choi et al. (2005). From these studies it becomes clear that different countries have different cultural backgrounds that strongly influence individual preferences and criteria regarding user-friendly interfaces. Significant differences exist between Western cultures that are based on individualism and Eastern cultures that focus more on collectivistic elements. A more detailed analysis of culture-specific influences on the interaction with recommender applications has first been presented in Chen and Pu (2008).

11

Recommender systems and the next-generation web

In recent years, the way we use the web has changed. Today's web surfers are no longer mere consumers of static information or users of web-enabled applications. Instead, they play a far more active role. Today's web users connect via social networks, they willingly publish information about their demographic characteristics and preferences, and they actively provide and annotate resources such as images or videos or share their knowledge in community platforms. This new way of using the web (including some minor technical innovations) is often referred to as *Web 2.0* (O'Reilly 2007).

A further popular idea to improve the web is to transform and enrich the information stored in the web so that machines can easily interpret and process the web content. The central part of this vision (called the Semantic Web) is to provide defined meaning (semantics) for information and web services. The Semantic Web is also vividly advertised, with slogans such as "enabling computers to read the web" or "making the web readable for computers". This demand for semantics stems from the fact that web content is usually designed to be interpreted by humans. However, the processing of this content is extremely difficult for machines, especially if machines must capture the intended semantics. Numerous techniques have been proposed to describe web resources and to relate them by various description methods, such as how to exchange data, how to describe taxonomies, or how to formulate complex relations among resources.

These recent developments also open new opportunities in the area of recommender systems. One of the basic challenges in many domains is the sparsity of the rating databases and the limited availability of user preference information. In Web 2.0 times, these valuable pieces of information are increasingly available through social networks, in which users not only exhibit their preferences to others but are also explicitly connected to possibly like-minded, trusted users. One research question, therefore, consists of finding ways to exploit this additional knowledge in the recommendation process.

Web 2.0 also brought up new types of public information spaces, such as web logs (blogs), wikis, and platforms for sharing multimedia resources. Because of the broad success and sheer size of many of those public platforms, finding interesting content becomes increasingly challenging for the individual user. Thus, these platforms represent an additional application area for recommender systems technology.

Furthermore, in knowledge-based recommendation the acquisition and maintenance of knowledge is an important precondition for a successful application. Semantic information offered by the Semantic Web allows us to apply knowledge-based recommender techniques more efficiently and to employ these techniques to new application areas. Semantic information about users and items also allows us to improve classical filtering methods.

All these new capabilities of Web 2.0 and the Semantic Web greatly influence the field of recommender systems. Although almost all areas of recommender technologies are touched by the introduction of new web technologies and the changed usage of the web, our goal in this chapter is to provide a central overview about recent developments triggered by these innovations.

For example, we can exploit information provided by online communities to build trust-aware recommender systems to avoid misuse or to improve recommendations (Section 11.1). Furthermore, the (free) annotation of web resources are forming so-called folksonomies (Section 11.2), which give additional valuable information to improve recommendations. If these annotations, which can be considered as additional characterization of items, are interpreted more formally (i.e., by a formal description of a taxonomy), then we receive some additional means to enhance recommender systems, as described in Section 11.3. On one hand, semantics are valuable for computing recommendations, but on the other hand, semantic information must be acquired. In Section 11.4 we discuss approaches as to how this task can be solved more efficiently in the context of recommender systems.

11.1 Trust-aware recommender systems

When we talked about "user communities" – for instance, in the context of collaborative filtering approaches – the assumption was that the community simply consists of all users of the online shop (and its recommender system) and that the members of the community are only implicitly related to each other through their co-rated items. On modern consumer review and price comparison platforms as well as in online shops, however, users are also given the opportunity to rate the reviews and ratings of other users. On the Epinions.com consumer review platform, for instance, users may not only express that they liked a particular item review, but they can also state that they generally trust specific other users and their opinions. By this means, users are thus becoming explicitly connected in a "web of trust". In the following section, we show how such trust networks can serve as a basis for a recommender system.

The phrase "trust in recommender systems" is interpreted in the research community in three different ways: first, in the sense of getting users to believe that the recommendations made by the system are correct and fair, such as with the help of suitable explanations; second, that recommender systems assess the "trustworthiness" of users to discover and avoid attacks on recommender systems. These system-user and user-system trust relationships were discussed in previous chapters. In this section, we focus on the third interpretation of trust, which is based on trust relationships between users – users put more trust in the recommendations of those users to which they are connected.

11.1.1 Exploiting explicit trust networks

In general, trust-enhanced nearest-neighbor recommender systems, such as the one presented by Massa and Avesani (2007), aim to exploit the information from trust networks to improve the systems "performance" in different dimensions. The hope is that the accuracy of the recommendations can be increased – for instance, by taking the opinions of explicitly trusted neighbors into account instead of using peers that are chosen based only on a comparison of the rating history. In particular, the goal here is also to alleviate the cold-start problem and improve on the user coverage measure – in other words, if no sufficiently large neighborhood can be determined based on co-rated items, the opinions of trusted friends can serve as a starting point for making predictions. In addition, one conjecture when using explicit trust networks is that they help make the recommender system more robust against attacks, because desired "incoming" trust relationships to a fake profile cannot easily be injected into a recommender database.

Next, we briefly sketch the general architecture of the "trust-aware recommender system" (TARS) proposed by Massa and Avesani (2007) as an example and summarize their findings of an evaluation on real-world datasets.

There are two inputs to the proposed recommender system: a standard rating database and the *trust network*, which can be thought of as a user-to-user matrix T of trust statements. The possible values in the matrix range from 0 (no trust)

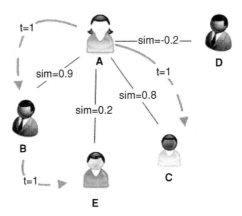

Figure 11.1. Similarity and trust network.

to 1 (full trust). A special null value such as ⊥ (Massa and Avesani 2004) is used to indicate that no trust statement is available.

In principle, the TARS system aims to embed the trust knowledge in a standard nearest-neighbor collaborative filtering method. Basically, the prediction computation is the same as the usual nearest-neighbor weighting scheme from Equation 11.1, which predicts the rating for a not-yet-seen item p based on the active user's average rating $\overline{r_a}$ and the weighted opinions of the N most similar neighbors.

$$pred(a, p) = \overline{r_a} + \frac{\sum_{b \in N} sim(a, b) * (r_{b,p} - \overline{r_b})}{\sum_{b \in N} sim(a, b)} \qquad (11.1)$$

In the TARS approach, the only difference to the original scheme is that the similarity of the active user a and neighbor b, $sim(a, b)$, is not calculated based on Pearson's correlation coefficient (or something similar). Instead, the trust value $T_{a,b}$ is used to determine the weights of the opinions of the neighbors.

Figure 11.1 shows an example of a situation in which the problem is to generate a recommendation for user A (Alice). Both the standard similarity measures (denoted as *sim*) and some trust statements (denoted as *t*) are shown in the figure: A has issued two explicit trust statements: she trusts users B and C, probably because in the past, she found the opinions of B and C valuable. Assume that in this setting A's actual rating history (similarity of 0.8 and 0.9) is also quite similar to the ones of B and C, respectively.

Furthermore, assume that in this example we parameterize a nearest-neighbor algorithm to make a prediction for A only if at least three peers with a similarity above the 0.5 level can be identified. With these parameters and the given trust network, no recommendation can be made. If we use the trust values instead of Pearson's similarity measure, the situation will not improve, as *A* has issued only two trust statements. The key point of trust-aware recommendation, however, is that we may assume that the trust-relationships are *transitive* – that is, we assume that if *A* trusts *B* and *B* trusts *E*, *A* will also trust *E*. Although *A* might not fully trust *E*, because of the missing direct experience, we may assume that *A* trusts *E* at least to some extent. In the example setting shown in Figure 11.1, a third peer, namely *E*, is therefore assumed trustworthy and a recommendation can be made if we again assume a lower bound of at least three neighbors.

It is relatively easy to see in the example how trust-aware recommenders can help alleviate the cold-start problem. Although the matrix of trust statements may also be sparse, as in a typical rating database, the transitivity of the trust relationship allows us to derive an indirect measure of trust in neighbors not directly known to the target user. In addition, if the trust matrix is very sparse, it can be combined in one way or another with the standard similarity measure and serve as an additional source of knowledge for the prediction.

An important property of trust relationships is that they are usually not assumed to be symmetric – that is, the fact that *A* trusts the opinion of *C* does not tell us that *C* also trusts *A*. This aspect helps us make trust networks more robust against attacks: although it is easy to inject profiles that express trust both to real users and other fake users, it is intuitively not easy to attract trust statements from regular users – that is, to connect the fake network to the network of the real users, which is important to propagate trust relationships to fake users. To limit the propagation of false trust values, the number of propagating steps can be limited or the trust values multiplied by some damping factor.

11.1.2 Trust metrics and effectiveness

How indirect trust and distrust values should be calculated between users that are not directly connected in a social network is a relatively new area of research. One typical option is to use a multiplicative propagation of trust values in combination with a maximum propagation distance and a minimum trust threshold (Massa and Avesani 2007, Golbeck 2006); these parameters, for instance, can be determined empirically. The algorithm implementations are relatively straightforward (Massa and Avesani 2007) and may differ in the way the graph formed by the social connections is explored.

These relatively simple metrics already lead to some measurable increase in recommendation accuracy in special cases. Still, various other schemes for trust calculation that, for instance, also take into account explicit distrust statements differently than "low trust" have been proposed. A broader discussion of this topic and more complex propagation schemes are discussed, for instance, by Guha et al. (2004), Ziegler and Lausen (2004, 2005), and Victor et al. (2006); proposals based on subjective logic or in the context of the Semantic Web are presented by Jøsang et al. (2006) and Richardson et al. (2003).

Although the aforementioned metrics calculate an estimate of the "local" trust between a source and a target user, it would be possible to use a "global" trust metric as well. A very simple metric would be to average the trust values received for every user to compute an overall user "reputation". Such an approach often can be found, for example, on online auction platforms. Google's PageRank algorithm (Brin and Page 1998) can also be seen as an example of a global trust metric, which, however, goes beyond simple averaging of the number of links pointing to a page.

To what extent the information in the trust networks can really help to improve the quality of recommendations has been evaluated by Massa and Avesani (2004) on the basis of a dataset from the Epinions.com platform. On this web site, users can not only review all types of items (such as cars, books, or movies) but also rate the reviewers themselves by adding them to their personal web of trust if they have the feeling that they "have consistently found [the reviews] to be valuable". Because no other information is available, adding a user to the web of trust can be interpreted as a direct trust rating of 1. On this platform, users may also put other users on a block list, which can be seen as a trust rating of 0. More fine-grained ratings, however, are not possible on Epinions.com.

Interestingly, the item rating behavior of Epinions.com users is reported to be different from that of MovieLens users. Compared to MovieLens, not only are the rating database and the trust matrix even sparser, but about half the ratings had the highest possible value (5) and another 30 percent had the second-highest value (4). Furthermore, more than half the users are cold starters who have voted for fewer than five items. Standard accuracy metrics such as mean absolute error do not reflect reality very well in such settings, because every prediction error will be weighed in the same way, although significant differences for heavy raters and cold starters are expected. Massa and Avesani (2007) therefore propose to use the mean absolute user error (MAUE) to ensure that all users have the same weight in the accuracy calculation. In addition, the usage of a *user coverage* metric is recommended, which measures the number of users for which a recommendation can be made – a problem that virtually does not appear in the MovieLens dataset, because it is guaranteed that every user has rated at least twenty items.

The results of an evaluation that used different algorithms (standard CF, unpersonalized, and trust-based ones) and views on the dataset to measure the impact on heavy raters, cold starters, niche items, and so forth can be summarized as follows:

- *Effectiveness of simple algorithms.* Given such a specific, but probably in practice not uncommon, distribution of ratings, simple algorithms such as "always predict value 5" or "always predict the mean rating value of a user" work quite well and are only slightly worse than standard CF algorithms with respect to MAE.

  Another simple technique is to predict the average item rating. In this application scenario, this unpersonalized algorithm even outperforms standard CF techniques, an effect that cannot be observed when using other datasets, such as the MovieLens database. With respect to rating coverage, the simple method is better, in particular, when it comes to the many cold-start users. When the different slices of the rating database are analyzed, it can be observed, however, that CF techniques are effective for controversial items for which the standard deviation is relatively large and an averaging technique does not help.

- *Using direct trust only.* In this setting, a trust-based technique is employed that uses only the opinions of users for which an explicit trust statement is available; no trust propagation over the network is done. Although the overall MAE of this method is between that of standard CF and the simple average technique, it works particularly well for cold-start users, niche items (items for which only very few ratings exist), and opinionated users (users having a high standard deviation in their ratings). When the MAUE measure is used, the "direct-trust" method called MT1 achieves the highest overall accuracy of all compared methods. With respect to coverage, it is finally observed that "MT1 is able to predict fewer ratings than CF but the predictions are spread more equally over the users (which can then be at least partially satisfied)" (Massa and Avesani 2007).

- *Trust propagation.* The evaluations with propagation levels of 2, 3, and 4 lead to the following observations: First, when the propagation level is increased, the trust network quickly grows from around ten directly trusted neighbors to nearly 400 in the second level, and several thousand in the third level. An increase in the propagation distance thus leads directly to an increase in rating coverage. At the same time, however, the prediction accuracy constantly decreases, because the opinions of the faraway neighbors are not the best predictors.

- *Hybrids.* In their final experiments, Massa and Avesani tried to calculate predictions based on both a standard similarity measure and the trust measure and combine the results as follows: when only one method was able to compute a weight, this value was taken. If both methods were applicable, a weighted average was computed. Although such a combination quite intuitively leads to increased coverage, the performance did not increase and typically fall between the CF and the trust-based algorithms.

To determine whether global trust metrics (in which every user gets the same trust value from everyone) work well, the performance of using an adapted version of PageRank to determine global trust weights was also evaluated. The experiments showed, however, that such nonpersonalized metrics cannot compete with the personalized trust metrics described earlier.

In summary, the evaluation showed two major things. First, the exploitation of existing trust relations between users can be very helpful for fighting the cold-start problem. Second, there are still many open questions in the context of the evaluation of recommender systems, most probably because research evaluations are often made only on the MovieLens dataset.

11.1.3 Related approaches and recent developments

The concept of trust in social networks has raised increased interest during the past few years. Among others, the following topics have also been covered in the area of recommender systems.

- *Similar approaches.* Several approaches to exploiting and propagating trust information have been proposed in the last years that are similar to the work by Massa and Avesani (2007). Golbeck (2005) and Golbeck and Hendler (2006), report on an evaluation of the trust-enhanced movie recommender FilmTrust. Although the size of the rating database was very small (only about 500 users), observations could be made that were similar to those with the larger Epinions.com dataset. The distribution of ratings in the database was not very broad, so a "recommend the average" approach generally worked quite well. Again, however, the trust-based method worked best for opinionated users and controversial ratings and outperformed the baseline CF method. The propagation method used in the FilmTrust system is similar to the one used by Massa and Avesani (2007); however, the trust ratings can be chosen from a scale of 1 to 10.

Another approach that exploits explicit trust statements and combines them with another measure of item importance was recently proposed by Hess et al. (2006). In their work, a trust network among scientific reviewers is combined with standard visibility measures for scientific documents to personalize the document visibility measure in a community. Although no real-world evaluation of this approach has been made, it can be counted as another interesting idea that shows how in the future trust information from various sources might be combined to generate personalized information services.

- *Implicit trust.* Although their paper is titled "Trust in Recommender Systems", the trust concept used by O'Donovan and Smyth (2005) is not based on direct trust statements as in the work described above. Instead, they propose new neighbor selection and weighting metrics that go beyond simple partner similarity. The real-life analogy on which they base their work is that one will typically ask friends who have similar overall tastes for a movie recommendation. However, not every one of those "general" neighbors might be a good advisor for every type of item/movie. In their work, therefore, trustworthiness is determined by measuring how often a user has been a reliable predictor in the past – either on average or for a specific item. In contrast to the work of Massa and Avesani, for instance, these trust values can be automatically extracted from the rating database.

  When these numbers are available, predictions can be computed in different ways: by using a weighted mean of the standard Pearson similarity and the trust measure, by using the trust value as a filter for neighbor selection, or by a combination of both. O'Donovan and Smyth evaluated their approach on the MovieLens dataset and showed that an accuracy improvement of more than 20 percent can be achieved when compared with the early CF approach of Resnick et al. Resnick et al. (1994).

  The usage of the term *trust* for this approach is not undisputed. The term *competence* or *reputation* would probably have been more suitable for this approach. Further approaches that automatically infer such "trust" relationships from existing rating data are those by Papagelis et al. (2005) and Weng et al. (2006).

- *Recommending new friends.* Before statements in an explicit trust network can be exploited, new users must be connected with a sufficient number of other members of the community – in other words, we face another form of cold-start problem here. Many of today's social web platforms aim to increase the connectivity of their members by suggesting other users as friends. The suggestions are based, for example, on existing relationships (friend-of-a-friend) or on a random selection.

However, because the number of other users a person will connect to is limited, it might be a good idea for the system to automatically recommend community members whose opinions are particularly valuable – the ones who, in our setting, are good predictors. How such users are characterized and how a good selection influences prediction quality is analyzed by Victor et al. (2008a, 2008b). In their work, the authors differentiate among the following particular user types (key figures): *mavens*, who write a lot of reviews; *frequent raters*, who rate many items; and *connectors*, who issue many trust statements and are trusted by many – that is, when trust is propagated over them, many other users are reached. To measure the impact of the opinion of one user on the other and to analyze the tradeoff between coverage and accuracy, several measures, such as accuracy change or betweenness (from the field of social network analysis; see Wasserman and Faust 1994), are used. Overall, Victor et al. can show – based again on an evaluation of an Epinions.com dataset – that the inclusion of key figures in the network leads to better accuracy and coverage results when compared with a situation in which users are initially connected to a randomly selected set of other users.

In summary, recent research has already shown that the exploitation of existing trust information can improve the accuracy of recommender systems. We believe, however, that this is only the beginning of a more comprehensive exploitation of all the different information available in today's and tomorrow's online social networks.

11.2 Folksonomies and more

Besides social networks such as Facebook.com, on which user communities willingly share many of their preferences and interests, platforms that support *collaborative tagging* of multimedia items have also become popular in Web 2.0. In contrast to classical keyword-assignment schemes that superimpose a defined classification hierarchy, so-called folksonomies (folk taxonomies) represent a far more informal way of allowing users to annotate images or movies with keywords. In these systems, users assign arbitrary tags to the available items; tags can describe several dimensions or aspects of a resource such as content, genre, or other metadata but also personal impressions such as *boring*. When annotating such tags, users frequently express their opinions about products and services. Therefore, it seems to be self-evident to exploit this information to provide recommendations.

The goals of the Semantic Web also include the semantic annotation of resources. In contrast to tagging systems, however, Semantic Web approaches postulate the usage of formal, defined, and machine-processable annotations. Although folksonomies are sometimes referred to as "lightweight ontologies", they are actually at the opposite spectrum of annotation options: although formal ontologies have the advantages of preciseness and definedness, they are hard to acquire. Folksonomies, on the other hand, not only can be used by everyone but also directly reflect the language and terms that are actually used by the community.

Recommender systems and folksonomies can be related to each other in different ways. First, in analogy to trust-enhanced approaches, one can try to exploit the information of how items are tagged by the community for predicting interesting items to a user. Second, because the arbitrary usage of tags also has its problems, recommender system technology can be used to recommend tags to users – for example, to narrow the range of used tags in the system. We will shortly summarize example systems of each type in the following.

11.2.1 Using folksonomies for recommendations

In collaborative tagging systems, users annotate resources with tags. The question in folksonomy-enhanced recommender approaches is how we can leverage this information for recommending items or improving prediction accuracy. Roughly speaking, there are two basic approaches to integrate tags in the recommendation process. Tags can be viewed as information about the content of items. Consequently, standard content-based methods are the starting point of extension. Tags can also be viewed as an additional dimension of the classical user–item matrix, and collaborative techniques serve as a basis for advancements.

In the following subsections, we introduce the basic ideas from both sides, using some recent works that clearly show the underlying considerations.

11.2.1.1 Folksonomies and content-based methods

We start our introduction by the exploitation of so-called tag clouds. Then we present two approaches to deal with the ambiguity and redundancy of free formulated tags. One approach tries to overcome this problem by linguistic methods, whereas the other direction is to exploit statistical information.

Recommendations based on tag clouds. Szomszor et al. (2007) exploits folksonomies for recommendations in the movie domain by combining the information from two data sources: the Netflix rating dataset and the *Internet*

Movie Database (IMDb). The Netflix database contains millions of ratings for several thousand movies. The IMDb database is a large collection of movie data and comprises nearly a million titles. In this experiment, only the community-provided keywords (tags) in the IMDb have been used: typically, several dozen keywords are associated with a popular movie. The two databases can be relatively easily combined in a single schema by matching the movie names.

Based on the ratings and keywords associated to films by various users, the recommendation method tries to estimate the future ratings of users. The central concept for tag-based recommendation by Szomszor et al. (2007) is what the authors call the user's *rating tag cloud*. The basic idea is to identify the keywords typically selected to annotate films that the user u has assigned a rating r. For instance, we might identify that users typically assign the keywords "army," "based-on-novel," or "blockbuster" to movies a specific user u usually rates with a rating value 5. Based on this information, we can search for movies not rated by u but annotated with keywords "army," "based-on-novel," or "blockbuster" and guess that these movies will also be rated 5 by the user u.

In particular, let a specific user be denoted by $u \in U$, where U is the set of all users and $m \in M$ is a movie, where M is the set of all available movies. A rating value is denoted by $r \in R$, where R is the set of possible rating values – for instance, $R = \{1, 2, 3, 4, 5\}$. The set of movies rated by user u is M_u. The rating value for a user $u \in U$ and a movie $m \in M_u$ is denoted by $f_u(m) \in R$.

K is the global set of keywords, and K_m is the set of keywords associated with movie m. N_k denotes the global frequency of occurrences of keyword $k \in K$ for all movies.

Based on this information, a rating tag-cloud $T_{u,r}$ is introduced by Szomszor et al. (2007) for a given user u and rating r. $T_{u,r}$ is defined as the set of -tuples $\langle k, n_k(u, r) \rangle$, where $k \in K$ refers to a keyword and $n_k(u, r)$ is the frequency of keyword k assigned to movies, which user u has rated with rating value r:

$$n_k(u, r) = |\{m|m \in M_u \wedge k \in K_m \wedge f_u(m) = r\}| \qquad (11.2)$$

In other words, given a user u, a rating value r, and a keyword k, $n_k(u, r)$ gives the number of movies annotated by keyword k that have been assigned a rating r by user u. The tag-cloud $T_{u,r}$ expresses the keywords assigned to movies that have been rated with r by user u and how often these keywords were used (which can be seen as a weight/significance factor).

In general, a tag cloud can serve as means of visually depicting the importance of individual tags: more frequent terms are accentuated by a larger font size (Figure 11.2). Figure 11.2 could, for instance, be the tag cloud of the movies that user A has rated with "5."

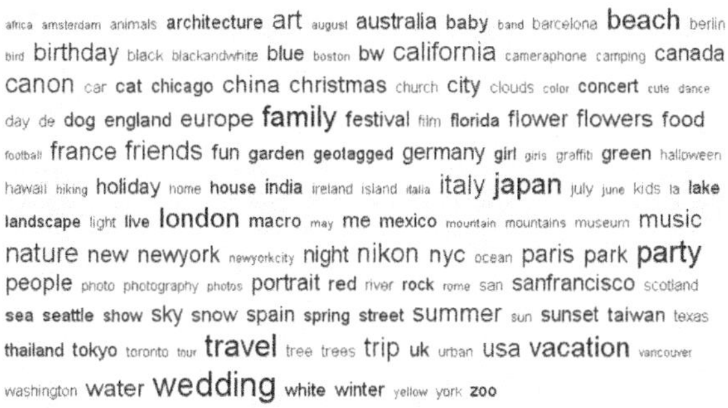

africa amsterdam animals architecture art august australia baby band barcelona beach berlin
bird birthday black blackandwhite blue boston bw california cameraphone camping canada
canon car cat chicago china christmas church city clouds color concert cute dance
day de dog england europe family festival film florida flower flowers food
football france friends fun garden geotagged germany girl girls graffiti green halloween
hawaii hiking holiday home house india ireland island italia italy japan july june kids la lake
landscape light live london macro may me mexico mountain mountains museum music
nature new newyork newyorkcity night nikon nyc ocean paris park party
people photo photography photos portrait red river rock rome san sanfrancisco scotland
sea seattle show sky snow spain spring street summer sun sunset taiwan texas
thailand tokyo toronto tour travel tree trees trip uk urban usa vacation vancouver
washington water wedding white winter yellow york zoo

Figure 11.2. Tag cloud.

Szomszor et al. (2007), use the rating tag clouds as the only basis for recommending items according to the following two schemes.

- *Simple tag cloud comparison.* This metric makes a prediction for an unseen movie m^* by comparing the keywords K_{m^*} associated with m^* with each of the active user's u rating tag clouds. In particular, for an unseen movie m^*, the rating r^* is guessed where the intersection of the tags of tag cloud T_{u,r^*} and the tags associated to movie m^* is maximal – that is, r^* is assigned to the r where $\sigma(u, m^*, r) = |\{(k, n_k) \in T_{u,r} | k \in K_{m^*}\}|$ is maximal.
- *Weighted tag cloud comparison.* This slightly more sophisticated metric not only measures keyword overlap, but also takes the weight of the tags in the clouds into account. This is done by defining a similarity measure, which follows the idea of TF-IDF weighting – the relative frequency of term occurrences and global keyword frequencies are considered when comparing a tag cloud and a keyword set.

Given a user u, a movie m^*, and a rating r^*, the appropriateness of r^* is estimated by:

$$\sigma(u, m^*, r^*) = \sum_{\{\langle k, n_k(u,r^*) \rangle \in T_{u,r^*} | k \in K_{m^*}\}} \frac{n_k(u, r^*)}{log(N_k)} \qquad (11.3)$$

In this measure, Szomszor et al. (2007) sum the frequency of all keywords K_{m^*} of movie m^* where user u has used these keywords to rate movies with rating value r^*. The frequencies are weighted by dividing them by the logarithm of the global frequency N_k of keyword k, as commonly done in term-weighting schemes. The weighted average for all possible rating values is defined by

$$\overline{\sigma}(u, m^*) = \frac{1}{S(u, m^*)} \sum_{r \in R} r \times \sigma(u, m^*, r) \qquad (11.4)$$

where $S(u, m^*) = \sum_{r \in R} \sigma(u, m^*, r)$ is a normalization factor.

In the experiments conducted by Szomszor et al. (2007) the estimation of r^* is based on combining $\overline{\sigma}(u, m^*)$ with the average rating of movie m^*. The average rating is simply defined as usual, where U_{m^*} is the set of users who rated m^*:

$$\overline{r}(m^*) = \frac{1}{|U_{m^*}|} \sum_{u \in U_{m^*}} f_u(m^*) \tag{11.5}$$

Finally, the weighted estimated rating value of a movie m^* for users u is computed by

$$\sigma^*(u, m^*) = 0.5\, \overline{r}(m^*) + 0.5\, \overline{\sigma}(u, m^*) \tag{11.6}$$

In the experiments, the accuracy of these two metrics is compared with an unpersonalized prediction method, which always predicts the average value of all ratings an item in question has received. What is shown in these preliminary evaluations is that recommendations can, in principle, be made solely based on the tag clouds and that the weighted approach performs better than the unweighted approach. For the Netflix dataset, a root mean squared error of roughly 0.96 was achieved. The analysis of the experiments showed that the proposed approach does pretty well for average ratings but has potential for improvements for extreme ratings. The work of Szomszor et al. (2007) is a first indication that the rating tag clouds can serve as an additional source of user profile information in a hybrid system.

Linguistic methods for tag-based recommendation. The work of de Gemmis et al. (2008) goes in a similar direction as the previously described approach; however, the main difference is that sophisticated techniques are exploited to identify the intended sense of a tag (keyword) associated with an item and to apply a variant of a naive Bayesian text classifier.

Basically, de Gemmis et al. implemented a content-based recommender system for recommending descriptions about paintings. They assumed that such descriptions are structured in slots – there are slots for the title, the painter, and a general painting description. These slots are called *static slots* because they do not change over time. Compared with plain content-based methods, the exploitation of slots is just an additional feature because the approach is also applicable if the content of slots are merged in just one slot.

The idea of de Gemmis et al. (2008) is to merge tags assigned by users to descriptions in special slots. These slots are called *dynamic slots* because they change as users add tags. In particular, given item I, the set of tags provided by all the users who rated I is called *SocialTags(I)* and set of tags provided by a specific user U for I is called *PersonalTags(U, I)*. For each set of tags, slots are added to the items.

Because tags may be formulated freely by users, the sense of tags can be ambiguous or tags may be a synonym of other tags. This problem must be addressed for content-based recommender systmes in general. To solve this problem, de Gemmis et al. (2008) propose the application of semantic indexing of documents. In particular, words in a slot are replaced by synsets using WORDNET. A *synset* (*synonym set*) is a structure containing sets of words with synonymous meanings, which represents a specific meaning of a word. This addresses the problem of synonyms. The problem of an ambiguous word sense is tackled by analyzing the semantic similarity of words in their context. A detailed description of word sense disambiguation (WSD) is presented by Semeraro et al. (2007). The result of applying WSD to the content of slots is that every slot contains a set of synsets, which are called *semantic tags*.

Following the common practice of content-based recommendation, the user classifies items he or she likes and items he or she dislikes. This classification information is exploited to compute the parameters of a naive Bayesian classifier – the conditional probability $P(c|d_j)$ is computed where c has the values *likes/dislikes* and d_j represents a specific item.

In order to apply Bayes' rule, the specialty of de Gemmis et al. (2008) is to compute the required conditional probability $P(d_j|c)$ by exploiting slots. Let M be the number of slots, s a slot, and d_j^s representing the slot s of document d_j; then according to the naive assumption of conditional independence, $P(d_j|c)$ is computed by

$$P(d_j|c) = \prod_{s=1}^{M} P(d_j^s|c) \tag{11.7}$$

Furthermore, $P(d_j^s|c)$ is estimated by applying a multivariate Poisson model in which the parameters of this model are determined by the average frequencies of tokens in the slots and in the whole collection of items. The method does not distinguish between various types of slots. Hence, static and dynamic slots – the slots representing the social and personal tags – are processed equally.

The evaluation by Gemmis et al. (2008) is based, as usual, on a k-fold cross-validation. Five configurations were explored: items were described (a) only with social tags, (b) with personal tags, (c) with social tags and static slots, (d) with personal tags and static slots, and (e) with static slots, neglecting dynamic slots. The best-performing combinations with respect to an F-measure are static slots combined with social or personal tags. However, it was observed that if only a few training examples are available, social tags without static tags performed best. The more examples that were available to assess the preferences of an individual, the better was the performance of a combination of static and dynamic slots.

Tag clustering. Although folksonomies provide many opportunities to improve recommendations, the free formulation of tags leads to unique challenges. Unsupervised tagging results in redundant, ambiguous, or very user-specific tags. To overcome this problem, Shepitsen et al. (2008) propose the clustering of tags.

The basic idea is to compute the interest of a user u in a resource r by the following formula:

$$I(u, r) = \sum_{c \in C} ucW(u, c) \times rcW(r, c) \qquad (11.8)$$

C is the set of all tag clusters; $ucW(u, c)$ is the user's interest in cluster c, calculated as the ratio of times user u annotated a resource with a tag from that cluster over the total annotations by that user; $rcW(r, c)$ determines the closeness of a resource to a cluster c by the ratio of times the resource was annotated with a tag from the cluster over the total number of times the resource was annotated.

This user interest in a resource is exploited to weight the similarity (denoted by $S(q, r)$) between a user query q and a resource r, where a user query corresponds to a tag. $S(q, r)$ is computed by the cosine similarity using term frequencies. In particular, the tag frequency $tf(t, r)$ for a tag t and a resource r is the number of times the resource has been annotated with the tag. T is the set of tags.

$$S(q, r) = cos(q, r) = \frac{tf(q, r)}{\sqrt{\sum_{t \in T} tf(t, r)^2}} \qquad (11.9)$$

This similarity $S(q, r)$ between a query q and a tag t is personalized by the interest of user u in a resource r resulting in similarities $S'(u, q, r)$ relating users, queries, and resources: $S'(u, q, r) = S(q, r) \times I(u, r)$.

For the computation of clusters of tags, tags are represented as a vector of weights over the set of resources. Both TF and TF-IDF can be applied; however, TF-IDF showed better results. The idea of clustering is that similar terms, such as *web design* and *design*, appear in similar clusters. Shepitsen et al. (2008) experimented with various cluster methods, showing in their experiments that agglomerative clustering leads to better improvements. Furthermore, Shepitsen et al. (2008) they propose a query-dependent cluster method. The idea is that given a query term (e.g., baseball), the clustering algorithm starts to build a cluster of tags around the query term.

In their evaluation, Shepitsen et al. (2008) showed the improvements compared with a standard recommendation technique based on cosine similarity exploiting test data from last.fm and del.icio.us; last.fm is a music community web site; del.icio.us is a social bookmarking web service. Using a query-dependent clustering technique showed higher improvements in the del.icio.us test domain compared with last.fm. The authors assume that this comes from the fact that the tags in last.fm have a higher density and are less ambiguous.

Comparison with classical collaborative methods. Sen et al. (2009) compare various algorithms that explore tags for recommendation tasks. This comparison is based on the ratings of MovieLens users. The algorithms evaluated are classified in two groups. *Implicit-only algorithms* do not explore an explicit rating of items, but examine only feedback from the users, such as clicks or searches. For example, if a user clicks on a movie, this could be interpreted as a sign of interest. *Explicit algorithms* exploit the rating of items provided by the user. Implicit-only methods are of special interest for domains in which the users cannot provide ratings.

For the evaluation, the performance of the algorithms with respect to the so-called recommendation task was measured. In particular, the task was to predict the top-five rated movies for a user. All these movies should be top-rated (4 or 5 stars in the MovieLens domain) by the users in the test set. Precision is applied as a metric. Twelve methods were compared, consisting of three naive baseline approaches, three standard CF methods, and six tag-based algorithms.

With respect to the recommendation task, the conclusion of this evaluation was that tag-based algorithms performed better than traditional methods and explicit algorithms showed better results than implicit-only methods. The best-performing approach was a combination of the best-performing explicit tag-based algorithm with Funk's value decomposition algorithm (described by Koren et al. 2009).

In addition to the recommendation task, the so-called prediction task was evaluated. In this task, the recommender system has to predict the rating of a user. Sen et al. (2009) use the MAE to assess the quality of algorithms. Two traditional CF methods, three baseline methods, and four tag-based algorithms were compared. In this evaluation, the tag-based methods did not show an improvement over the traditional CF methods. However, the evaluation was conducted on one particular domain.

11.2.1.2 Folksonomies and collaborative filtering

We now present two approaches that extend collaborative filtering techniques by tag information. The first one follows the memory-based method (Herlocker et al. 1999), whereas the second one extends probabilistic matrix factorization (Koren et al. 2009). Finally, we show how CF methods can be applied for retrieving tagged items, given a query formulated as a set of tags.

Extensions of classical collaborative filtering methods. In contrast to a content-based view of tags, recently some researchers viewed tags as additional information for discovering similarities between users and items in the sense of CF.

In particular, Tso-Sutter et al. (2008) viewed tags as additional attributes providing background knowledge. Although there is a reasonable amount of work on integrating such additional information into CF, there are some important differences. *Attributes* in the classical sense are global descriptions of items. However, when tags are provided by users for items, the usage of tags may change from user to user. Consequently, tag information is local and three-dimensional: tags, items, and users.

The basic idea of Tso-Sutter et al. (2008) is to combine user-based and item-based CF. In user-based CF, usually the k most similar users are selected to compute a recommendation. The similarity of users is based on their ratings of items. Likewise, this similarity among users is influenced if users assign the same (or similar) tags. Consequently, the tags are just viewed as additional items. Therefore, the user–item rating matrix is extended by tags. The entries of this matrix are just Boolean values indicating whether a user is interested in an item or used a specific tag.

Similarly, item-based CF usually exploits the k most similar items rated by a user. Tags are viewed as additional users and the user–item matrix is extended by new users. The entries for these new users (i.e., the tags) are set to true if an item was labeled by the tag.

The ratings of unrated items i for user u is computed by the following formulas proposed by Tso-Sutter et al. (2008). The rating matrix $O_{x,y}$ has entry 1 or 0 for users x and items y. The dimensions of this matrix differ, depending on whether we apply user-based or item-based CF. Let N_u be the k most similar neighbors based on some traditional CF method. The user-based prediction value is computed by:

$$p^{\text{ucf}}(O_{u,i} = 1) := \frac{|\{v \in N_u | O_{v,i} = 1\}|}{|N_u|} \tag{11.10}$$

For the computation of an item-based prediction value, the k most similar items are exploited. N_i denotes these items, which are computed by applying some standard CF similarity function. The similarity between items i and j is denoted by $w(i, j)$. The item-based prediction value is computed by:

$$p^{\text{icf}}(O_{u,i} = 1) := \sum_{j \in N_i \cap O_{u,j} = 1} w(i, j) \tag{11.11}$$

Because the ranges of p^{ucf} and p^{icf} are different, they are normalized in the following combination formula. The parameter λ adjusts the significance of the two predictions and must be adjusted for a specific application domain.

The combined prediction value is computed by

$$p^{\text{iucf}}(O_{u,i} = 1) := \lambda \frac{p^{\text{ucf}}(O_{u,i} = 1)}{\sum_i p^{\text{ucf}}(O_{u,i} = 1)} + (1 - \lambda) \frac{p^{\text{icf}}(O_{u,i} = 1)}{\sum_i p^{\text{icf}}(O_{u,i} = 1)}$$

$$\tag{11.12}$$

Based on these combined prediction values, the unrated items of user u can be ranked and the N top-ranked items are displayed to the user as a recommendation.

Tso-Sutter et al. (2008) evaluated this method based on the data of last.fm. Best results were achieved with $\lambda = 0.4$ and a neighborhood size of 20. The evaluation showed a significant improvement when tag information was exploited. Interestingly, this improvement could be shown only if item-based and user-based CF methods were combined.

The CF approach of Tso-Sutter et al. (2008) is a memory-based method. The second main approach for CF is the so-called model-based method, which tries to learn a model by employing statistical learning methods. Zhen et al. (2009) extended probabilistic matrix factorization (PMF) to exploit tag information. The key idea is to use tagging information to regularize the matrix factorization procedure of PMF.

Tag-based collaborative filtering and item retrieval. In contrast to the previously described approaches, social ranking – as introduced by Zanardi and Capra (2008) – is a method that aims to determine a list of potentially interesting items in the context of a user query. This query can either consist of a list of words provided by the user in a search engine scenario or be implicitly derived in one way or another from the user profile in a more classical recommendation scenario. Standard Web 2.0 search engines of this style use relatively simple metrics that combine a measure of overlap of search keywords and resource tags (ensuring accuracy) with a measure of how many users employed the particular tags for annotating the item (ensuring high confidence). Such measures, however, are good at retrieving popular items but not the so-called long tail of not-so-popular items.

In particular, we can distinguish between the long tail of tags and the long tail of items. The long tail of tags refers to the phenomenon that most of the tags are used only by a small subset of the user population. In the domain studied by Zanardi and Capra (2008), roughly 70 percent of the tags were used by twenty or fewer users, which represents roughly 0.08 percent of the whole user set. This suggests that standard keyword search will fail because of the small overlap of item tags and tags contained in a user query.

The long tail of items refers to the observation that most items are tagged by a small portion of the user population – 85 percent of the items are tagged by five or fewer users, as reported by Zanardi and Capra (2008). This suggests that standard recommender techniques fall short because of the almost empty overlap of user profiles.

Social ranking aims to overcome this problem by applying traditional CF ideas in a new way – by using user and tag similarities to retrieve a ranked list of items for a given user query. The similarity between users is determined based on an analysis of their tagging behavior. A simple definition of similarity is employed that states that users are considered similar if they have used the same set of tags (ignoring on which resources these tags have been put). When this information is encoded as a vector of tag usage counts, cosine similarity can be used as a metric to quantify the similarity. Similarly, tag similarity is calculated based on the co-occurrence of tags for a resource. Again, the cosine similarity measure can be used.

The calculated similarities are then used in the subsequent, two-step query phase as follows. To cope with the problem of free formulated tags, the original query u is expanded with a set of similar tags to a larger query u^*. The main idea here is to improve on the coverage measure, as users do not always use the same set of keywords to describe items. An expansion based on tag similarity will thus help to overcome this problem.

Next, all items that are tagged with at least one tag of u^* are retrieved and ranked. The ranking process is based on a function that consists of a combination of (a) the relevance of the tags with respect to the query and (b) the similarity of taggers to the active user who issued the query. The ranking $R(p)$ of item p is computed by summing up for every user u_i the similarities of the tags t_x user u_i used to tag p with the tags t_j contained in the expanded query q^* posed by the active user $\overline{u}$. This sum is amplified if the active user $\overline{u}$ is similar to user u_i:

$$R(p) = \sum_{u_i} \left(\sum_{\{t_x | u_i \text{ tagged } p \text{ with } t_x\}, t_j \in q^*} sim(t_x, t_j) \right) \times (sim(\overline{u}, u_i) + 1) \quad (11.13)$$

The approach was evaluated on the CiteULike social bookmarking data. Different views of the data have been used (compare with Massa and Avesani 2007), including heavy, medium, and low taggers, as well as popular and unpopular items. As a baseline, a standard method was used, as described above. It showed that social ranking consistently led to better results when it came to the long tail – that is, when searches for medium- or low-taggers or unpopular items were made. The query expansion method, quite intuitively, also helps improve on the coverage measure and thereby reduces the number of unsuccessful searches. Although the results are promising, further improvements seem possible – for example, through the usage of other methods for finding similar users or through the exploitation of external, semantic information (e.g., from WordNet) to make the query expansion more precise (Zanardi and Capra 2008).

As a side point, traditional recommender system algorithms (as well as standard quality measures; see Huang et al. 2008 and Firan et al. 2007) have been designed to work mostly on user-item matrices or content information. In social web and Web 2.0 scenarios, additional sources of information (such as trust relationships or tags) are available, however. Therefore, it can be expected that additional ways of exploiting these knowledge sources in an integrated way in Web 2.0 recommendation scenarios will be required in the future.

11.2.2 Recommending tags

As we see from these methods, one of the problems of folksonomies is that users differ in the way they annotate resources, so some fuzziness is introduced in the tags. This, in turn, hampers the computerized use of this information.

One way of alleviating this problem could be to enhance such a Web 2.0 system – such as a media-sharing platform – with an intelligent agent that supports the user in selecting and assigning tags to resources. This could not only help to align the sets of user tags but also serve as a motivator for end users to annotate the resources. Some of today's image and bookmark sharing, as well as music, platforms already provide some user guidance and tag recommendations. How these tagging suggestions are generated is, however, not published. It can be assumed that tag usage frequencies are at the core of these metrics.

The question arises of whether CF techniques can be a means for recommending a set of tags. As mentioned earlier, the problem setting is a bit different in folksonomies, as we do not have a two-dimensional user–item rating matrix but rather ternary relationships among users, resources, and tags. Jäschke et al. (2007) propose applying a nearest-neighbor approach, as follows: from the original ternary relation (called Y), two different binary projections, the user resource table $\pi_{UR}Y$ and the user-tag table $\pi_{UT}Y$, can be derived. The values in these projections are binary. $(\pi_{UR}Y)_{u,r}$ is, for instance, set to 1 if there exists an element (u, t, r) in Y – that is, when the user u has rated resource r with an arbitrary tag. Based on one of these projections, we can compute N_u^k, the set of k most similar neighbors for a user u. The cosine similarity measure can be used for comparing two rows of the matrix.

After the set of nearest neighbors N_u^k is determined, and given a resource r and user u, we can compute the top n recommended tags $\tilde{T}(u, r)$ from the set of all tags T for user u as follows:

$$\tilde{T}(u, r) := argmax_{t \in T}^n \sum_{v \in N_u^k} sim(\vec{x}_u, \vec{x}_v)\delta(v, t, r) \qquad (11.14)$$

where $\delta(v, t, r) := 1$ if $(v, t, r) \in Y$ and 0 otherwise. $\vec{x}_u$ and $\vec{x}_v$ are the rows of $\pi_{UT}Y$ or $\pi_{UR}Y$, depending on which of the possible projections was used.

As an alternative to this neighbor-based method, the same authors propose a recommendation technique based on FolkRank in Hotho et al. (2007). FolkRank is a graph-based search and ranking method for folksonomies. As the name suggests, it is inspired by PageRank, the underlying idea being that when resources are annotated with "important" tags by "influential" users, the resource will also become important and should be highly ranked in a search result. Because a direct application of PageRank is not possible because of the ternary nature of the relationships and the nondirectional relationships, an alternative form of weight-spreading is proposed by Hotho et al. (2007).

Jäschke et al. (2007) report the results of an evaluation of different approaches based on datasets from the popular social bookmarking systems Bibsonomy and last.fm. To determine the prediction accuracy, a variant of the standard leave-one-out method was chosen – that is, for a given resource the system should predict the tags a certain user would assign. Besides the CF variants and the graph-based FolkRank approach, an unpersonalized counting metric (*most popular tag by resource*) was also evaluated. For this metric, it was first counted how often every tag was associated with a certain resource. The tags that occurred most often with a given resource r were then used as a recommendation. The evaluation showed that FolkRank led to the best results with respect to both the recall and precision measure. The CF method performed slightly better when the user-tag projection was used; when the user-rating projection served as a basis, the performance of the CF algorithm was similar to the *most popular by resource* metric. The baseline method *recommend most popular tags* was outperformed by all other algorithms.

In summary, it can be seen that better prediction accuracy can be achieved with the computationally more expensive method based on FolkRank, and that an unpersonalized and easy-to-compute popularity-based method can already lead to relatively good results when compared with CF-based methods.

A new method that follows neither the PageRank nor the standard recommendation approach is proposed by Krestel et al. (2009). The authors employ the so-called latent Dirichlet allocation (LDA) to recommend tags. The basic idea is that various topics (e.g., *photography* or *how-to*) are hidden in the set of resources. Given a set of resources, tags, and users, LDA computes the probability that a tag t_i is related to a resource d by

$$P(t_i|d) = \sum_{j=1}^{Z} P(t_i|z_i = j)P(z_i = j|d) \tag{11.15}$$

where z_i represents a topic, $P(t_i|z_i = j)$ is the conditional probability that tag t_i is related to topic j, and $P(z_i = j|d)$ is the conditional probability that topic j is related to resource d. The number of topics Z is given as input. By exploiting the Gibbs sampling method, these conditional probabilities are estimated. As a result, by applying LDA, a numerical weight, expressing the association between tags and resources, can be computed. This weight is subsequently exploited to recommend tags for resources. Krestel et al. (2009) compare their approach to methods based on mining of association rules showing better results for LDA.

Krestel and Fankhauser (2009) extend this approach to tag recommendation by also considering the content of resources and by using a mixture of the unfactorized unigram language models with latent topic models. The evaluation on the bibsonomy dataset, provided by the discovery challenge at European Conference on Machine Learning and Principles and Practice of Knowledge Discovery in Databases (ECML PKDD) (Buntine et al. 2009), showed that on this dataset, language models slightly outperform latent topic models, but the mixture of both models achieves the best accuracy. Moreover, combining content with tags also yielded significant improvements.

A different method for generating tag recommendations on the popular image-sharing platform Flickr.com is proposed by Sigurbjörnsson and van Zwol (2008). In contrast to the aforementioned work, their recommendation system is strongly based on "tag co-occurrence" and different aggregation/ranking strategies. The method starts with a small set of given user-defined tags for an image and bases the recommendation on the aggregation of tags that are most frequently used together with the start set. Instead of simply using the "raw" frequencies, some normalization is required (similar to the TF-IDF method) to take the overall frequency of tag usages into account. The candidate tags are then ranked (different strategies are possible) and a top-n list is presented to the user.

Besides this ranking technique, Sigurbjörnsson and van Zwol (2008) also introduce the idea of tag "promotion". An analysis of "how users tag photos" in the Flickr.com dataset showed that the tag frequency distribution follows the power law and that this distribution can be used to determine a set of the most promising candidates for recommendation. Very popular tags are too general to be useful; rarely used words, on the other hand, are unstable predictors. The evaluation of different promotion techniques based on a large extract of the Flickr.com dataset showed that good improvements with respect to precision can be achieved.

In contrast to the FolkRank method described above, the work of Sigurbjörnsson and van Zwol (2008) is based solely on data of an image-sharing platform. To what extent their findings can also be transferred to other social web platforms and other types of resources has not been analyzed so far. Examples for other recent collaborative, content-based, or probabilistic methods for tag recommendation can be found in the work of Xu et al. (2006), Basile et al. (2007), Byde et al. (2007), and Mishne (2006).

11.2.3 Recommending content in participatory media

Besides resource tagging and participation in social networks, a third aspect of the second-generation web is the phenomenon of the increasing importance of *participatory media*, in which users contribute the content – for instance, in the form of blog messages.

On such media platforms – and in particular, on popular ones that have a broad community of contributors – the problem of filtering out the interesting messages quickly arises. At first glance, this problem is very similar to the classical news filtering problem discussed in Chapters 2 and 3. This time, however, we can hope to have more information available than just the active user's reading history and preference statements; we can additionally try to exploit the information in the social network to judge whether the message is important. Messages can, for instance, be judged to be important because other users rated them highly, which is the standard CF setting. We have also seen that explicit trust statements can help improve the accuracy of such estimates of interest. Being allowed to issue only one trust statement per person might, however, be too simplistic. In reality, we typically do not trust the opinion of a friend on every topic. Moreover, we are often embedded in a social environment, in which there are particularly important persons whose opinions we generally believe.

Given that more of this social information is available in Web 2.0, it would be desirable to take these pieces of information properly into account when recommending messages to a user. A first approach to incorporating such phenomena that appear in real-world social relationships into a computerized system has been made by Seth et al. (2008). In their work, they aim to develop a metric of credibility that takes into account various factors that should implicitly help to determine whether a recently posted message is credible (i.e., possibly relevant and important) to a user.

One of the basic ideas of their approach is that their multidimensional credibility measure is subjective – that is, different users may judge the credibility of a posting differently, possibly depending on their context and their community. In addition, Seth et al. postulate that credibility must be topic-specific – that is, one can be a trusted expert in one field and not trusted when it comes to a different subject area.

The proposed metric is based on various insights from the fields of media studies, political science, and social networks and combines different aspects that contribute to a person's credibility. These aspects range from direct experiences to the credibility we attribute to someone because of his or her role or local community, and including, finally, the general opinion of the public about a certain user. In Seth et al.'s model, each of the individual credibility measures are captured in a real number [0 . . . 1] and combined in a Bayesian network to an overall credibility estimate.

The Bayesian model is trained based on stochastic methods using the following data, which are assumed to be available:

- Messages are labeled with their authors (or posters) and a set of ratings that describe the supposed credibility of the message. These ratings are, as usual, provided by the users of the community.
- Users are explicitly connected with their "friends".
- Every user can declare a list of topics in which he or she is interested. Based on this, a set of topic-specific social network graphs can be derived, in which clusters of users and links can be identified with the help of community identification algorithms. From this, strong and weak ties between users can be identified.

On arrival of a new message, the learned model can be used to make a probabilistic prediction as to whether a user will find this message credible.

For evaluation purposes, a prepared data set from the digg.com knowledge sharing platform was used. The digg.com web site allows users to post articles, rate other items and also to connect to other users. Unfortunately, the first reported measurements by Seth et al. (2008) are not yet fully conclusive. Still, one can see this work as a further step toward the more extensive and integrated usage of information in social web platforms, with the ultimate goal of filtering interesting items more precisely in the future.

The idea to differentiate between users is also followed by Guy et al. (2009). Basically, the authors build on the insight to distinguish between people who are similar to a user and people who are familiar with a user.

The computation of a familiarity score between users is based on organizational charts, direct connections in a social network system, tagging of persons, and coauthorship of various items, such as papers, wikis, and patents. The computation of a similarity score is based on the co-usage of the same tag, co-bookmarking of the same web page, and co-commenting on the same blog entry. All these pieces of information are exploited to compute an overall score. Based on these scores, information items such as web pages, blog entries, and communities are recommended to users.

Furthermore, for a recommendation of an item, the names of the most related persons are displayed, serving as a kind of explanation why this item is recommended.

Guy et al. (2009) compared recommendations based exclusively on either the familiarity, similarity, or overall score. Test persons were asked to classify recommended items as *interesting*, *not interesting*, and *already known*. The evaluation showed that recommendations based on the familiarity score were classified as significantly more interesting than recommendations based on the similarity score. In addition, it was shown that providing explanations as described here results in a significant increase in the number of items classified as interesting.

The study was conducted in a work environment and did not ask the utility of the presented information objects for the users' tasks. Therefore, it is open whether the observed effect contributes rather to a persuasion or leads to the discovery of information objects that indeed support the work process.

The problem of filtering information in Web 2.0 collaborative media platforms is already in the focus of providers of commercial portal software. Nauerz et al. (2008), for instance, report on the implementation of a "recommender of background information" in the IBM WebSphere portal system. In their work, the selection of similar documents is based on automatically extracted or manually assigned meta-tags. Although details of the recommendation algorithms are not given and a straightforward method can be assumed, this example shows the practical relevance of providing additional support in Web 2.0 collaborative systems.

11.3 Ontological filtering

As mentioned earlier, a fundamental building block of the Semantic Web is the description of web resources by languages that can be interpreted by software systems. In particular, the idea is to better locate information on the web by such descriptions. Indeed, parts of the Semantic Web community deal with the classification of web content that best matches the information need of users by exploiting machine interpretable information – for example, web content is annotated and a formal description language (e.g., OWL) is applied to deduce classifications. This task shares many similarities with recommender systems that aim at the classification of items that best fulfill some user needs.

In particular, one central idea of the Semantic Web is to formulate a domain ontology (i.e., a logical theory) that is exploited to describe web resources and to reason about the properties of these resources by inference systems. Consequently, various researchers have applied ontologies to improve filtering techniques of recommender systems. Of course, one can argue that, in fact, long-known knowledge-based techniques have been applied, such as simple inheritance taxonomies and other forms of logical description of items and their relations. Therefore, these recommender systems are actually hybrid systems leveraging their capabilities by knowledge-based methods.

Most of the research in this area, however, was published in the context of the Semantic Web umbrella, and therefore we prefer to mirror this originally intended classification.

11.3.1 Augmentation of filtering by taxonomies

Assume that there is an ontology describing the domain by a super-/subconcept taxonomy with the meaning that every member of the subconcept is also a member of the superconcept, but not necessarily vice versa. For example, news can be classified by such an ontology, saying that news about "world soccer tournaments" is more specific than news about "soccer", which is more specific than news about "sport". In addition, news about "baseball" is different from news about "soccer" but more specific than "sport". In the following, we distinguish between parent relations (e.g., "sport" is a parent of "soccer") and grandparent relations (e.g., "sport" is a grandparent of "world soccer tournaments").

Based on this hierarchical ontology, items (e.g., pieces of information on the web) can be associated with such concepts. The set of concepts associated to an item is called the *item profile*. For example, a news item can be annotated by the label "soccer". Conversely, information about the interests of users can be provided by associating concepts to individual users and annotating these associations by the strength of their interests. The set of concepts associated to a user is called the *user profile*. For example, a user is strongly interested in "soccer" but not in "baseball". Information about the interests of users can be either provided directly by the users or acquired by indirect means, such as observing which items are clicked. By such observations, the interests of users can be continuously enhanced.

Given these three pieces of information (item profile, user profile, and the domain taxonomy) Maidel et al. (2008) propose a similarity function between items and user interests, in which the taxonomy is exploited.

Five different cases of matches were considered, in which each concept c_i in the item profile is assigned to a matching score depending on the user profile.

Case a: The perfect match. The item concept c_i is also contained in the user profile – for example, both the user and item profiles contain "soccer".

Case b: The item concept c_i is a parent of a concept contained in the user profile – for instance, the item concept is "sport" and the user profile contains "soccer".

Case c: The item concept c_i is a child of a concept contained in the user profile – for example, the item concept is "soccer" and the user profile contains "sport".

Case d: The item concept c_i is a grandparent of a concept contained in the user profile – for instance, the item concept is "sport" and the user profile contains "world soccer tournaments".

Case e: The item concept c_i is a grandchild of a concept contained in the user profile – for example, the item concept is "world soccer tournaments" and the user profile contains "sport".

For all the these cases, matching scores must be determined. Whereas the matching score of case a (the perfect match) is 1 (the maximum), the score values of the other cases must be determined. Intuitively, matching scores for cases d and e, which are defined by grandparent/grandchild relations, are lower than scores for cases b and c, which are specified by parent/child relations.

The degree of similarity between an item and a user is based on the profiles, matching scores of the concepts in the two profiles, and on the weights of the concepts in the user profile. The overall similarity score between item and user is defined by

$$IS = \frac{\sum_{c_i \in I} N_{c_i} S_{c_i}}{\sum_{c_j \in U} N_{c_j}} \tag{11.16}$$

where I is the item profile, U is the user profile, c_i is a concept in the item profile, c_j is a concept in the user profile, S_{c_i} is the matching score of concept c_i, and N_{c_j} is the number of clicks on items containing concept c_j, representing the weight of concept c_j for the user. This weight of concepts for a user may be given implicitly by monitoring the user or may be specified explicitly by the user.

Formula 11.13 may be extended by weights representing the importance of concepts in items (concept/item weights). In particular, the matching scores S_{c_i} can be multiplied by the weight of c_i in the item. However, the evaluation by Maidel et al. (2008) showed no significant improvements if concept/item weights are considered. Furthermore, Maidel et al. (2008) experimented systematically with different settings of matching scores. It turned out that the best choice is around $a = 1$, $b = 0.8$, $c = 0.4$, $d = 0$, and $e = 0.2$. Consequently, after the perfect match, the next important parameter is b, representing the match in which the item's concept is a more general parent concept of a user's concept. The next important parameter is c, in which a user's concept is a more general parent concept of an item's concept. The different weights depend on the direction of the generalization. The parameters d and e, in which item and user concepts are in a grandparent/grandchild relation, turned out to be less important.

The extensive evaluation presented by Maidel et al. (2008) showed that if weights of concepts representing the interests of users are given explicitly, much better recommendation results can be achieved compared with the case in which the interest of users is acquired during a session. Moreover, it was shown that if the taxonomy is not considered in the overall similarity score IS (i.e., only perfect matches are counted), the recommendation quality significantly drops. Consequently, the utility of exploiting taxonomies in matching user and item profiles for improving the error in this type of recommender systems was shown.

The approach of Maidel et al. (2008) was applied to recommend news, and follows the idea of Middleton et al. (2004) to take advantage of the information contained in taxonomies. Unfortunately, a comparison with Middleton et al. (2004) is missing.

The underlying recommendation method of Middleton et al. (2004) is a combination of content-based and collaborative techniques. The goal of the system is to recommend "interesting" research papers to computer scientists. The interests of a computer scientist are described by a set of computer science topics (such as hypermedia, artificial intelligence, or agents); each topic is weighted by a numerical value. These topics are organized in a subconcept-superconcept hierarchy, forming a simple ontology of research fields and their subfields. For example, papers in the field of recommender agents are papers in the agents field, which are papers in the artificial intelligence field. Computer scientists in a department read articles, and these articles form an article repository.

The idea of the Foxtrot system is based on the following information pieces; given that we know the topics in which a researcher is interested and the topic for each research paper, then we can estimate the interest a researcher possibly has in a particular paper. To estimate the topics of research papers, the Foxtrot system starts with a set of manually classified documents (the training set), in which their topic is known. The topic of unclassified research papers is estimated by classical content-based techniques. In particular, a research paper is represented by its term frequency, in which the set of terms is reduced by standard techniques, such as stemming and the application of a stop list to remove common words.

The interests of researchers are determined by various data sources. Researchers can explicitly declare in which topics they are interested. In addition, the Foxtrot system monitors which papers a researcher browsed. Because the topic of papers is known, it is assumed that the researcher is interested in this topic. Based on this information, a *topic interest value* is computed. This topic interest value is a summation of various single interest values. Given a user u and a specific topic t, the topic interest value $i(u, t)$ is increased for each paper browsed by user u if the paper's topic is classified to be t. Browsing papers that are recommended by the system is given a higher weight. Furthermore, the interest values of papers are weighted by time. The more recently a paper was browsed, the higher its weight. In addition, the topic interest value is increased or decreased by a certain amount if the user explicitly declared his or her interest or disinterest. The most important fact of the computation of the topic interest value is that the topic interest value of topic t is increased by 50 percent of the topic interest values of its subtopics.

Papers are ranked for a specific user by their recommendation value. The recommendation value for a paper p and a user u is computed, as usual, by summing over all topics t and multiplying the classification confidence (i.e., a measure of how confident we are that paper p deals with topic t) and the topic interest value of the user u in that topic t.

In an experimental evaluation, it was shown that the association of topics to users is more accurate if the ontology of topics is employed compared to the case in which only a flat list of topics was exploited. In addition, the recommendation accuracy could be improved by using an ontology.

The cold-start problem is addressed by Middleton et al. (2004) by using previously published papers of researchers. These unclassified papers are compared with the classified papers of the repository to estimate their topic classification. The interest values of a user are computed based on the topics of his or her published papers, weighting previously published papers higher. In addition, the ontology of topics is exploited as described – the interest in a topic is increased by some fraction depending on the interest values of the subtopics. Furthermore, if the system is up and running and a new user is added, this new user is compared with already participating users. The interest values of a new user are adjusted depending on the interest values of similar users.

In contrast to the work of Middleton et al. (2004), the work of Ziegler et al. (2004) follows a different strategy of propagated interests in a topic along the taxonomy. Furthermore, a different strategy for computing recommendations is developed that also addresses the problem of topic diversification – recommending not always more of the same, but pinpointing different but still interesting items.

The basic idea of Ziegler et al. (2004) for incorporating taxonomic information is to base the recommendation algorithm on the user's interest in categories of products. Consequently, user similarity is determined by common interests in categories and not by common interests in items. The goal of this approach is to reduce problems of collaborative filtering methods if user ratings are sparse. The rationale is that although users have not rated the same item, they may show common interest in a category – for instance, even though different books were bought, these books can belong to the same topic.

In particular, Ziegler et al. (2004) assume a set of products $B = \{b_1, \ldots, b_m\}$ and a set of user ratings R_i for every user u_i where $R_i \subseteq B$. In this approach, a user is either interested in a product if this product is in R_i, or we have no information about his or her interest in a product if this product is not in R_i. This is the typical case of e-commerce applications, in which interests in products are implicitly rated by purchase data or product mentions. In addition, a set of product categories $D = \{d_1, \ldots, d_l\}$ is given, representing topics into which products may fall. These product topics are organized in a tree representing subconcept-superconcept relations between them. For every product, a descriptor assignment function is defined that associates to each product b_k a subset $D_k \subseteq D$ of topics. Product b_k is a member of each topic in D_k.

Instead of representing a user's interest by a vector of dimension $|B|$ (i.e., for each product the vector contains an entry), the user's interests are characterized by a vector of dimension $|D|$, in which each entry represents the interest of a user in a topic.

Based on the user rating R_i, the set of products in which the user is interested can be determined. By the descriptor assignment function, the topics can be computed in which a user is interested. The interest in a topic is propagated from the subtopics to the supertopics. The propagating depends on the number of siblings a topic has. The fewer siblings a topic possesses, the more interest is assigned to its supertopic. The results of this computation are user interest vectors containing scores for topics. These user interest vectors are exploited to compute the similarities of users based on Pearson correlation. Finally, as usual, for each user the k nearest neighbors are determined.

The generation of recommendations combines two proximity values. *User proximity* takes into account how similar two users are. If user u_j recommends an item to user u_i, this recommendation receives more weight the closer the interest profiles of u_i and u_j are. The second proximity exploited is called *product proximity*. Here the idea is that the closer the topics of a product are to the topics in which a user is interested, the higher the weight for recommending this product. Based on these proximity values, weights for products are computed, and the ordered set of the top N products represents the recommendation.

Finally, Ziegler et al. (2004) propose a method for topic diversification to avoid the phenomenon that only products of the most interesting topic are recommended. Here the idea is that the recommendation list is incrementally expanded. The topics that have not been covered in the recommendation list so far receive higher weights. These weights are increased for a topic d_j depending on the length of the list – the longer the list that does not contain an item of d_j, the higher the weights.

Evaluations show the advantage of this approach compared with standard CF approaches and hybrid approaches combining content-based and collaborative filtering.

11.3.2 Augmentation of filtering by attributes

Mobasher et al. (2004) exploited semantic information to enhance item-based CF. The basic idea is to use semantic information about items to compute similarities between them. These semantic similarities are combined with similarities based on past user ratings to estimate future user ratings. In particular, it is assumed that an ontology describes a domain, such as movies. This ontology describes the attributes used to characterize items. For example, in the movie domain, these attributes are genre, actors, director, and name. In a further step, the method assumes the instantiation of the ontology by items. Mobasher et al. (2004) accomplished this instantiation process by web mining techniques.

To support the computation of item similarities, the instances are converted into a vector representation. This conversion includes normalization and discretization of continuous attributes. The process also results in the addition of new attributes, such as representing different intervals in a continuous range or representing each unique discrete value for categorial attributes. The outcome of this process is a $n \times d$ matrix $S_{n \times d}$, where n is the number of items and d is the number of unique semantic attributes. Matrix S is called the *semantic attribute matrix*.

In a further step, singular value decomposition (SVD) is applied to reduce the number of attributes of the matrix. SVD is a well-known technique of latent semantic indexing (Deerwester et al. 1990), which has been shown to improve accuracy of information retrieval. Each dimension in the reduced space is a latent variable representing groups of highly correlated attributes. Reducing the dimensionality of the original matrix reduces noise of the data and its sparsity, thus addressing inherent problems of filtering techniques.

Based on the semantic attribute matrix similarities, $SemSim(i_p, i_q)$ for all pairs of items i_p and i_q are computed. For this computation the standard vector-based cosine similarity is applied. In addition, the usual item similarities $RateSim(i_p, i_q)$ are computed based on the ratings of users. Finally, these two similarities are combined by a linear function: $CombinedSim(i_p, i_q) = \alpha \cdot SemSim(i_p, i_q) + (1 - \alpha) \cdot RateSim(i_p, i_q)$. The best value for α depends on the domain and is determined by a sensitivity analysis. The predictions of ratings for a user u_a regarding item i_t is realized by a weighted sum approach exploiting the combined similarity values of k nearest neighbors (see Section 2.1). Mobasher et al. (2004) reported results of an evaluation that show that the proposed approach improves the prediction accuracy and that the application of SVD results in further enhancements. Furthermore, the experiments show that the approach produces reasonably accurate predictions of user ratings for new items, thus alleviating the "new item problem" of CF techniques.

11.4 Extracting semantics from the web

Semantic information can provide valuable means for improving recommendations. However, where does this information come from, and how costly and reliable is the acquisition process?

To address this problem, we can distinguish two approaches to generate semantic information. The first approach assumes that humans are providing semantics by annotating content and by declaring logical sentences. The second approach is to develop software systems that are able to generate semantics with little or no human intervention. Given the lessons learned in applying knowledge-based systems, this second path is particularly attractive, as it reduces the needed development and maintenance efforts.

In the work described by Shani et al. (2008), the basic idea is to generate the information needed for CF systems through web mining. They developed two different methods, *WebCount* and *WebCrawl*. WebCount is based on the cosine score $cosine(i_1, i_2)$ for binary ratings (i.e., the user expresses only that he or she likes or does not like an item) where i_1 and i_2 are items, and $count(i_1, i_2)$ is the number of users who liked both item i_1 and i_2. $count(i)$ is the number of users who just liked i.

$$cosine(i_1, i_2) = \frac{count(i_1, i_2)}{count(i_1)count(i_2)} \qquad (11.17)$$

Based on these scores, item-based recommendations can be computed as described in Section 2.2. Shani et al. (2008) propose to use the number of pages that list an item i as an approximation of the count of i. Similarly, $count(i_1, i_2)$ is estimated by the number of pages that mention both items. The simplest way to acquire these numbers is to input the names of items into a web search engine and to approximate the count by the number of hits returned – for instance, in the movie domain, the movie names are exploited.

Obviously, this approach gives a rough estimation, which could be improved by more sophisticated query and mining techniques. For the movie domain, a simple improvement extends the query with additional keywords and filters, such as "movie recommendations" OR "recommended movies" OR "related movies" (Shani et al. 2008).

The WebCrawl method applies a more sophisticated strategy that, however, may not be generally applicable. The idea is that in web systems, which host a web community, the members of such a community provide data for item ratings in their profiles. In particular, in some web communities (such as MySpace), members declare the list of movies or musicians they like. Each page in WebCrawl is treated as a user, and the items recognized on this web page are counted as a positive rating. Obviously, such a method could be easily enhanced by more sophisticated web crawling techniques such as detecting if the item is really a positive rating or rather negative.

Shani et al. (2008) describe a comparison of WebCount and WebCrawl with a standard approach. This approach exploits the Netflix dataset, in which users explicitly rated movies. This comparison showed that WebCrawl provided the best recommendations. Ratings based on Netflix were close behind, with WebCount somewhat worse. Given that the methods for mining the web could be easily enhanced for both WebCount and WebCrawl, it seems reasonable that web mining techniques will play an important role in improving the data collection for CF techniques. The results suggest that in the movie domain, the user–item matrix generated by crawling MySpace has similar or better quality compared with the one derived from the Netflix ratings. In addition, the results of the simple WebCount method were surprisingly good, as more than

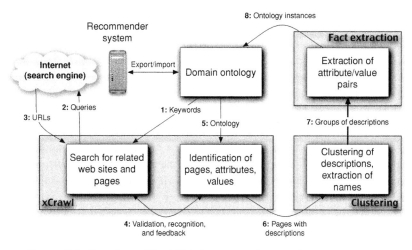

Figure 11.3. Workflow of the AllRight ontology instantiation system.

70 percent of the recommendation lists computed by WebCount were classified as reasonable by the users.

As shown by the contribution of Shani et al. (2008), CF systems can take advantage of the huge amount of information on the web. However, the exploitation of the web to enhance recommendation technology is not limited to collaborative methods. In fact, web mining techniques have a high potential to reduce the efforts for implementing and maintaining the knowledge bases of knowledge-based recommender systems.

In particular, knowledge-based recommenders require a description of all available products and their attributes. Acquiring and maintaining this data could be a costly and error-prone task. Jannach et al. (2009) describe the AllRight system, which instantiates an ontology by exploiting tabular web sources.

The basic idea of the AllRight system is to search the web for tables describing products (e.g., digital cameras), which are the information sources to populate product ontologies. For this task, numerous problems must be solved, as the web content is designed to be easily comprehended by humans, but not by machines. Figure 11.3 depicts the workflow of the AllRight system, which also shows the major challenges.

In a first step, the knowledge engineer must specify the set of attributes that describe the products of a domain (e.g., digital cameras). This description includes the domains of the attributes and their units. In addition, the knowledge engineer can associate keywords to attributes and units reflecting the fact that various names are used to name attributes and units. As shown in Figure 11.3, the system exploits these keywords (1) to crawl the web by xCrawl, by posting queries (2) and downloads (3) all web pages that describe products of the domain. The downloaded pages are analyzed by the Identification component (4) to check if the pages contain the desired product descriptions. This is supported by the domain ontology, which provides constraints exploited by the validation. To correct errors, it is desirable to have many redundant descriptions of products, as this allows us to clean data by statistical methods. The identified pages containing product information are forwarded to a module that clusters the pages so each cluster describes one product (7). In a further step, attribute/value pairs are extracted from these clusters, which serve as an input to create instances of the domain ontology (8). The result is a fact extraction system that achieves an F-measure between 0.8 and 0.9 for digital cameras and laptops, which served as test domains.

11.5 Summary

In this chapter we have shown the opportunities, current methods, and realizations of Web 2.0 and the Semantic Web for the field of recommender systems. With these new developments of the web, we are able to exploit additional information so users can be better served by recommender technology. We have shown how this information can contribute to more trustworthy and qualitative enhanced recommendations satisfying information needs. In particular, we started with approaches exploiting very little semantics (lightweight ontologies) and moved to semantically richer domains. Finally, we pointed out how semantics can be extracted from the web.

We must acknowledge, however, that the advancements of the web are still tremendously fast. Consequently, we have provided the current state, which outlines the fundamental development paths of web-based recommender technology. Both Web 2.0 and the Semantic Web in combination not only drive new technologies but, maybe even more important, also have huge impacts on society regarding the communication and interaction patterns of humans. This can be impressively observed through the success and growth rate of various web-based service and communication platforms. Technological as well as social developments provide, on one hand, new means to improve recommendation methods, but on the other hand, generate new demands for supporting humans with advice in a more complex and faster moving world. Consequently, we will see many new developments and high-impact applications of recommender technology in the context of Web 2.0 and the Semantic Web, boosted by a growing interest of major Internet players.

12

Recommendations in ubiquitous environments

In previous sections we restricted our discussion of the application and use of recommender systems to the context of traditional websites. When information systems extend their reach to offer access and interaction opportunities virtually anywhere, however, the so-called ubiquitous environments become application domains for recommender systems.

In this chapter, we therefore discuss the idiosyncrasies of recommending in ubiquitous environments compared with traditional web applications. First we reflect on the evolution of mobile systems and the associated technological issues in a short introductory statement. Second, we focus on the challenges and proposed algorithms for introducing additional context data, such as location. Finally, we provide an overview of selected application domains and related work.

12.1 Introduction

Mobile applications have always been a domain for recommendation because small display sizes and space limitations naturally require access to personalized information, on one hand, and location provides an additional exploitable source of user feedback, on the other hand. Since the end of the 1990s, research into mobile applications has focused heavily on adaptivity with regards to heterogenous hardware and software standards (see Miller et al. 2003). Therefore, most proposed mobile applications have remained in a prototypical state and have been evaluated only in small field trials with a limited scope for usage. One exception in this respect is the ClixSmart system (Smyth and Cotter 2002), which personalizes users' navigation on mobile portals and has been evaluated and fielded in real-world scenarios. For most scientific prototypes, however, wider productive use has been hindered for reasons such as restrictive hardware requirements or client-side software installations.

Nevertheless, some of these limitative circumstances are now starting to disappear. For instance, the latest generation of smart phones (and also netbooks) not only has more powerful CPUs and better displays than previous generations, but many of these devices also come with built-in GPS modules that can be used to determine geographical position, which can, in turn, be used as contextual knowledge. In addition, modern wireless broadband data transfer standards and low network access prices make this technology affordable for a broader user community. Therefore, location-aware mobile applications, such as Google Maps, have already become common among early technology adopters. Subsequently, research is beginning to focus on user experience and interaction design. In a current study conducted by Jannach and Hegelich (2009), the impact of different algorithms recommending games were compared. However, only 2 percent of all mobile users participating in this field experiment actually rated at least one of the proposed items explicitly. Thus, sparsity of user feedback in particular must be addressed when building recommender systems in the mobile context. Further research questions that come up in ubiquitous application domains are, for instance

- What are the specific goals of recommender systems in a mobile context? Do users expect serendipitous recommendations, or is it more important to be pointed to things that are close to one's current position?
- What are the implications of contextual parameters such as localization for the design of recommendation algorithms? Is location just another preference, a requirement that is always strictly enforced, or something in between?
- Is there something such as a mobile application domain, or are there plenty of different scenarios that only partially share common characteristics, such as city and museum guides, recommender systems for tourists and visitors, or ad-hoc work networks, to name a few?
- What role does the modality of interaction play when addressing users "on the go"? Pushing information can be useful to draw recipients' attention to "windows of opportunity" close to them, but the users' permission is surely needed. Comparable to permission-based marketing, how should permission protocols for "push" recommendations function?

Although these questions remain essentially unanswered because of the lack of results from case studies and surveys, we discuss current research into the context awareness of recommendation algorithms, present selected examples of pilot systems in different application domains (in Subsection 12.3), and conclude with a short summary.

12.2 Context-aware recommendation

Context awareness is a requirement for recommender systems that is particularly relevant in ubiquitous domains. Whereas some researchers denote virtually any domain aspect as context, we will denote as context only situation parameters that can be known by the system and may have an impact on the selection and ranking of recommendation results. Shilit et al. (1994) name the most important aspects of context as where you are, who you are with, and what resources are nearby. Exploiting the current location of the user, his or her companions, and the availability of resources in his or her surroundings can considerably increase the perceived usefulness of a mobile application. Dix et al. (2000) discuss awareness of space and location with respect to interactive mobile systems from a design perspective that also includes virtual worlds, although we will focus only on physical worlds in our discussion. Ranganathan and Campbell (2003) see context as "any information about the circumstances, objects or conditions surrounding a user that is considered relevant to the interaction between the user and the ubiquitous computing environment". Thus, context denotes additional information to what is traditionally represented in a user model, such as demographics or interests, and refers to "physical contexts (e.g., location, time), environmental contexts (weather, light and sound levels), informational contexts (stock quotes, sports scores), personal contexts (health, mood, schedule, activity), social contexts (group activity, social activity, whom one is in a room with), application contexts (emails, websites visited) and system contexts (network traffic, status of printers)" (Ranganathan and Campbell 2003). As becomes obvious from this enumeration, the border between user model and context is not well defined. In particular, differentiating between ephemeral and short-term user interests, with the latter constituting largely what is also considered as personal or application context, has always been the focus of user modeling and personalization research.

A context-aware user model for personalization is also sketched by Anand and Mobasher (2007). A person buying and rating books might do this in different situations. Sometimes the person buys fiction books for himself or herself, but sometimes books are work related or for children. Thus, Anand and Mobasher (2007) argue that aggregating all the information in a simple, not context-aware, user profile is suboptimal, and they propose a recommendation method that can take this contextual information better into account. Their approach relies on a more complex user model that has both long-term and short-term memories, supporting the automated generation of "contextual cues" from the short-term memory.

Thus, even early recommendation systems such as Fab (Balabanović and Shoham 1997) that differentiated between short- and long-term interest profiles can be seen as implementing some form of context awareness. However, here we denote only approaches that focus on impersonal context parameters, such as location, as context-aware.

Schwinger et al. (2005) give an overview of the different levels of context awareness implemented by mobile tourism guides, and Höpken et al. (2008) present a two-dimensional framework that matches contextual dimensions with the adaptation space of a mobile tourist guide. They argue that a mobile guide possesses several dimensions according to which its functionality and appearance can be adapted, such as content elements (e.g., topic, textual content, images); interface design issues such as modality, layout and structure, or navigation options, and behavioral and interactivity aspects. Therefore, they propose that a change in a specific context dimension such as the client technology of the device can have implications for several of the adaptation dimensions. Whereas Höpken et al. (2008) address adaptation aspects in general of ubiquitous applications with web interfaces, in this section we focus only on the implications of context-awareness for the function of recommendation systems themselves.

At a minimum, most systems filter the presented information content according to users' current location and consider additional preferences (e.g., "display only objects from category A"). However, such approaches are quite static and do not include machine learning aspects such as content-based or collaborative filtering. Lee et al. (2006), for instance, first mine relevant personal factors from historic data that seem to influence users' choice for restaurants and produce a recommendation list based on the user's personal preferences. Second, restaurant recommendations considering only their proximity to the user's current location are computed, and finally a weighted list of both is presented to the requestor.

Adomavicius et al. (2005) consider the notion of context in recommendation by proposing a multidimensional approach. They formalize contextual information in a general way by encompassing additional data dimensions. Adomavicius and Tuzhilin (2005) traditionally understand recommendation as a two-dimensional function $rec : U \times I \mapsto R$ that maps a user (U) and an item dimension (I) onto a utility score R, as already discussed in Chapter 5. Consequently, the multidimensional approach defines the rating function rec_{md} over an arbitrary n-dimensional space $D_1 \times \cdots \times D_n$:

$$rec_{md} : D_1 \times \cdots \times D_n \mapsto R \tag{12.1}$$

The domain space D_i can, for instance, be *location, time,* or *companion.* To derive predictions from such multidimensional ratings data, a reduction-based approach can be employed that restricts the ratings matrix to entries that conform to the context criteria. For instance, when one wants to compute whether a user will like a specific restaurant that is situated in a city, only ratings of restaurants in the city and none located in the countryside will be considered. Such an approach makes sense only if the quality of recommendations is improved when the ratings input is reduced by a specific situational context such as location. In addition, exploiting only a limited segment of ratings data based on some contextual parameters sharply aggravates the cold-start problems mentioned in Chapter 2. In particular, reduction-based approaches that consider several contextual dimensions in parallel become rather impractical for applications with a relatively small-scale user base. Adomavicius et al. (2005) therefore propose additional enhancements such as aggregating several contextual segments and combining the approach with traditional two-dimensional recommendation as a fallback scenario. An obvious example is the aggregation of ratings from Monday to Friday as weekday ratings, in which an aggregation function such as *average* is employed to resolve potential conflicts when the same item is rated by the same user at different time points.

Adomavicius et al. (2005) experimentally evaluated their approach in the movie domain, in which for instance they employed the place where the movie was watched (home or theatre), the time (weekday or weekend), the type of friends who were present, as well as release information about the movie indicating its novelty as contextual data dimensions. Their approach was able to outperform a traditional two-dimensional collaborative filtering recommender system in terms of accuracy on an historical dataset. However, they observed that not all contextual segments positively contribute to recommendation results, and therefore they employed a preselection mechanism that identifies segments that reach significant improvements.

Another approach, presented by Bohnert et al. (2008), studied the sequence patterns of visitor locations in museums and developed interest and transition models to predict a visitor's next locations. They collected a dataset from tracking real visitors in a museum that contains the sequences of exhibitions they observed, as well as their interest profiles. The latter were derived from content descriptions of exhibits and by interpreting relative viewing times of exhibits as implicit ratings. The *interest model* considers only the visitor's relative interest and does not take the order of visits into account. In contrast, the *transition model* reflects the probabilities of transitions between two exhibitions (i.e., locations) that allows the algorithm to predict the next locations by finding a maximum probability sequence of k unvisited exhibitions. Although their findings need to be considered preliminary because of the small size of the dataset, they observed that the transition model significantly outperformed the interest model and that a hybrid exploiting both models could provide only minor additional improvements. In any case, this is an additional argument in favor of considering the location context in ubiquitous recommendation applications.

Ranganathan and Campbell (2003) applied first-order predicate calculus to ensure a transparent line of reasoning on context models. Contexts are thus first-order predicates, and logical rules allow higher-level concepts to be derived from low-level sensor information. Thus the proposed system represents a deductive user model for context management that can be queried by recommendation algorithms in order to receive input for personalization.

Having discussed the different viewpoints of context-awareness and hinted at a few computation schemes that can reason on context, we now examine the variety of different systems and prototypes that have been constructed.

12.3 Application domains

Mobile recommendation applications have been shown to be a very active area, and applications have been fielded in domains such as tourism, cultural heritage, or commerce in general.

M-Commerce. M-commerce refers to monetary transactions that are conducted via wireless networks. The adoption of context awareness for m-commerce applications is crucial for their success. Tarasewich (2003) distinguished between the context of the participant, the environment he or she is in, and the activities currently being carried out. Ngai and Gunasekaran (2007) provide meta-research classifying published work on m-commerce. Both motivate the necessity of information reduction and context sensitivity for this application domain, although recommendation is not explicitly mentioned. However, recommender system research is particularly active in the following subfield of m-commerce.

Tourism and visitor guides. The tourism industry as one of the biggest economic sectors worldwide, together with the fact that travelers have specific information needs, makes this domain a natural choice for mobile information systems. Kabassi (2010) provides a coherent overview of recommendation applications that also includes a comparative analysis with respect to contextual aspects such as weather, season, or distance. Cyberguide, an experimental tour guide that provides location-aware services (Abowd et al. 1997), was one of the pioneers in this field. It requires, however, a specific personal digital assistant

(PDA) hardware platform. In comparison, the GUIDE system (Cheverst et al. 2002b) is a context-aware mobile guide for visitors to the city of Lancaster, requiring an end system with a touch screen and the ability to run Java applications. The system offers its users adaptive pull-based access to information that builds on existing wireless networking infrastructure. In addition, Cheverst et al. (2002a) explored the appropriateness of information push for this application domain in a small field trial. One of the findings was that people showed enthusiasm for push-based guidance information, but context-aware support requires very fine-grained location information. For instance, users want to be notified when they take a wrong turn, or an attraction should be announced when it first comes into the visitor's field of vision. However, such detailed context information requires specific hardware that cannot be assumed to be available in a usage scenario with a wider scope.

Ardissono et al. (2005) presented the interactive tourist information guide INTRIGUE, which was developed on the basis of a multiagent infrastructure for the personalization of web-based systems. It not only personalizes the content presented by ranking items according to assumed user interest but also customizes the information according to the display capabilities of the user device (Ardissono et al. 2003). INTRIGUE incorporates a fuzzy utility-based personalization approach for ranking results, and it also supports group recommendation.

The Dynamic Tour Guide (DTG) is a mobile agent that supports visitors in locating attractions of interest and proposes personalized tour plans for the German city of Goerlitz-Zittau (Kramer et al. 2006). It implements a semantic match algorithm to determine the user's assumed interest for city attractions and computes a personalized tour plan. Comparable to INTRIGUE, the DTG system also needs initial acquisition of personalization knowledge. The system is aware of the location and time context, for instance, and the description of an attraction depends on the user's position. The tour plan is rearranged based on the progress made and the remaining time. However, the system is not specifically a tour guide with detailed background knowledge about sights and attractions; rather, the system focuses on offering a wide range of useful information during a visit to the historic city.

The COMPASS application (van Setten et al. 2004) is a context-aware mobile personal assistant based on 3G network services. It uses a recommendation service to offer its users interactive maps with a set of nearby points of interest. The system integrates different types of context awareness, such as location, time, or weather, with recommendation functionality. Interestingly, one of the findings of van Setten et al. (2004) is that a large share of users want to decide for themselves which contextual factors should be taken into account and which should not. Nguyen and Ricci (2007a) developed a critique-based mobile recommender system that recommends restaurants and enables its users not only to specify their initial preferences but also to critique recommended items. Thus users can not only reject a proposal but can also give reasons why. Their user study showed that the additional constraints acquired from users during the interactive process lead to more satisfied users.

Adaptation to technical restrictions such as display size and personalization of presented content, as addressed by some of the aforementioned systems, are provided by the innsbruck.mobile system (Höpken et al. 2006). Its focus lies on widespread and actual use among tourists, and therefore avoids client-side installation requirements. One of its novelties is the support of two different types of communication paradigms. First, information seekers have personalized browsing access to categories such as events, sights, restaurants, or accommodations. However, in addition to its web-based information pull service, the system also supports context-aware information push (Beer et al. 2007) that regularly provides, for instance, weather and news messages as well as security warnings. It is configured by event-condition-action rules, in which events trigger the evaluation of the subsequent conditions. Examples of possible events are rapid weather changes, time points, or intervals, as well as users entering specific location areas. An outstanding pecularity of innsbruck.mobile is that prior to its development, substantial empirical research was performed into the usage intentions of Tyrolean tourists. Rasinger et al. (2007) asked tourists if they would use, for instance, a sightseeing or event guide, and what type of features, such as search and browse functionality, recommendation, or push services, would be most useful. Interestingly, users identified weather and news, transport and navigation, and security to be the most important mobile information services for tourists.

SPETA (Garcia-Crespo et al. 2009) is a recently proposed social pervasive e-tourism advisor that combines Semantic Web techniques, geographic information system (GIS) functionality, social networks features, and context awareness. Its recommendation functionality consists of of a hybridization component that combines several filters that reduce the set of potential recommendations based on context, domain knowledge, and collaborative filtering in parallel.

Cultural heritage and museum guides. Mobile guides for archeological sites or museums providing multimedia services, such as Archeoguide (Vlahakis et al. 2002) or MobiDENK (Kroesche et al. 2004), typically impose specific hardware requirements on their potential users. MobiDENK runs on a PDA and is a location-aware information system for historic sites. It displays multimedia background information on monuments of historic significance. Archeoguide goes a step further and reconstructs ruined sites and simulates ancient life in an augmented reality tour with a head-mounted display. As already outlined in Section 12.2, Bohnert et al. (2008) analyzed the sequence patterns of museum visitors' locations to predict their next locations. The museum guide LISTEN, presented by Zimmermann et al. (2005), generates a personalized three-dimensional audio experience based on the assumed interests of users because of their walking patterns and their specific location. For determining the latter, sensors of the ubiquitous environment provide input to the system, and actuators display visual and acoustic objects. Pilot applications for such ubiquitous technologies are also popular for the domain of home computing and consumer electronics in general, as the reader will see next.

Home computing and entertainment. Nakajima and Satoh (2006) present, for instance, software infrastructure that supports spontaneous and personalized interaction in home computing. Their notion of "personalized" basically means that users are able to personally configure and adapt smart devices in their environment based on their preferences and on specific situations. In this respect, "portable personalities" denotes the existence of distributed user models for the same user, each of which holds only partial information. Thus merging, harmonization, and reuse of these models becomes necessary when personalization and recommendation are applied in scenarios such as those discussed by Uhlmann and Lugmayr (2008).

12.4 Summary

Rapid technical advancements toward ever more powerful mobile devices and their fast market penetration are reality. Therefore, mobile applications – and ubiquitous applications in general – constitute a promising application domain for different types of personalization and recommendation. The context awareness of applications is thereby a necessity, as they have to coexist with activities such as walking, driving, or communicating. This is the main difference from traditional web applications that may assume the undivided attentiveness of their users.

When analyzing the distribution of research work on recommendation in mobile and ubiquitous environments, it is obvious that the tourism application domain is by far the most active field. Despite this, not many applications have been evaluated in broad field studies and involving not only students, perhaps with the exceptions of Rasinger et al. (2007) and Jannach and Hegelich

(2009), for instance. One of the challenges for wide-scale application of tourism recommenders is the availability of extensive and accurate resource data. For instance, a mobile restaurant recommender requires not only the positioning coordinates of all restaurants within a specific region but also some additional qualitative data, such as the type of food served, the atmosphere perceived by guests, or the business hours. As acquisition and maintenance of product data are quite cost-intensive, only widespread use and acceptance of mobile recommendation applications by end users will justify the development effort. An approach that partially addresses this problem by automated generation of additional semantic knowledge from geocoded information objects is presented by Zanker et al. (2009). It derives qualitative evidence for a given object, such as proximity to beaches or aptness for specific sports activities, from the fact that other geocoded objects that are known to be beach resorts or sports facilities are nearby. They apply this approach to the tourism domain to help users identify regions that match their holiday preferences and interests.

One even more fundamental bottleneck will have to be resolved before recommendation applications can become successful in ubiquitous environments: technical interoperability between ubiquitous devices themselves (Shacham et al. 2007) and the privacy concerns of users.

When seen only from the viewpoint of recommendation technology, in most practical applications support of context basically denotes the filtering out of inappropriate items based on context parameters such as location. Consequently, more research with respect to context awareness in recommender systems in the sense of Adomavicius et al. (2005) will be needed.

13

Summary and outlook

13.1 Summary

Recommender systems have their roots in various research areas, such as information retrieval, information filtering, and text classification, and apply methods from different fields, such as machine learning, data mining, and knowledge-based systems. With this book, we aimed to give the reader a broad overview and introduction to the area and to address the following main topics:

- *Basic recommendation algorithms*: We discussed collaborative and content-based filtering as the most popular recommendation technologies. In addition, the basic recommendation schemes, as well as different optimizations, limitations, and recent approaches, were presented.
- *Knowledge-based and hybrid approaches*: As the value of exploiting additional domain knowledge (in addition to user ratings or item "content") for improving a recommender system's accuracy is undisputed, two chapters were devoted to knowledge-based and hybrid recommender systems. We discussed both knowledge-based recommendation schemes, such as constraint and utility-based recommendation, as well as possible hybridization strategies.
- *Evaluation of recommender systems and their business value*: In most cases, recommender systems are e-commerce applications. As such, their business value and their impact on the user's decision-making and purchasing behavior must be analyzed. Therefore, this book summarized the standard approaches and metrics for determining the predictive accuracy of such systems in the chapter on recommender systems evaluation. A further chapter was devoted to the question of how recommender systems can influence the decision-making processes of online users. Finally, a comprehensive

299

case study demonstrated that recommender systems can help to measurably increase sales.

- *Recent research topics*: Further chapters of the book were devoted to research areas at the forefront of the field. Topics included the opportunities for employing recommendation technology in Web 2.0, ubiquity aspects for recommendations, and the question of how to prevent attacks on recommender systems.

Even though this book covered a broad range of topics, it cannot cover every possible technique or algorithm optimization in detail owing to its introductory nature and the speed of development within the field. Therefore, selections and compromises had to be made, for instance, with respect to the presented application domains. Examples were given for both classical domains, such as books or movies, and less obvious application fields, such as financial products. A discussion of further application domains, such as music recommendation or recommenders for the tourism domain (see, e.g., Staab et al. 2002 or Jannach et al. 2007), which also have their own peculiarities and may require specific recommendation techniques, is beyond the scope of this book.

13.2 Outlook

We conclude the book with a subjective selection of current developments, as well as emerging future problem settings in recommender systems research, many of which were not or were only partially covered in this work.

Improved collaborative filtering techniques. Although a massive number of algorithms and optimizations for the basic CF schemes have been developed over the past fifteen years, new techniques or combinations of CF methods are continually being proposed. A recent example for a rather simple approach is described by Zhang and Pu (2007), whose basic idea is to recursively make predictions for neighbors of the active user u who are very similar to u but have not rated the target item, to improve recommendation accuracy. In addition to such improvements, the Netflix Prize (Bell et al. 2009) competition[1] gave CF research an additional boost recently: in 2006, the online DVD rental company Netflix announced a US $1 million prize for the team whose recommendation system could improve the accuracy of their existing recommendation algorithm (measured in terms of the root mean square error [RMSE]) by 10 percent. The competition was won in 2009 in a joint effort by

---

[1] http://www.netflixprize.com.

four different competitors. The accuracy improvements were reached by using a combination of various techniques (Töscher et al. 2008). The main aspects were the inclusion of the time aspect as a third aspect beside movies and users, the calculation and combination of various predictors, new techniques such as "neighborhood-aware matrix factorization", and the automatic fine-tuning of parameters.

Context awareness. Time aspects – as mentioned above – are only one of many additional pieces of information that could be taken into account in recommendation tasks. Recent works, for instance, have also tried to take additional context aspects of the user into account. Under the term *context*, various aspects of the user's situation can be subsumed:

- A rather simple form of contextual information, which was discussed in Chapter 12, is the user's current geographical location, which might be exploited not only for ubiquitous or mobile recommendation scenarios but also in the context of the geospatial web in general.
- Time, in the sense of weekday or current time of the day, can also be seen as contextual information. Users might, for instance, be interested in different news topics at different times of the week.
- The emotional context of the user is another interesting dimension that will surely have an impact in some domains. For instance, in the classic movie domain, the user's mood will obviously affect how much users like movies of specific genres. González et al. (2007) present an approach that captures this emotional context of users.
- Accompanying persons represent another context dimension that is surely relevant for making recommendations. The term *group recommendations* has become popular in this regard and is used, for instance, by McCarthy et al. (2006).

When the context of the user's decision process is captured explicitly, recommender systems may exploit *multicriteria ratings* containing this contextual information as an additional source of knowledge for improving the accuracy of recommendations. In contrast with classical settings, which allow each user issues exactly one rating per item, multicriteria ratings would, for example, permit the user to evaluate a movie along different dimensions, such as plot, actors, and so forth. Initial promising methods for exploiting this additional information are reported by Adomavicius and Kwon (2007) and Lee and Teng (2007). One practical example for the use of multicriteria ratings is the e-tourism platform tripadvisor.com. There, users can rate hotels along dimensions such as value, price, or location. Unfortunately, however, no standard datasets for multicriteria ratings are freely available yet, which makes the development and comparison of different methods extremely difficult.

Recommendation on algorithms and techniques. Since the first algorithms and systems that have now become known as recommender systems were developed, myriad new techniques and improvements have been proposed. Thus, from the perspective of design science (Hevner et al. 2004) the research community was very productive in coming up with more scalable and more accurate algorithms. From the viewpoint of behavioral science, however, more research efforts are needed before we can develop an explanatory theory as to why and how recommender systems affect users' decision processes. In addition, prescriptive theory is required to guide practitioners in terms of which domains and situations are suitable for which recommendation algorithms. Some initial thoughts toward *recommending recommenders* were made by Ramezani et al. (2008), but clearly more research in this direction will take place.

User interaction issues/virtual advisors. More elaborate user interaction models are relevant not only in mobile recommender systems but also in classical web-based systems, in particular where additional knowledge sources, such as explicit user preferences, can be leveraged to improve the recommendation process. The provision of better explanations or the use of "persuasive" technologies are examples of current research in the area. In addition, we believe that more research is required in the area of conversational user interaction – for example, into the development of dialog-based systems for interactive preference elicitation. Also, the use of natural language processing techniques, as well as multimodal, multimedia-enhanced rich interfaces, is, in our opinion, largely unexplored, although it is an important step in the transition between classical recommender systems and "virtual advisors" (Jannach 2004).

Such next-generation recommenders might someday be able to simulate the behavior of an experienced salesperson. Instead of only filtering and ranking items from a given catalog, future advisory systems could, for instance, help the user interactively explore the item space, conduct a dialog that takes the customer's skill level into account, help the user make compromises if no item satisfies all of his or her requirements, give intuitive explanations why certain items are recommended, or provide personalized arguments in favor of one particular product.

Recommendation techniques will merge into other research fields. Over the past few years, the research field has experienced considerable growth; the annual ACM conference series on recommender systems attracted more than

200 submissions in 2010. We expect that additional interest in the basic building blocks of recommendation systems, such as user modeling and personalized reasoning, will come from neighboring and related fields such as information retrieval. The personalization of search results might transform search engines into context-aware recommender systems in the future. Moreover, the enormous growth of freely available user-generated multimedia content, thanks to Web 2.0, will lead to emphasis on personalized retrieval results (Sebe and Tian 2007). Analogously, recommending appropriate items could also be of interest in other fast growing areas such as the "Internet of Things" or the "Web of Services". Thus the authors are convinced that knowledge of recommendation techniques will be helpful for many application areas.

Bibliography

G. D. Abowd, C. G. Atkeson, J. Hong, S. Long, R. Kooper, and M. Pinkerton, *Cyberguide: A mobile context-aware tour guide*, Wireless Networks **3** (1997), no. 5, 421–433.

G. Adomavicius and Y. O. Kwon, *New recommendation techniques for multicriteria rating systems*, Intelligent Systems, IEEE **22** (2007), no. 3, 48–55.

G. Adomavicius, R. Sankaranarayanan, S. Sen, and A. Tuzhilin, *Incorporating contextual information in recommender systems using a multidimensional approach*, ACM Transactions on Information Systems **23** (2005), no. 1, 103–145.

G. Adomavicius and A. Tuzhilin, *Toward the next generation of recommender systems: A survey of the state-of-the-art and possible extensions*, IEEE Transactions on Knowledge and Data Engineering **17** (2005), no. 6, 734–749.

R. Agrawal and R. Srikant, *Fast algorithms for mining association rules*, Proceedings of the 20th International Conference on Very Large Data Bases (VLDB'94), Morgan Kaufmann, 1994, pp. 487–499.

J. Allan, J. Carbonell, G. Doddington, J. Yamron, and Y. Yang, *Topic detection and tracking pilot study final report*, Proceedings of the DARPA Broadcast News Transcription and Understanding Workshop, 1998, pp. 194–218.

S. S. Anand and B. Mobasher, *Intelligent techniques for web personalization*, Lecture Notes in Computer Science, vol. 3169, Springer, Acapulco, Mexico, 2005, pp. 1–36.

S. S. Anand and B. Mobasher, *Contextual recommendation*, From Web to Social Web: Discovering and Deploying User and Content Profiles: Workshop on Web Mining (Berlin, Germany), Springer, 2007, pp. 142–160.

L. Ardissono, A. Felfernig, G. Friedrich, A. Goy, D. Jannach, G. Petrone, R. Schäfer, and M. Zanker, *A Framework for the Development of Personalized, Distributed Web-Based Configuration Systems*, AI Magazine **24** (2003), no. 3, 93–110.

L. Ardissono, A. Goy, G. Petrone, and M. Segnan, *A multi-agent infrastructure for developing personalized web-based systems*, ACM Transactions on Internet Technology **5** (2005), no. 1, 47–69.

L. Ardissono, A. Goy, G. Petrone, M. Segnan, and P. Torasso, *Intrigue: Personalized recommendation of tourist attractions for desktop and handset devices*, Applied Artificial Intelligence **17** (2003), no. 8–9, 687–714.

E. Aronson, T. Wilson, and A. Akert, *Social psychology*, 6th ed., Pearson Prentice Hall, 2007.

R. Baeza-Yates and B. Ribeiro-Neto, *Modern information retrieval*, Addison-Wesley, 1999.

M. Balabanović and Y. Shoham, *Fab: content-based, collaborative recommendation*, Communications of the ACM **40** (1997), no. 3, 66–72.

P. Basile, D. Gendarmi, F. Lanubile, and G. Semeraro, *Recommending smart tags in a social bookmarking system*, Proceedings of the International Workshop Bridging the Gap between Semantic Web and Web 2.0 at ESWC 2007 (Innsbruck, Austria), 2007, pp. 22–29.

C. Basu, H. Hirsh, and W. Cohen, *Recommendation as classification: using social and content-based information in recommendation*, in Proceedings of the 15th National Conference on Artificial Intelligence (AAAI'98) (Madison, WI), American Association for Artificial Intelligence, 1998, pp. 714–720.

G. Beenen, K. Ling, X. Wang, K. Chang, D. Frankowski, P. Resnick, and R. Kraut, *Using social psychology to motivate contributions to online communities*, Proceedings of the 2004 ACM Conference on Computer Supported Cooperative Work (CSCW '04) (Chicago), 2004, pp. 212–221.

T. Beer, M. Fuchs, W. Höpken, J. Rasinger, and H. Werthner, *Caips: A context-aware information push service in tourism*, Proceedings of the 14th International Conference on Information and Communication Technologies in Tourism 2007 (ENTER) (Ljubljana, Slovenia), Springer, January 2007, pp. 129–140.

F. Belanger, *A conjoint analysis of online consumer satisfaction*, Journal of Electronic Commerce Research **6** (2005), 95–111.

R. Bell, J. Bennett, Y. Koren, and C. Volinsky, *The million dollar programming prize*, IEEE Spectrum (2009), 28–33.

R. M. Bell, Y. Koren, and C. Volinsky, *The BellKor solution to the Netflix Prize*, Tech. Report http://www.netflixprize.com/assets/ProgressPrize2007_KorBell.pdf, AT&T Labs Research, 2007.

A. Bellogín, I. Cantador, P. Castells, and A. Ortigosa, *Discerning relevant model features in a content-based collaborative recommender system*, Preference Learning (J. Fürnkranz and E. Hüllermeier, eds.), Springer, 2010.

L. D. Bergman, A. Tuzhilin, R. Burke, A. Felfernig, and L. Schmidt-Thieme (eds.), *Proceedings of the 2009 ACM Conference on Recommender Systems (RecSys '09)*, New York, 2009.

S. Berkovsky, Y. Eytani, T. Kuflik, and F. Ricci, *Enhancing privacy and preserving accuracy of a distributed collaborative filtering*, Proceedings of the 2007 ACM Conference on Recommender Systems (RecSys '07) (Minneapolis), ACM, 2007, pp. 9–16.

J. Bettman, M. Luce, and J. Payne, *Constructive consumer choice processes*, Journal of Consumer Research **25** (1998), no. 3, 187–217.

R. Bhaumik, R. D. Burke, and B. Mobasher, *Crawling attacks against web-based recommender systems*, Proceedings of the 2007 International Conference on Data Mining (DMIN '07) (Las Vegas) (Robert Stahlbock, Sven F. Crone, and Stefan Lessmann, eds.), June 2007, pp. 183–189.

R. Bhaumik, C. Williams, B. Mobasher, and R. Burke, *Securing collaborative filtering against malicious attacks through anomaly detection*, Proceedings of the 4th Workshop on Intelligent Techniques for Web Personalization (ITWP '06) (Boston), July 2006.

D. Billsus and M. Pazzani, *User modeling for adaptive news access*, User Modeling and User-Adapted Interaction: The Journal of Personalization Research **10** (2000), no. 2–3, 147–180.

D. Billsus and M. J. Pazzani, *Learning collaborative information filters*, Proceedings of the 15th International Conference on Machine Learning (ICML'98), Morgan Kaufmann, San Francisco, 1998, pp. 46–54.

D. Billsus and M. J. Pazzani, *A personal news agent that talks, learns and explains*, Proceedings of the 3rd Annual Conference on Autonomous Agents (AGENTS'99) (Seattle), ACM, 1999, pp. 268–275.

D. Billsus and M. J. Pazzani, *Adaptive news access*, The Adaptive Web (Peter Brusilovsky, Alfred Kobsa, and Wolfgang Nejdl, eds.), Lecture Notes in Computer Science, vol. 4321, Springer, 2007, pp. 550–570.

D. Billsus, M. J. Pazzani, and J. Chen, *A learning agent for wireless news access*, Proceedings of the 5th International Conference on Intelligent User Interfaces (IUI '00) (New Orleans), ACM, 2000, pp. 33–36.

F. Bohnert, I. Zukerman, S. Berkovsky, T. Baldwin, and L. Sonenberg, *Using interest and transition models to predict visitor locations in museums*, AI Communications **21** (2008), no. 2–3, 195–202.

C. Bomhardt, *NewsRec, a SVM-driven Personal Recommendation System for News Websites*, Proceedings of the 2004 IEEE/WIC/ACM International Conference on Web Intelligence (WI '04) (Washington, DC), IEEE Computer Society, 2004, pp. 545–548.

S. Botti and S. Iyengar, *The psychological pleasure and pain of choosing: When people prefer choosing at the cost of subsequent outcome satisfaction*, Journal of Personality and Social Psychology **87** (2004), no. 3, 312–326.

A. Bouza, G. Reif, A. Bernstein, and H. Gall, *Semtree: Ontology-based decision tree algorithm for recommender systems*, International Semantic Web Conference (Posters and Demos), CEUR Workshop Proceedings, vol. 401, CEUR-WS.org, 2008.

J. S. Breese, D. Heckerman, and C. M. Kadie, *Empirical analysis of predictive algorithms for collaborative filtering*, Proceedings of the 14th Conference on Uncertainty in Artificial Intelligence (Madison, WI) (Gregory F. Cooper and Serafín Moral, eds.), Morgan Kaufmann, 1998, pp. 43–52.

W. F. Brewer, C. A. Chinn, and A. Samarapungavan, *Explanation in scientists and children*, Minds and Machines **8** (1998), no. 1, 119–136.

D. Bridge, M. Göker, L. McGinty, and B. Smyth, *Case-based recommender systems*, Knowledge Engineering Review **20** (2005), no. 3, 315–320.

S. Brin and L. Page, *The anatomy of a large-scale hypertextual web search engine*, Computer Networks and ISDN Systems **30** (1998), no. 1–7, 107–117.

B. G. Buchanan and E. H. Shortliffe, *Rule-based expert systems: The Mycin experiments of the Stanford Heuristic Programming Project (the Addison-Wesley series in artificial intelligence)*, Addison-Wesley Longman, Boston, 1984.

C. Buckley, G. Salton, and J. Allan, *The effect of adding relevance information in a relevance feedback environment*, Proceedings of the 17th Annual International ACM SIGIR Conference on Research and Development in Information Retrieval (SIGIR'94) (Dublin), Springer, 1994, pp. 292–300.

W. L. Buntine, M. Grobelnik, D. Mladenic, and J. Shawe-Taylor, *European conference on machine learning and principles and practice of knowledge discovery in databases*,

ECML/PKDD (1) (Bled, Slovenia), Lecture Notes in Computer Science, vol. 5781, Springer, September 2009.

R. Burke, *Knowledge-based recommender systems*, Encyclopedia of Library and Information Science **69** (2000), no. 32, 180–200.

———, *Interactive critiquing for catalog navigation in e-commerce*, Artificial Intelligence Review **18** (2002a), no. 3–4, 245–267.

R. Burke, *The wasabi personal shopper: A case-based recommender system*, Proceedings of the 16th National Conference on Artificial Intelligence and the 11th Innovative Applications of Artificial Intelligence Conference Innovative Applications of Artificial Intelligence (AAAI'99/IAAI'99) (Orlando, FL), AAAI Press, 1999, pp. 844–849.

———, *Hybrid recommender systems: Survey and experiments*, User Modeling and User-Adapted Interaction **12** (2002b), no. 4, 331–370.

R. Burke, P. Brusilovsky and A. Kobsa and W. Nejdl, *Hybrid web recommender systems*, The Adaptive Web: Methods and Strategies of Web Personalization, Springer, Heidelberg, Germany, 2007, pp. 377–408.

R. Burke, K. Hammond, and B. Young, *The findme approach to assisted browsing*, IEEE Expert **4** (1997), no. 12, 32–40.

R. Burke, K. J. Hammond, and B. C. Young, *Knowledge-based navigation of complex information spaces*, Proceedings of the 13th National Conference on Artificial Intelligence (AAAI '96) (Portland, OR), AAAI Press, 1996, pp. 462–468.

R. Burke, B. Mobasher, and R. Bhaumik, *Limited knowledge shilling attacks in collaborative filtering systems*, in Proceedings of the 3rd IJCAI Workshop in Intelligent Techniques for Personalization (Edinburgh, Scotland), 2005, pp. 17–24.

A. Byde, H. Wan, and S. Cayzer, *Personalized tag recommendations via tagging and content-based similarity metrics*, Proceedings of the International Conference on Weblogs and Social Media (ICWSM '07), poster session (Boulder, CO), 2007.

J. Canny, *Collaborative filtering with privacy*, Proceedings of the 2002 IEEE Symposium on Security and Privacy (SP '02) (Washington, DC), IEEE Computer Society, 2002a, pp. 45–57.

———, *Collaborative filtering with privacy via factor analysis*, Proceedings of the 25th Annual International ACM SIGIR Conference on Research and Development in Information Retrieval (SIGIR '02) (Tampere, Finland), ACM, 2002b, pp. 238–245.

G. Carenini and J. Moore, *Generating and evaluating evaluative arguments*, Artificial Intelligence **170** (2006), 925–952.

G. Carenini and J. D. Moore, *An empirical study of the influence of user tailoring on evaluative argument effectiveness*, Proceedings of the 17th International Joint Conference on Artificial Intelligence (IJCAI '01) (Seattle) (Bernhard Nebel, ed.), Morgan Kaufmann, August 2001, pp. 1307–1312.

G. Carenini and R. Sharma, *Exploring more realistic evaluation measures for collaborative filtering*, Proceedings of the 19th National Conference on Artificial Intelligence (AAAI) (San Jose, CA), AAAI Press, 2004, pp. 749–754.

O. Celma and P. Herrera, *A new approach to evaluating novel recommendations*, Proceedings of the 2008 ACM Conference on Recommender Systems (RecSys '08) (Lausanne, Switzerland), ACM Press, 2008, pp. 179–186.

S. Chakrabarti, *Mining the web: Discovering knowledge from hypertext data*, Science and Technology Books, 2002.

P. Chau, M. Cole, A. Massey, M. Montoya-Weiss, and R. O'Keefe, *Cultural differences in the online behavior of consumers*, Communications of the ACM **10** (2002), no. 45, 138–143.

S. H. S. Chee, J. Han, and K. Wang, *Rectree: An efficient collaborative filtering method*, Proceedings of the 3rd International Conference on Data Warehousing and Knowledge Discovery (Munich), 2001, pp. 141–151.

K. Chellapilla and P. Y. Simard, *Using machine learning to break visual human interaction proofs (HIPS)*, Proceedings of the 18th Annual Conference on Neural Information Processing Systems (NIPS '04), 2004, pp. 265–272.

L. Chen and P. Pu, *Trust building in recommender agents*, 1st International Workshop on Web Personalisation, Recommender Systems and Intelligent User Interfaces (WPRSIUI '05) (Reading, UK), 2005, pp. 135–145.

———, *A cross-cultural user evaluation of product recommender interfaces*, Proceedings of the 2008 ACM Conference on Recommender Systems (RecSys '08) (Lausanne, Switzerland), ACM, 2008, pp. 75–82.

Y.-H. Chen and E. I. George, *A Bayesian model for collaborative filtering*, in Proceedings of Uncertainty 99: The Seventh International Workshop on Artificial Intelligence and Statistics (Fort Lauderdale, FL), January 1999.

J. A. Chevalier and D. Mayzlin, *The effect of word of mouth on sales: Online book reviews*, Journal of Marketing Research **43** (2006), no. 9, 345–354.

K. Cheverst, K. Mitchel, and N. Davies, *Exploring context-aware information push*, Personal and Ubiquitous Computing **6** (2002a), no. 4, 276–281.

———, *The role of adaptive hypermedia in an context-aware tourist guide*, Communications of the ACM **45** (2002b), no. 5, 47–51.

D. Chickering, D. Heckerman, and C. Meek, *A Bayesian approach to learning Bayesian networks with local structure*, Proceedings of the 13th Annual Conference on Uncertainty in Artificial Intelligence (UAI '97) (San Francisco), Morgan Kaufmann, 1997, pp. 80–89.

P.-A. Chirita, W. Nejdl, and C. Zamfir, *Preventing shilling attacks in online recommender systems*, Proceedings of the 7th Annual ACM International Workshop on Web Information and Data Management (WIDM '05) (Bremen, Germany), ACM, 2005, pp. 67–74.

Y. H. Cho, C. Y. Kim, and D.-H. Kim, *Personalized image recommendation in the mobile internet*, Proceedings of 8th Pacific Rim International Conference on Artificial Intelligence (PRICAI '04) (Auckland, New Zealand), Lecture Notes in Computer Science, vol. 3157, Springer, 2004, pp. 963–964.

B. Choi, I. Lee, J. Kim, and Y. Jeon, *A qualitative cross-national study of cultural influences on mobile data*, Proceedings of the SIGCHI Conference on Human Factors in Computing Systems (CHI '05) (Portland, OR), ACM, 2005, pp. 661–670.

M. Claypool, A. Gokhale, T. Miranda, P. Murnikov, D. Netes, and M. Sartin, *Combining content-based and collaborative filters in an online newspaper*, Proceedings of the ACM SIGIR Workshop on Recommender Systems: Algorithms and Evaluation (Berkeley, CA), 1999.

W. Cohen, *Learning rules that classify e-mail*, Proceedings of the AAAI Symposium on Machine Learning in Information Access (Stanford, CA) (Marti Hearst and Haym Hirsh, eds.), 1996, pp. 18–25.

W. W. Cohen, *Fast effective rule induction*, Proceedings of the 12th International Conference on Machine Learning (ICML '95) (Tahoe City, CA) (Armand Prieditis and Stuart Russell, eds.), Morgan Kaufmann, July 1995, pp. 115–123.

L. Console, D. T. Dupre, and P. Torasso, *On the relationship between abduction and deduction*, Journal of Logic and Computation **1** (1991), no. 5, 661–690.

D. Cosley, S. Lam, I. Albert, J. Konstan, and J. Riedl, *Is seeing believing? How recommender system interfaces affect users' opinions*, Proceedings of the SIGCHI Conference on Human Factors in Computing Systems (CHI '03) (Fort Lauderdale, FL), 2003, pp. 585–592.

P. Cotter and B. Smyth, *PTV: Intelligent personalised tv guides*, Proceedings of the 17th National Conference on Artificial Intelligence and 12th Conference on Innovative Applications of Artificial Intelligence, AAAI Press/MIT Press, 2000, pp. 957–964.

J. W. Creswell, *Research design: Qualitative, quantitative and mixed methods approaches*, 3rd ed., SAGE Publications, 2009.

A. S. Das, M. Datar, A. Garg, and S. Rajaram, *Google news personalization: scalable online collaborative filtering*, Proceedings of the 16th International Conference on World Wide Web (WWW '07) (New York), ACM Press, 2007, pp. 271–280.

S. Deerwester, S. T. Dumais, G. W. Furnas, T. K. Landauer, and R. Harshman, *Indexing by latent semantic analysis*, Journal of the American Society for Information Science **41** (1990), 391–407.

M. de Gemmis, P. Lops, G. Semeraro, and P. Basile, *Integrating tags in a semantic content-based recommender*, Proceedings of the 2008 ACM Conference on Recommender Systems (RecSys '08), ACM, Lausanne, Switzerland, 2008, pp. 163–170.

C. Dellarocas, *Strategic manipulation of internet opinion forums: Implications for consumers and firms*, Management Science **52** (2006), no. 10, 1577–1593.

F. H. del Olmo and E. Gaudioso, *Evaluation of recommender systems: A new approach*, Expert Systems with Applications **35** (2008), no. 3, 790–804.

A. P. Dempster, N. M. Laird, and D. B. Rubin, *Maximum likelihood from incomplete data via the EM algorithm*, Journal of the Royal Statistical Society, Series B **39** (1977), no. 1, 1–38.

J. Demšar, *Statistical comparisons of classifiers over multiple data sets*, Journal of Machine Learning Research **7** (2006), 1–30.

M. Deshpande and G. Karypis, *Item-based top-n recommendation algorithms*, ACM Transactions on Information Systems **22** (2004), no. 1, 143–177.

M. B. Dias, D. Locher, M. Li, W. El-Deredy, and P. J. G. Lisboa, *The value of personalised recommender systems to e-business: a case study*, Proceedings of the 2008 ACM Conference on Recommender Systems (RecSys '08) (Lausanne, Switzerland), 2008, pp. 291–294.

A. Dix, T. Rodden, N. Davies, J. Trevor, A. Friday, and K. Palfreyman, *Exploiting space and location as a design framework for interactive mobile systems*, ACM Transactions on Computer-Human Interaction **7** (2000), no. 3, 285–321.

P. Domingos and M. J. Pazzani, *Beyond independence: Conditions for the optimality of the simple Bayesian classifier*, Proceedings of the 13th International Conference on Machine Learning (ICML '96) (Bari, Italy), 1996, pp. 105–112.

_____, *On the optimality of the simple Bayesian classifier under zero-one loss*, Machine Learning **29** (1997), no. 2-3, 103–130.

P. Duttweiler, *The internal control index: A newly developed measure of locus of control*, Educational and Psychological Measurement **44** (1984), 209–221.

B. Fasolo, G. McClelland, and P. Todd, *Escaping the tyranny of choice: When fewer attributes make choice easier*, Marketing Theory **7** (2007), no. 1, 13–26.

A. Felfernig and R. Burke, *Constraint-based recommender systems: technologies and research issues*, Proceedings of the 10th International Conference on Electronic Commerce (ICEC '08) (Innsbruck, Austria), ACM, 2008, pp. 1–10.

A. Felfernig, G. Friedrich, B. Gula, M. Hitz, T. Kruggel, R. Melcher, D. Riepan, S. Strauss, E. Teppan, and O. Vitouch, *Persuasive recommendation: Exploring serial position effects in knowledge-based recommender systems*, Proceedings of the 2nd International Conference of Persuasive Technology (Persuasive '07) (Stanford, California), vol. 4744, Springer, 2007a, pp. 283–294.

A. Felfernig, G. Friedrich, M. Schubert, M. Mandl, M. Mairitsch, and E. Teppan. Plausible Repairs for Inconsistent Requirements, Proceedings of the 21st International Joint Conference on Artificial Intelligence (IJCAI '09), Pasadena, California, USA, pp. 791–796, 2009.

A. Felfernig, G. Friedrich, D. Jannach, and M. Stumptner, *Consistency-based diagnosis of configuration knowledge bases*, Artificial Intelligence **152** (2004), no. 2, 213–234.

A. Felfernig, B. Gula, G. Leitner, M. Maier, R. Melcher, S. Schippel, and E. Teppan, *A dominance model for the calculation of decoy products in recommendation environments*, Proceedings of the AISB Symposium on Persuasive Technologies (Aberdeen, Scotland), vol. 3, University of Aberdeen, 2008a, pp. 43–50.

A. Felfernig, B. Gula, and E. Teppan, *Knowledge-based recommender technologies for marketing and sales*, International Journal of Pattern Recognition and Artificial Intelligence **21** (2006), no. 2, 1–22.

A. Felfernig, K. Isak, K. Szabo, and P. Zachar, *The VITA financial services sales support environment*, Proceedings of the 22nd National Conference on Artificial Intelligence (AAAI '07), AAAI, 2007b, pp. 1692–1699.

A. Felfernig, M. Mairitsch, M. Mandl, M. Schubert, and E. Teppan, *Utility-based repair of inconsistent requirements*, 22nd International Conference on Industrial, Engineering and Other Applications of Applied Intelligent Systems, IEA/AIE (Tainan, Taiwan), 2009, pp. 162–171.

A. Felfernig and K. Shchekotykhin, *Debugging user interface descriptions of knowledge-based recommender applications*, Proceedings of the 11th International Conference on Intelligent User Interfaces (IUI '06) (Sydney, Australia), ACM Press, 2006, pp. 234–241.

A. Felfernig and E. Teppan, *The asymmetric dominance effect and its role in e-tourism recommender applications*, Proceedings of Wirtschaftsinformatik 2009, Austrian Computer Society, 2009, pp. 791–800.

A. Felfernig, E. Teppan, G. Leitner, R. Melcher, B. Gula, and M. Maier, *Persuasion in knowledge-based recommendation*, Proceedings of the 2nd International Conference on Persuasive Technologies (Persuasive '08) (Oulu, Finland), vol. 5033, Springer, 2008c, pp. 71–82.

A. Felfernig, G. Friedrich, D. Jannach, and M. Zanker, *An integrated environment for the development of knowledge-based recommender applications*, International Journal of Electronic Commerce **11** (2006–07), no. 2, 11–34.

A. Felfernig and B. Gula, *An empirical study on consumer behavior in the interaction with knowledge-based recommender applications*, Proceedings of the 8th IEEE International Conference on E-Commerce Technology (CEC '06)/3rd IEEE International Conference on Enterprise Computing, E-Commerce and E-Services (EEE '06) (Palo Alto, CA), 2006, p. 37.

C. S. Firan, W. Nejdl, and R. Paiu, *The benefit of using tag-based profiles*, Proceedings of the 2007 Latin American Web Conference (LA-WEB '07) (Washington, DC), IEEE Computer Society, 2007, pp. 32–41.

D. M. Fleder and K. Hosanagar, *Recommender systems and their impact on sales diversity*, Proceedings of the 8th ACM Conference on Electronic Commerce (EC '07) (San Diego, California, USA), 2007, pp. 192–199.

B. J. Fogg, *Persuasive technologies*, Communications of the ACM **42** (1999), no. 5, 26–29.

———, *Persuasive technology – using computers to change what we think and do*, Morgan Kaufmann, 2003.

P. W. Foltz and S. T. Dumais, *Personalized information delivery: an analysis of information filtering methods*, Communications of the ACM **35** (1992), no. 12, 51–60.

J. H. Friedman, *On bias, variance, 0/1–loss, and the curse-of-dimensionality*, Data Mining and Knowledge Discovery **1** (1997), no. 1, 55–77.

G. Friedrich, *Elimination of spurious explanations*, Proceedings of the 16th European Conference on Artificial Intelligence (ECAI '04), including Prestigious Applicants of Intelligent Systems (PAIS '04) (Valencia, Spain) (Ramon López de Mántaras and Lorenza Saitta, eds.), IOS Press, August 2004, pp. 813–817.

X. Fu, J. Budzik, and K. J. Hammond, *Mining navigation history for recommendation*, Proceedings of the 5th International Conference on Intelligent User Interfaces (IUI '00) (New Orleans), ACM, 2000, pp. 106–112.

A. Garcia-Crespo, J. Chamizo, I. Rivera, M. Mencke, R. Colomo-Palacios, and J. M. Gómez-Berbís, *Personalizing recommendations for tourists*, Telematics and Informatics **26** (2009), no. 3, 306–315.

S. Garcìa and F. Herrera, *An extension on "statistical comparisons of classifiers over multiple data sets" for all pairwise comparisons*, Journal of Machine Learning Research **9** (2008), 2677–2694.

T. George and S. Merugu, *A scalable collaborative filtering framework based on co-clustering*, Proceedings of the 5th IEEE International Conference on Data Mining (ICDM '05) (Washington, DC), IEEE Computer Society, 2005, pp. 625–628.

F. Gershberg and A. Shimamura, *Serial position effects in implicit and explicit tests of memory*, Journal of Experimental Psychology: Learning, Memory, and Cognition **20** (1994), no. 6, 1370–1378.

G. Gigerenzer, *Bauchentscheidungen*, Bertelsmann Verlag, March 2007.

J. Golbeck, *Semantic web interaction through trust network recommender systems*, End User Semantic Web Interaction Workshop at the 4th International Semantic Web Conference, Galway, Ireland, 2005.

———, *Generating predictive movie recommendations from trust in social networks*, Proceedings of the 4th International Conference on Trust Management (iTrust '06) (Pisa, Italy), May 2006, pp. 93–104.

J. Golbeck and J. Hendler, *Inferring binary trust relationships in web-based social networks*, ACM Transactions Internet Technology **6** (2006), no. 4, 497–529.

D. Goldberg, D. Nichols, B. M. Oki, and D. Terry, *Using collaborative filtering to weave an information tapestry*, Communications of the ACM **35** (1992), no. 12, 61–70.

K. Goldberg, T. Roeder, D. Gupta, and C. Perkins, *Eigentaste: A constant time collaborative filtering algorithm*, Information Retrieval **4** (2001), no. 2, 133–151.

N. Golovin and E. Rahm, *Reinforcement learning architecture for web recommendations*, Proceedings of the International Conference on Information Technology: Coding and Computing (ITCC '04) (Las Vegas), vol. 2, 2004, pp. 398–402.

G. Golub and W. Kahan, *Calculating the singular values and pseudo-inverse of a matrix*, Journal of the Society for Industrial and Applied Mathematics, Series B: Numerical Analysis **2** (1965), no. 2, 205–224.

G. Gonzalez, B. Lopez, and J. D. L. Rosa, *The emotional factor: An innovative approach to user modelling for recommender systems*, Proceedings of AH2002 Workshop on Recommendation and Personalization in e-Commerce (Malaga, Spain), 2002, pp. 90–99.

G. González, J. L. de la Rosa, and M. Montaner, *Embedding emotional context in recommender systems*, Proceedings of the 20th International Florida Artificial Intelligence Research Society Conference (Key West, FL), AAAI Press, 2007, pp. 454–459.

S. Grabner-Kräuter and E. A. Kaluscha, *Empirical research in on-line trust: a review and critical assessment*, International Journal of Human-Computer Studies **58** (2003), no. 6, 783–812.

D. Grether and C. Plott, *Economic theory of choice and the preference reversal phenomenon*, American Economic Review **69** (1979), no. 4, 623–638.

U. Gretzel and D. Fesenmaier, *Persuasion in recommender systems*, International Journal of Electronic Commerce **11** (2006), no. 2, 81–100.

I. Grigorik, *SVD recommendation system in ruby*, web blog, 01 2007, http://www.igvita.com/2007/01/15/svd-recommendation-system-in-ruby/ [accessed March 2009].

R. Guha, R. Kumar, P. Raghavan, and A. Tomkins, *Propagation of trust and distrust*, Proceedings of the 13th International Conference on World Wide Web (WWW '04) (New York), ACM, 2004, pp. 403–412.

I. Guy, N. Zwerdling, D. Carmel, I. Ronen, E. Uziel, S. Yogev, and S. Ofek-Koifman, *Personalized recommendation of social software items based on social relations*, Proceedings of the 2009 ACM Conference on Recommender Systems (RecSys '09), New York, 2009, pp. 53–60.

G. Häubl and K. Murray, *Preference construction and persistence in digital marketplaces: The role of electronic recommendation agents*, Journal of Consumer Psychology **13** (2003), 75–91.

———, *Double agents*, MIT Sloan Management Review **47** (2006), no. 3, 7–13.

G. Häubl and V. Trifts, *Consumer decision making in online shopping environments: The effects of interactive decision aids*, Marketing Science **19** (2000), no. 1, 4–21.

K. Hegelich and D. Jannach, *Effectiveness of different recommender algorithms in the mobile internet: A case study*, Proceedings of the 7th Workshop on Intelligent Techniques for Web Personalization and Recommender Systems (ITWP) at IJCAI '09 (Pasadena, CA), 2009, pp. 41–50.

J. Herlocker, J. A. Konstan, and J. Riedl, *An empirical analysis of design choices in neighborhood-based collaborative filtering algorithms*, Information Retrieval **5** (2002), no. 4, 287–310.

J. L. Herlocker, J. A. Konstan, et al., *An Algorithmic Framework for Performing Collaborative Filtering*, Proceedings of the 22nd Annual International ACM SIGIR Conference, ACM Press, 1999, pp. 230–237.

J. L. Herlocker, J. A. Konstan, and J. Riedl, *Explaining collaborative filtering recommendations*, Proceedings of the 2000 ACM Conference on Computer Supported Cooperative Work (CSCW '00) (Philadelphia), ACM, 2000, pp. 241–250.

J. L. Herlocker, J. A. Konstan, L. G. Terveen, and J. T. Riedl, *Evaluating collaborative filtering recommender systems*, ACM Transactions on Information Systems (TOIS) **22** (2004), no. 1, 5–53.

A. Herrmann, M. Heitmann, and B. Polak, *The power of defaults*, Absatzwirtschaft **6** (2007), 46–47.

C. Hess, K. Stein, and C. Schlieder, *Trust-enhanced visibility for personalized document recommendations*, Proceedings of the 2006 ACM Symposium on Applied Computing (SAC '06) (Dijon, France) (Hisham Haddad, ed.), ACM, 2006, pp. 1865–1869.

A. R. Hevner, S. T. March, J. Park, and S. Ram, *Design science in information systems research*, MIS Quarterly **28** (2004), no. 1, 75–105.

W. Hill, L. Stead, M. Rosenstein, and G. Furnas, *Recommending and evaluating choices in a virtual community of use*, Proceedings of the SIGCHI Conference on Human Factors in Computing Systems (CHI '95) (Denver), 1995, pp. 194–201.

W. Höpken, M. Fuchs, M. Zanker, T. Beer, A. Eybl, S. Flores, S. Gordea, M. Jessenitschnig, T. Kerner, D. Linke, J. Rasinger, and M. Schnabl, *etPlanner: An IT framework for comprehensive and integrative travel guidance*, Proceedings of the 13th International Conference on Information Technology and Travel and Tourism (ENTER) (Lausanne, Switzerland), 2006, pp. 125–134.

T. Hofmann, *Probabilistic latent semantic indexing*, Proceedings of the 22nd annual International ACM SIGIR Conference on Research and Development in Information Retrieval (SIGIR '99) (Berkeley, CA), 1999, pp. 50–57.

_____ , *Latent semantic models for collaborative filtering*, ACM Transactions on Information Systems **22** (2004), no. 1, 89–115.

T. Hofmann and J. Puzicha, *Latent class models for collaborative filtering*, Proceedings of the 16th International Joint Conference on Artificial Intelligence (IJCAI '99) (San Francisco), 1999, pp. 688–693.

W. Höpken, M. Scheuringer, D. Linke, and M. Fuchs, *Context-based adaptation of ubiquitous web applications in tourism*, Information and Communication Technologies in Tourism (ENTER) (Innsbruck, Austria), 2008, pp. 533–544.

A. Hotho, R. Jäschke, C. Schmitz, and G. Stumme, *Information retrieval in folksonomies*, Proceedings of the European Semantic Web Conference 2006 (Budva, Montenegro), Lecture Notes in Computer Science, vol. 4011, Springer, 2007, pp. 411–426.

E. M. Housman and E. D. Kaskela, *State of the art in selective dissemination of information*, IEEE Transactions on Engineering Writing and Speech **13** (1970), no. 2, 78–83.

C.-N. Hsu, H.-H. Chung, and H.-S. Huang, *Mining skewed and sparse transaction data for personalized shopping recommendation*, Machine Learning **57** (2004), no. 1–2, 35–59.

Y.-C. Huang, J. Y. jen Hsu, and D. K.-C. Wu, *Tag-based user profiling for social media recommendation*, Workshop on Intelligent Techniques for Web Personalization and

Recommender Systems (ITWP) at AAAI '08 (Chicago), AAAI Press, 2008, pp. 49–55.

Z. Huang, H. Chen, and D. Zeng, *Applying associative retrieval techniques to alleviate the sparsity problem in collaborative filtering*, ACM Transactions on Information Systems **22** (2004), no. 1, 116–142.

J. Huber, W. Payne, and C. Puto, *Adding asymmetrically dominated alternatives: Violations of regularity and the similarity hypothesis*, Journal of Consumer Research **9** (1982), 90–98.

C. Huffman and B. Kahn, *Variety for sale: Mass customization or mass confusion*, Journal of Retailing **74** (1998), no. 4, 491–513.

N. J. Hurley, M. P. O'Mahony, and G. C. M. Silvestre, *Attacking recommender systems: A cost-benefit analysis*, IEEE Intelligent Systems **22** (2007), no. 3, 64–68.

J. Hutchinson, *Is more choice always desirable? Evidence and arguments from leks, food, selection, and environmental enrichment*, Biological Reviews **80** (2005), 73–92.

I. Im and A. Hars, *Does a one-size recommendation system fit all? the effectiveness of collaborative filtering based recommendation systems across different domains and search modes*, ACM Transactions on Information Systems **26** (2007), no. 1, 4.

S. Iyengar, R. Wells, and B. Schwartz, *Doing better but feeling worse: Looking for the best job undermines satisfaction*, Psychological Science **17** (2006), no. 2, 143–150.

D. Jannach, *Advisor suite – a knowledge-based sales advisory system*, Proceedings of European Conference on Artificial Intelligence (Valencia, Spain) (R. Lopez de Mantaras and L. Saitta, eds.), IOS Press, 2004, pp. 720–724.

———, *Finding preferred query relaxations in content-based recommenders*, Proceedings of IEEE Intelligent Systems Conference (IS '2006) (Westminster, UK), IEEE Press, 2006a, pp. 355–360.

———, *Techniques for fast query relaxation in content-based recommender systems*, Proceedings of the 29th German Conference on Artificial Intelligence (KI '06) (Bremen, Germany) (C. Freksa, M. Kohlhase, and K. Schill, eds.), Lecture Notes in Artificial Intelligence, vol. 4314, Springer, 2006b, pp. 49–63.

D. Jannach and K. Hegelich, *A case study on the effectiveness of recommendations in the mobile internet*, Proceedings of the 2009 ACM Conference on Recommender Systems (RecSys '09) (New York), 2009, pp. 41–50.

D. Jannach, K. Shchekotykhin, and G. Friedrich, *Automated ontology instantiation from tabular web sources – the Allright system*, Web Semantics: Science, Services and Agents on the World Wide Web **7** (2009a), no. 3, 136–153.

D. Jannach, M. Zanker, and M. Fuchs, *Constraint-based recommendation in tourism: A multi-perspective case study*, Information Technology and Tourism **11** (2009b), no. 2, 139–156.

D. Jannach, M. Zanker, M. Jessenitschnig, and O. Seidler, *Developing a Conversational Travel Advisor with ADVISOR SUITE*, Information and Communication Technologies in Tourism (ENTER '07) (Ljubljana, Slovenia) (Marianna Sigala, Luisa Mich, and Jamie Murphy, eds.), Springer, 2007, pp. 43–52.

S. Jarvenpaa and P. Todd, *Consumer reactions to electronic shopping on the world wide web*, International Journal of Electronic Commerce **1** (1996), no. 2, 59–88.

S. Jarvenpaa, N. Tractinsky, and M. Vitale, *Consumer trust in an internet store*, Information Technology and Management **1** (2000), no. 1–2, 45–71.

R. Jäschke, L. Marinho, A. Hotho, L. Schmidt-Thieme, and G. Stumme, *Tag recommendations in folksonomies*, Knowledge Discovery in Databases: PKDD 2007 (Warsaw), Lecture Notes in Computer Science, vol. 4702, Springer, 2007, pp. 506–514.

Z. Jiang, W., and I. Benbasat, *Multimedia-based interactive advising technology for online consumer decision support*, Communications of the ACM **48** (2005), no. 9, 92–98.

R. Jin, L. Si, and C. Zhai, *A study of mixture models for collaborative filtering*, Information Retrieval **9** (2006), no. 3, 357–382.

T. Joachims, *Text categorization with support vector machines: learning with many relevant features*, Proceedings of the 10th European Conference on Machine Learning (ECML-98) (Chemnitz, Germany) (Claire Nédellec and Céline Rouveirol, eds.), no. 1398, Springer Verlag, Heidelberg, Germany, 1998, pp. 137–142.

T. Joachims, L. Granka, B. Pan, H. Hembrooke, and G. Gay, *Accurately interpreting clickthrough data as implicit feedback*, Proceedings of the 28th Annual International ACM SIGIR Conference (SIGIR '05) (Salvador, Brazil), ACM, 2005, pp. 154–161.

A. Jøsang, S. Marsh, and S. Pope, *Exploring different types of trust propagation*, Proceedings of the 4th International Conference on Trust Management (iTrust '06) (Pisa, Italy), Lecture Notes in Computer Science, vol. 3986, Springer, 2006, pp. 179–192.

U. Junker, *QUICKXPLAIN: Preferred explanations and relaxations for over-constrained problems*, Proceedings of the 19th National Conference on Artificial Intelligence (AAAI '04) (San Jose, CA), AAAI, 2004, pp. 167–172.

K. Kabassi, *Personalizing recommendations for tourists*, Telematics and Informatics **27** (2010), no. 1, 51–66.

S. Kaplan, H. Reneau, and S. Whitecotton, *The effects of predictive ability information, locus of control, and decision maker involvement on decision aid reliance*, Journal of Behavioral Decision Making **14** (2001), 35–50.

J. W. Kim, B. H. Lee, M. J. Shaw, H.-L. Chang, and M. Nelson, *Application of decision-tree induction techniques to personalized advertisements on internet storefronts*, International Journal of Electronic Commerce **5** (2001), no. 3, 45–62.

J. Koenemann and N. J. Belkin, *A case for interaction: a study of interactive information retrieval behavior and effectiveness*, Proceedings of the SIGCHI conference on Human Factors in Computing Systems (CHI '96) (Vancouver, BC), ACM, 1996, pp. 205–212.

J. Konstan, B. Miller, D. Maltz, J., L. Gordon, and J. Riedl, *Grouplens: applying collaborative filtering to usenet news*, Communications of the ACM **40** (1997), no. 3, 77–87.

I. Koprinska, J. Poon, J. Clark, and J. Chan, *Learning to classify e-mail*, Information Sciences **177** (2007), no. 10, 2167–2187.

Y. Koren, R. Bell, and C. Volinsky, *Matrix factorization techniques for recommender systems*, Computer **42** (2009), no. 8, 30–37.

R. Kramer, M. Modsching, and K. ten Hagen, *A city guide agent creating and adapting individual sightseeing tours based on field trial results*, International Journal of Computational Intelligence Research **2** (2006), no. 2, 191–206.

R. Krestel and P. Fankhauser, *Tag recommendation using probabilistic topic models*, ECML/PKDD Discovery Challenge (DC '09), Workshop at ECML/PKDD 2009)

(Bled, Slovenia) (Folke Eisterlehner, Andreas Hotho, and Robert Jäschke, eds.), September 2009, pp. 131–141.

R. Krestel, P. Fankhauser, and W. Nejdl, *Latent dirichlet allocation for tag recommendation*, Proceedings of the 2009 ACM Conference on Recommender Systems (RecSys '09) (New York), 2009, pp. 61–68.

V. Krishnan, P. K. Narayanashetty, M. Nathan, R. T. Davies, and J. A. Konstan, *Who predicts better?: Results from an online study comparing humans and an online recommender system*, Proceedings of the 2008 ACM Conference on Recommender Systems (RecSys '08) (Lausanne, Switzerland), ACM, 2008, pp. 211–218.

J. Krösche, J. Baldzer, and S. Boll, *MobiDENK – mobile multimedia in monument conservation*, IEEE Multimedia **11** (2004), no. 2, 72–77.

A. Kruglanski, D. Webster, and A. Klem, *Motivated resistance and openness to persuasion in the presence or absence of prior information*, Journal of Personality and Social Psychology **65** (1993), no. 5, 861–876.

B. Krulwich and C. Burkey, *The infofinder agent: Learning user interests through heuristic phrase extraction*, IEEE Expert: Intelligent Systems and Their Applications **12** (1997), no. 5, 22–27.

S. K. Lam and J. Riedl, *Shilling recommender systems for fun and profit*, Proceedings of the 13th International Conference on World Wide Web (WWW '04) (New York), ACM, 2004, pp. 393–402.

K. Lang, *Newsweeder: learning to filter netnews*, Proceedings of the 12th International Conference on Machine Learning (ICML '95) (Tahoe City, CA), 1995, pp. 331–339.

N. Lathia, S. Hailes, and L. Capra, *Private distributed collaborative filtering using estimated concordance measures*, Proceedings of the 2007 ACM Conference on Recommender Systems (RecSys '07) (Minneapolis), ACM, 2007, pp. 1–8.

B.-H. Lee, H.-N. Kim, J.-G. Jung, and G.-S. Jo, *Location-based service with context data for a restaurant recommendation*, Proceedings of the 17th International Conference on Database and Expert Systems Applications (DEXA '06) (Krakow, Poland), 2006, pp. 430–438.

H.-H. Lee and W.-G. Teng, *Incorporating multi-criteria ratings in recommendation systems*, Proceedings of IEEE International Conference on Information Reuse and Integration (IRI '07) (Las Vegas), 2007, pp. 273–278.

H. Lee, A. F. Smeaton, N. E. O'Connor, and B. Smyth, *User evaluation of físchlár-news: An automatic broadcast news delivery system*, ACM Transactions on Information Systems **24** (2006), no. 2, 145–189.

D. Lemire and A. Maclachlan, *Slope one predictors for online rating-based collaborative filtering*, Proceedings of the 5th SIAM International Conference on Data Mining (SDM '05) (Newport Beach, CA), 2005, pp. 471–480.

I. Levin, S. Schneider, and G. Gaeth, *All frames are not created equal: A typology and critical analysis of framing effects*, Organizational Behavior and Human Decision Processes **76** (1998), 90–98.

D. D. Lewis, R. E. Schapire, J. P. Callan, and R. Papka, *Training algorithms for linear text classifiers*, Proceedings of the 19th Annual International ACM SIGIR Conference on Research and Development in Information Retrieval (SIGIR '96) (Zurich, Switzerland), ACM, 1996, pp. 298–306.

Q. Li, C. Wang, and G. Geng, *Improving personalized services in mobile commerce by a novel multicriteria rating approach*, Proceedings of the 17th International Conference on World Wide Web (WWW '08) (Beijing), 2008, pp. 1235–1236.

T. Li, M. Ogihara, and Q. Li, *A comparative study on content-based music genre classification*, Proceedings of the 26th Annual International ACM SIGIR Conference on Research and Development in Information Retrieval (SIGIR '03) (Toronto), ACM, 2003, pp. 282–289.

W. Lin, *Association rule mining for collaborative recommender systems*, Master's thesis, Worcester Polytechnic Institute, May 2000.

W. Lin, S. A. Alvarez, and C. Ruiz, *Efficient adaptive-support association rule mining for recommender systems*, Data Mining and Knowledge Discovery **6** (2002), no. 1, 83–105.

G. Linden, B. Smith, and J. York, *Amazon.com recommendations: item-to-item collaborative filtering*, Internet Computing, IEEE **7** (2003), no. 1, 76–80.

C. Ling and C. Li, *Data mining for direct marketing: Problems and solutions*, Proceedings of the 4th International Conference on Knowledge Discovery and Data Mining (KDD '98) (New York), 1998, pp. 73–79.

B. Logan, *Music recommendation from song sets*, Proceedings of 5th International Conference on Music Information Retrieval (ISMIR '04) (Barcelona, Spain), 2004, pp. 425–428.

F. Lorenzi and F. Ricci, *Case-based recommender systems: A unifying view*, Intelligent Techniques for Web Personalisation, Lecture Notes in Artificial Intelligence, vol. 3169, Springer, 2005, pp. 89–113.

Z. Ma, G. Pant, and O. R. L. Sheng, *Interest-based personalized search*, ACM Transactions on Information Systems **25** (2007), no. 1, 5.

T. Mahmood and F. Ricci, *Learning and adaptivity in interactive recommender systems*, Proceedings of the 9th International Conference on Electronic Commerce (ICEC '07), ACM, 2007, pp. 75–84.

V. Maidel, P. Shoval, B. Shapira, and M. Taieb-Maimon, *Evaluation of an ontology-content based filtering method for a personalized newspaper*, Proceedings of the 2008 ACM Conference on Recommender Systems (RecSys '08) Lawsanne, Switzerland) (Pearl Pu, Derek Bridge, Bamshad Mobasher, and Francisco Ricci, eds.), ACM, 2008, pp. 91–98.

O. Maimon and L. Rokach (eds.), *The data mining and knowledge discovery handbook*, Springer, 2005.

N. Mandel and E. Johnson, *Constructing preferences online: can web pages change what you want?*, Unpublished manuscript (Wharton School, University of Pennsylvania), 1999.

C. D. Manning, P. Raghavan, and H. Schütze, *Introduction to information retrieval*, Cambridge University Press, 2008.

B. M. Marlin and R. S. Zemel, *Collaborative prediction and ranking with non-random missing data*, Proceedings of the 3rd ACM Conference on Recommender Systems (RecSys '09) (New York), ACM, 2009, pp. 5–12.

P. Massa and P. Avesani, *Trust-aware collaborative filtering for recommender systems*, Springer, Lecture Notes in Computer Science, vol. 3290, 2004, pp. 492–508.

———, *Trust-aware recommender systems*, Proceedings of the 2007 ACM Conference on Recommender Systems (RecSys '07) (Minneapolis, MN), ACM, 2007, pp. 17–24.

E. Maylor, *Serial position effects in semantic memory: reconstructing the order of verses of hymns*, Psychonomic Bulletin and Review **9** (2002), no. 4, 816–820.

D. Mayzlin, *Promotional chat on the internet*, Marketing Science **25** (2006), no. 2, 155–163.

A. McCallum and K. Nigam, *A comparison of event models for naive bayes text classification*, In AAAI-98 Workshop on Learning for Text Categorization (Madison, WI), 1998.

K. McCarthy, J. Reilly, B. Smyth, and L. McGinty, *Generating diverse compound critiques*, Artificial Intelligence Review **24** (2005), no. 3–4, 339–357.

K. McCarthy, M. Salamó, L. Coyle, L. McGinty, B. Smyth, and P. Nixon, *Group recommender systems: a critiquing based approach*, Proceedings of the 11th International Conference on Intelligent User Interfaces (IUI '06) (Sydney, Australia), ACM, 2006, pp. 267–269.

P. McCrae and P. Costa, *The neo personality inventory: Using the five-factor model in counseling*, Journal of Counseling and Development **69** (1991), 367–372.

D. McFadden, *Rationality for economists?*, Journal of Risk and Uncertainty **19** (1999), no. 1, 73–105.

L. McGinty and B. Smyth, *On the role of diversity in conversational recommender systems*, Proceedings of the International Conference on Case-Based Reasoning Research and Development (ICCBR '03), 2003, pp. 276–290.

M. R. McLaughlin and J. L. Herlocker, *A collaborative filtering algorithm and evaluation metric that accurately model the user experience*, Proceedings of the 27th Annual International ACM SIGIR Conference on Research and Development in Information Retrieval (SIGIR '04) (Sheffield, OK), ACM, 2004, pp. 329–336.

T. McNamara, *Theories of priming II: Types of primes*, Journal of Experimental Psychology: Learning, Memory, and Cognition **20** (1994), no. 3, 507–520.

S. M. McNee, J. Riedl, and J. A. Konstan, *Being accurate is not enough: how accuracy metrics have hurt recommender systems*, Extended Abstracts on Human Factors in Computing Systems (CHI '06) (Montréal), ACM, 2006, pp. 1097–1101.

D. McSherry, *Similarity and compromise*, Proceedings of the 5th International Conference on Case-Based Reasoning (ICCBR '03) (Trondheim, Norway), 2003a, pp. 291–305.

———, *Incremental relaxation of unsuccessful queries*, Proceedings of the European Conference on Case-based Reasoning (P. Funk and P. A. Gonzalez Calero, eds.), Lecture Notes in Artificial Intelligence, vol. 3155, Springer, 2004, pp. 331–345.

D. McSherry, *Similarity and compromise*, Proceedings of the 5th International Conference on Case-Based Reasoning (ICCBR '03) (Trondheim, Norway) (Kevin D. Ashley and Derek G. Bridge, eds.), LNCS, vol. 2689, Springer, June 2003b, pp. 291–305.

———, *Explanation in recommender systems*, Artificial Intelligence Review **24** (2005), no. 2, 179–197.

P. Melville, R. J. Mooney, and R. Nagarajan, *Content-Boosted Collaborative Filtering for Improved Recommendations*, Proceedings of the 18th National Conference on Artificial Intelligence (AAAI) (Edmonton, Alberta, Canada), 2002, pp. 187–192.

S. E. Middleton, N. R. Shadbolt, and D. C. De Roure, *Ontological user profiling in recommender systems*, ACM Transactions on Information Systems **22** (2004), no. 1, 54–88.

M. B. Miles and A. M. Huberman, *Qualitative data analysis – an expanded sourcebook*, 2nd ed., SAGE Publications, 1994.

B. N. Miller, I. Albert, S. K. Lam, J. A. Konstan, and J. Riedl, *MovieLens unplugged: experiences with an occasionally connected recommender system*, Proceedings of the 8th International Conference on Intelligent User Interfaces (IUI '03) (Miami, FL), 2003, pp. 263–266.

B. N. Miller, J. A. Konstan, and J. Riedl, *Pocketlens: Toward a personal recommender system*, ACM Transactions on Information Systems **22** (2004), no. 3, 437–476.

G. Mishne, *Autotag: a collaborative approach to automated tag assignment for weblog posts*, Proceedings of the 15th International Conference on World Wide Web (WWW '06) (Edinburgh, Scotland), ACM, 2006, pp. 953–954.

K. Miyahara and M. J. Pazzani, *Collaborative filtering with the simple bayesian classifier*, Pacific Rim International Conference on Artificial Intelligence (Melbourne, Australia), 2000, pp. 679–689.

D. Mladenic, *Personal webwatcher: Design and implementation*, Technical Report IJS-DP-7472 (Pittsburgh, PA), 1996.

D. Mladenic, *Text-learning and related intelligent agents: A survey*, IEEE Intelligent Systems **14** (1999), no. 4, 44–54.

B. Mobasher, R. Bhamik, and C. Williams, *Effective attack models for shilling item-based collaborative filtering systems*, Proceedings of the 2005 WebKDD Workshop, held in Conjuction with ACM SIGKDD '05 (Chicago, IL), 2005, pp. 13–23.

B. Mobasher, R. Burke, R. Bhaumik, and C. Williams, *Toward trustworthy recommender systems: An analysis of attack models and algorithm robustness*, ACM Transactions on Internet Technology **7** (2007), no. 4, 23.

B. Mobasher, R. D. Burke, and J. J. Sandvig, *Model-based collaborative filtering as a defense against profile injection attacks*, Proceedings of National Conference on Artificial Intelligence and the 18th Innovative Applications of Artificial Intelligence Conference (Boston), AAAI Press, 2006, pp. 1388–1393.

B. Mobasher, H. Dai, T. Luo, and M. Nakagawa, *Effective personalization based on association rule discovery from web usage data*, Proceedings of the 3rd International Workshop on Web Information and Data Management (WIDM '01) (Atlanta), ACM, 2001, pp. 9–15.

B. Mobasher, X. Jin, and Y. Zhou, *Semantically enhanced collaborative filtering on the web*, Web Mining: From Web to Semantic Web (B. Berendt et al., ed.), Lecture Notes in Computer Science, vol. 3209, Springer, 2004, pp. 57–76.

R. J. Mooney and L. Roy, *Content-based book recommending using learning for text categorization*, Proceedings of the Fifth ACM Conference on Digital Libraries, San Antonio, TX, pp. 195–204, June 2000.

J. Murphy, C. Hofacker, and R. Mizerski, *Primacy and recency effects on clicking behavior*, Journal of Computer-Mediated Communication **11** (2006), no. 2, 522–535.

T. Nakajima and I. Satoh, *A software infrastructure for supporting spontaneous and personalized interaction in home computing environments*, Personal Ubiquitous Computing **10** (2006), no. 6, 379–391.

T. Nathanson, E. Bitton, and K. Goldberg, *Eigentaste 5.0: constant-time adaptability in a recommender system using item clustering*, Proceedings of the 2007 ACM Conference on Recommender Systems (RecSys '07) (Minneapolis, MN), ACM, 2007, pp. 149–152.

A. Nauerz, B. König-Ries, and M. Welsch, *Recommending background information and related content in web 2.0 portals*, Proceedings Adaptive Hypermedia (Hannover, Germany), Lecture Notes in Computer Science, vol. 5149, Springer, 2008, pp. 366–69.

E. W. T. Ngai and A. Gunasekaran, *A review for mobile commerce research and applications*, Decision Support Systems **43** (2007), no. 1, 3–15.

Q. N. Nguyen and F. Ricci, *Acquiring and revising preference in a critique-based mobile recommender system*, IEEE Intelligent Systems **22** (2007), May/Jun, 22–29.

_____ , *Replaying live-user interactions in the off-line evaluation of critique-based mobile recommendations*, Proceedings of the 2007 ACM Conference on Recommender Systems (RecSys '07) (Minneapolis, MN), ACM, 2007, pp. 81–88.

_____ , *Long-term and session-specific user preferences in a mobile recommender system*, Proceedings of the 13th International Conference on Intelligent User Interfaces (IUI '08) (Gran Canaria, Spain), 2008, pp. 381–384.

D. Nichols, *Implicit rating and filtering*, Proceedings of 5th DELOS Workshop on Filtering and Collaborative Filtering (Budapest), ERCIM, 1998, pp. 31–36.

K. Nigam, A. K. McCallum, S. Thrun, and T. M. Mitchell, *Learning to classify text from labeled and unlabeled documents*, Proceedings of the 15th Conference of the American Association for Artificial Intelligence (AAAI '98) (Madison, WI), 1998, pp. 792–799.

D. Nikovski and V. Kulev, *Induction of compact decision trees for personalized recommendation*, Proceedings of the 2006 ACM Symposium on Applied Computing (SAC '06) (Dijon, France), 2006, pp. 575–581.

K. Oatley and J. Jenkins, *Understanding emotions*, Blackwell, 1996.

J. O'Donovan and B. Smyth, *Trust in recommender systems*, Proceedings of the 10th International Conference on Intelligent User Interfaces (IUI '05) (San Diego, CA), ACM, 2005, pp. 167–174.

M. O'Mahony, N. Hurley, N. Kushmerick, and G. Silvestre, *Collaborative recommendation: A robustness analysis*, ACM Transactions on Internet Technology **4** (2004), no. 4, 344–377.

M. P. O'Mahony, N. J. Hurley, and G. C. M. Silvestre, *Recommender systems: Attack types and strategies*, Proceedings of the 20th National Conference on Artificial Intelligence and the 17th Innovative Applications of Artificial Intelligence Conference (Pittsburgh, PA) (Manuela M. Veloso and Subbarao Kambhampati, eds.), July 2005, pp. 334–339.

T. O'Reilly, *What is Web 2.0: Design patterns and business models for the next generation of software*, Communictions and Strategies, International Journal of Digital Economics **65** (2007), 17–37.

B. O'Sullivan, A. Papadopoulos, B. Faltings, and P. Pu, *Representative explanations for over-constrained problems*, Proceedings of the 22nd National Conference on Artificial Intelligence (AAAI '07), Vancouver, British Columbia, Canada, 2007, pp. 323–328.

D. O'Sullivan, B. Smyth, D. C. Wilson, K. McDonald, and A. Smeaton, *Improving the quality of the personalized electronic program guide*, User Modeling and User-Adapted Interaction **14** (2004), no. 1, 5–36.

M. Papagelis, D. Plexousakis, and T. Kutsuras, *Alleviating the sparsity problem of collaborative filtering using trust inferences*, Proceedings of the 3rd International

Conference on Trust Management (iTrust '05) (Paris), Lecture Notes in Computer Science, vol. 3477, Springer, 2005, pp. 224–239.

W. Parrot, *Emotions in social psychology*, Taylor and Francis, 2001.

J. Payne, J. Bettman, and E. Johnson, *The adaptive decision maker*, Cambridge University Press, 1993.

M. Pazzani, *A framework for collaborative, content-based and demographic filtering*, Artificial Intelligence Review **13** (1999), no. 5–6, 393–408.

M. J. Pazzani, *A framework for collaborative, content-based and demographic filtering*, Artificial Intelligence Review **13** (1999), no. 5–6, 393–408.

———, *Commercial applications of machine learning for personalized wireless portals*, Proceedings of the 7th Pacific Rim International Conference on Artificial Intelligence (PRICAI '02) (Tokyo), 2002, pp. 1–5.

M. Pazzani, J. Muramatsu, and D. Billsus, *Syskill & Webert: Identifying interesting web sites*, Proceedings of the 13th National Conference on Artificial Intelligence (Portland, OR), 1996, pp. 54–61.

M. Pazzani and D. Billsus, *Learning and revising user profiles: The identification of interesting web sites*, Machine Learning **27** (1997), no. 3, 313–331.

M. J. Pazzani and D. Billsus, *Content-based recommendation systems*, The Adaptive Web (Peter Brusilovsky, Alfred Kobsa, and Wolfgang Nejdl, eds.), Lecture Notes in Computer Science, vol. 4321, Springer, 2007, pp. 325–341.

E. J. Pedhazur and L. P. Schmelkin, *Measurement, design and analysis: An integrated approach*, Lawrence Erlbaum Associates, 1991.

D. Pennock, E. Horvitz, S. Lawrence, and C. L. Giles, *Collaborative filtering by personality diagnosis: A hybrid memory- and model-based approach*, Proceedings of the 16th Conference on Uncertainty in Artificial Intelligence (UAI '00) (Stanford, CA), 2000, pp. 473–480.

R. Picard, *Affective computing*, MIT Press, Cambridge, 1997.

R. Plutchik and R. Hope, *Circumplex models of personality and emotions*, American Psychological Association, 1997.

H. Polat and W. Du, *Privacy-preserving collaborative filtering using randomized perturbation techniques*, Third IEEE International Conference on Data Mining (ICDM '03) (Melbourne, FL), November 2003, pp. 625–628.

H. Polat and W. Du, *SVD-based collaborative filtering with privacy*, Proceedings of the 2005 ACM Symposium on Applied Computing (SAC '05) (Santa Fe, NM), ACM, 2005, pp. 791–795.

M. F. Porter, *An algorithm for suffix stripping*, Program **14** (1980), no. 3, 130–137.

P. Pu and L. Chen, *Trust-inspiring explanation interfaces for recommender systems*, Knowledge-based Systems **20** (2007), no. 6, 542–556.

P. Pu, D. G. Bridge, B. Mobasher, and F. Ricci (eds.), *Proceedings of the 2008 ACM Conference on Recommender Systems (RecSys '08)*, Lausanne, Switzerland, ACM, October 2008.

P. Pu, L. Chen, and P. Kumar, *Evaluating product search and recommender systems for e-commerce environments*, Electronic Commerce Research **8** (2008), no. 1–2, 1–27.

P. Pu, B. Faltings, and P. Kumar, *User-involved tradeoff analysis in configuration tasks*, The 3rd International Workshop on User-Interaction in Constraint Satisfaction at CP'2003 (Kinsale, Ireland), 2003, pp. 85–102.

J. R. Quinlan, *C4.5: programs for machine learning*, Morgan Kaufmann, San Francisco, 1993.

M. Ramezani, L. Bergman, R. Thompson, R. Burke, and B. Mobasher, *Selecting and applying recommendation technology*, Proceedings of International Workshop on Recommendation and Collaboration, in conjunction with the 2008 International ACM Conference on Intelligent User Interfaces (IUI '08) (Gran Canaria, Spain), ACM, January 2008.

A. Ranganathan and R. H. Campbell, *An infrastructure for context-awareness based on first order logic*, Personal Ubiquitous Computing **7** (2003), no. 6, 353–364.

A. Rashid, I. Albert, D. Cosley, S. Lam, S. McNee, J. Konstan, and J. Riedl, *Getting to know you: Learning new user preferences in recommender systems*, Proceedings of the 7th International Conference on Intelligent User Interfaces (IUI '02) (San Francisco), ACM, 2002, pp. 127–134.

J. Rasinger, M. Fuchs, W. Höpken, and T. Beer, *Information search with mobile tourist guides: A survey of usage intention*, Information Technology and Tourism **9** (2007), no. 3/4, 177–194.

J. Reilly, K. McCarthy, L. McGinty, and B. Smyth, *Dynamic critiquing*, Proceedings of the 7th European Conference on Case-Based Reasoning (ECCBR '04), 2004, pp. 763–777.

———, *Incremental critiquing*, Knowledge-Based Systems **18** (2005), no. 4–5, 143–151.

J. Reilly, J. Zhang, L. McGinty, P. Pu, and B. Smyth, *A comparison of two compound critiquing systems*, Proceedings of the 12th International Conference on Intelligent User Interfaces (IUI '07) (Honolulu), ACM, 2007, pp. 317–320.

———, *Evaluating compound critiquing recommenders: a real-user study*, Proceedings of the 8th ACM Conference on Electronic Commerce (EC '07) (San Diego, CA), ACM, 2007, pp. 114–123.

J. Reilly, K. McCarthy, L. McGinty, and B. Smyth, *Explaining compound critiques*, Artificial Intelligence Review **24** (2005), no. 2, 199–220.

R. Reiter, *A theory of diagnosis from first principles*, Artificial Intelligence **32** (1987), no. 1, 57–95.

P. Resnick, N. Iacovou, M. Suchak, P. Bergstorm, and J. Riedl, *Grouplens: An open architecture for collaborative filtering of netnews*, Proceedings of the 1994 ACM Conference on Computer Supported Cooperative Work (CSCW'94) (Chapel Hill, NC), ACM, 1994, pp. 175–186.

P. Resnick and H. R. Varian, *Recommender systems*, Communications of the ACM **40** (1997), no. 3, 56–58.

F. Ricci, L. Rokach, B. Shapira, and P. B. Kantor (eds.), *Recommender Systems Handbook*, Springer, 2010.

F. Ricci and Q. Nguyen, *Acquiring and revising preferences in a critique-based mobile recommender system*, IEEE Intelligent Systems **22** (2007), no. 3, 22–29.

M. Richardson, R. Agrawal, and P. Domingos, *Trust management for the semantic web*, The SemanticWeb – ISWC 2003 (Sanibel Island, FL), Lecture Notes in Computer Science, vol. 2870, Springer, 2003, pp. 351–368.

I. Ritov and J. Baron, *Status-quo and omission biases*, Journal of Risk and Uncertainty **5** (1992), no. 2, 49–61.

R. Roe, J. Busemeyer, and T. Townsend, *Multialternative decision field theory: A dynamic connectionist model of decision making*, Psychological Review **108** (2001), no. 2, 370–392.

M. Salamo, J. Reilly, L. McGinty, and B. Smyth, *Knowledge discovery from user preferences in conversational recommendation*, Proceedings of the 9th European Conference on Principles and Practice of Knowledge Discovery in Databases (PKDD '05) (Porto, Portugal), Lecture Notes in Computer Science, vol. 3721, Springer, 2005, pp. 228–239.

G. Salton, *The SMART retrieval system – experiments in automatic document processing*, Prentice-Hall, 1971.

G. Salton, A. Wong, and C.S. Yang, *A vector space model for information retrieval*, Journal of the American Society for Information Science **18** (1975), no. 11, 613–620.

G. Salton and C. Buckley, *Improving retrieval performance by relevance feedback*, pp. 355–364, Morgan Kaufmann, San Francisco, 1997.

G. Salton and C. Buckley, *Term-weighting approaches in automatic text retrieval*, Information Processing and Management **24** (1988), no. 5, 513–523.

W. Samuelson and R. Zeckhauser, *Status quo bias in decision making*, Journal of Risk and Uncertainty **108** (1988), no. 2, 370–392.

J. J. Sandvig, B. Mobasher, and R. Burke, *Robustness of collaborative recommendation based on association rule mining*, Proceedings of the 2007 ACM Conference on Recommender Systems (RecSys '07) (Minneapolis, MN), ACM, 2007, pp. 105–112.

J. J. Sandvig, B. Mobasher, and R. Burke, *A survey of collaborative recommendation and the robustness of model-based algorithms*, IEEE Data Engineering Bulletin **31** (2008), no. 2, 3–13.

B. Sarwar, G. Karypis, J. Konstan, and J. Riedl, *Application of dimensionality reduction in recommender systems – a case study*, Proceedings of the ACM WebKDD Workshop (Boston), 2000.

B. M. Sarwar, G. Karypis, J. A. Konstan, and J. Riedl, *Incremental singular value decomposition algorithms for highly scalable recommender systems*, Proceedings of the 5th International Conference on Computer and Information Technology (ICCIT '02), 2002, pp. 399–404.

B. Sarwar, G. Karypis, J. Konstan, and J. Riedl, *Analysis of recommendation algorithms for e-commerce*, Proceedings of the 2nd ACM Conference on Electronic Commerce (EC '00) (Minneapolis, MN), ACM, 2000, pp. 158–167.

————, *Item-based collaborative filtering recommendation algorithms*, Proceedings of the 10th International Conference on World Wide Web (WWW '01) (Hong Kong), ACM, 2001, pp. 285–295.

K. Satoh, A. Inokuchi, K. Nagao, and T. Kawamura (eds.), *New Frontiers in Artificial Intelligence, JSAI 2007 Conference and Workshops, Revised Selected Papers*, Lecture Notes in Computer Science, vol. 4914, Miyazaki, Japan, Springer, June 2007.

J. B. Schafer, D. Frankowski, J. Herlocker, and S. Sen, *Collaborative filtering recommender systems*, The Adaptive Web: Methods and Strategies of Web Personalization (P. Brusilovsky, A. Kobsa, and W. Nejdl, eds.), Lecture Notes in Computer Science, vol. 4321, Springer, Berlin Heidelberg, 2006, pp. 291–324.

B. Schwartz, A. Ward, J. Monterosso, S. Lyubomirsky, W. White, K. White, and R. Lehman, *Maximizing versus satisficing: Happiness is a matter of choice*, Journal of Personality and Social Psychology **83** (2002), no. 5, 1178–1197.

W. Schwinger, C. Grün, B. Pröll, W. Retschitzegger, and A. Schauerhuber, *Context-awareness in mobile tourism guides*, Tech. Report TR-05-04, JKU Linz, Austria, Institute of Bioinformatics, 2005.

N. Sebe and Q. Tian, *Personalized multimedia retrieval: the new trend?*, Proceedings of the International Workshop on Multimedia Information Retrieval (MIR '07) (Augsburg, Bavaria, Germany), ACM, 2007, pp. 299–306.

G. Semeraro, M. Degemmis, P. Lops, and P. Basile, *Combining learning and word sense disambiguation for intelligent user profiling*, Proceedings of the 20th International Joint Conference on Artifical Intelligence (IJCAI '07) (Hyderabad, India) (Manuela M. Veloso, ed.), January 2007, pp. 2856–2861.

S. Sen, J. Vig, and J. Riedl, *Tagommenders: connecting users to items through tags*, Proceedings of the 18th International Conference on the World Wide Web (WWW '09) (Madrid) (Juan Quemada, Gonzalo León, Yoëlle S. Maarek, and Wolfgang Nejdl, eds.), 2009, pp. 671–680.

S. Senecal and J. Nantel, *The influence of online product recommendations on consumers' online choices*, Journal of Retailing **80** (2004), no. 2, 159–169.

A. Seth, J. Zhang, and R. Cohen, *A subjective credibility model for participatory media*, Workshop Intelligent Techniques for Web Personalization and Recommender Systems (ITWP) at AAAI '08 (Chicago), AAAI Press, 2008, pp. 66–77.

R. Shacham, H. Schulzrinne, S. Thakolsri, and W. Kellerer, *Ubiquitous device personalization and use: The next generation of IP multimedia communications*, ACM Transactions on Multimedia Computing and Communication Applications **3** (2007), no. 2, 12.

G. Shani, R. I. Brafman, and D. Heckerman, *An MDP-based recommender system*, Journal of Machine Learning Research **6** (2002), 453–460.

G. Shani, D. M. Chickering, and C. Meek, *Mining recommendations from the web*, Proceedings of the 2008 ACM Conference on Recommender Systems (Lawrence, Switzerland) (Pearl Pu, Derek G. Bridge, Bamshad Mobasher, and Francesco Ricci, eds.), ACM, 2008, pp. 35–42.

U. Shardanand and P. Maes, *Social information filtering: Algorithms for automating "word of mouth"*, Proceedings of the ACM Conference on Human Factors in Computing Systems (CHI '95), vol. 1, 1995, pp. 210–217.

K. Shchekotykhin, D. Jannach, G. Friedrich, and O. Kozeruk, *Allright: Automatic ontology instantiation from tabular web documents*, Proceedings of the 6th International Conference on Semantic Web Conference (ISWC '07) (Busan, South Korea), 2007, pp. 466–479.

J. Shen, B. Cui, J. Shepherd, and K.-L. Tan, *Towards efficient automated singer identification in large music databases*, Proceedings of the 29th Annual International ACM SIGIR Conference on Research and Development in Information Retrieval (SIGIR '06) (Seattle), ACM, 2006, pp. 59–66.

A. Shepitsen, J. Gemmell, B. Mobasher, and R. D. Burke, *Personalized recommendation in social tagging systems using hierarchical clustering*, Proceedings of the 2008 ACM Conference on Recommender Systems (Lawrence, Switzerland) (Pearl Pu, Derek G. Bridge, Bamshad Mobasher, and Francesco Ricci, eds.), ACM, 2008, pp. 259–266.

B. Shilit, N. Adams, and R. Want, *Context-aware computing applications*, Proceedings of the 1994 First Workshop on Mobile Computing Systems and Applications (WMCSA '94) (Santa Cruz, CA) (Maria Sigala et al., ed.), IEEE Computer Society, 1994, pp. 85–90.

H. Shimazu, *Expertclerk: A conversational case-based reasoning tool for developing salesclerk agents in e-commerce webshops*, Artificial Intelligence Review **18** (2002), no. 3–4, 223–244.

E. H. Shortliffe, *A rule-based computer program for advising physicians regarding antimicrobial therapy selection*, Proceedings of the 1974 annual ACM conference (ACM'74) (New York), ACM, 1974, p. 739.

M.-L. Shyu, C. Haruechaiyasak, S.-C. Chen, and N. Zhao, *Collaborative filtering by mining association rules from user access sequences*, Proceedings of the International Workshop on Challenges in Web Information Retrieval and Integration (WIRI '05) (Washington, DC), IEEE Computer Society, 2005, pp. 128–135.

B. Sigurbjörnsson and R. van Zwol, *Flickr tag recommendation based on collective knowledge*, Proceeding of the 17th International Conference on World Wide Web (WWW '08) (Beijing), ACM, 2008, pp. 327–336.

H. Simon, *A behavioral model of choice*, Quarterly Journal of Economics **69** (1955), no. 1, 99–118.

I. Simonson and A. Tversky, *Choice in context: Tradeoff contrast and extremeness aversion*, Journal of Marketing Research **29**, (1992), no. 3, 281–295.

B. Smyth and P. Cotter, *Personalized adaptive navigation for mobile portals*, Proceedings of the 15th European Conference on Artificial Intelligence (ECAI '02) (Lyon, France), 2002, pp. 608–612.

B. Smyth, P. Cotter, and S. Oman, *Enabling intelligent content discovery on the mobile internet*, Proceedings of the 19th National Conference on Innovative Applications of Artificial Intelligence (IAAI '07) (Vancouver, BC), 2007, pp. 1744–1751.

F. Sørmo, J. Cassens, and A. Aamodt, *Explanation in case-based reasoning-perspectives and goals*, Artificial Intelligence Review **24** (2005), no. 2, 109–143.

S. Staab, H. Werthner, F. Ricci, A. Zipf, U. Gretzel, D. R. Fesenmaier, C. Paris, and C. A. Knoblock, *Intelligent systems for tourism*, IEEE Intelligent Systems **17** (2002), no. 6, 53–64.

R. E. Stake, *The art of case study research*, SAGE Publications, 1995.

X.-F. Su, H.-J. Zeng, and Z. Chen, *Finding group shilling in recommendation system*, Special Interest Tracks and Posters of the 14th International Conference on World Wide Web (WWW '05) (Chiba, Japan), ACM, 2005, pp. 960–961.

K. Swearingen and R. Sinha, *Beyond algorithms: An HCI perspective on recommender systems*, Workshop on Recommender Systems held in Conjunction with SIGIR '01 (New Orleans), 2001.

M. Szomszor, C. Cattuto, H. Alani, K. O'Hara, A. Baldassarri, V. Loreto, and V. D.P. Servedio, *Folksonomies, the semantic web, and movie recommendation*, Workshop on Bridging the Gap between Semantic Web and Web 2.0 at ESWC 2007 (Innsbruck, Austria), 2007, pp. 71–84.

P.-N. Tan, M. Steinbach, V. Kumar, *Introduction to data mining*, Addison Wesley, 2006.

P. Tarasewich, *Designing mobile commerce applications*, Communications of the ACM **46** (2003), no. 12, 57–60.

E. Teppan and A. Felfernig, *Minimization of product utility estimation errors in recommender result set evaluations*, Proceedings of the 2009 IEEE/WIC/ACM International Joint Conference on Web Intelligence and Intelligent Agent Technology (WI-IAT '09) (Milano, Italy), 2009a, pp. 20–27.

E. Teppan and A. Felfernig, *Asymmetric dominance and compromise effects in the financial services domain*, Proceedings of the 11th IEEE International Conference on Commerce and Enterprise Computing (CEC '09) (Vienna, Austria), 2009b, pp. 57–64.

C. Thompson, M. Göker, and P. Langley, *A personalized system for conversational recommendations*, Journal of Artificial Intelligence Research **21** (2004), 393–428.

T. Thompson and Y. Yeong, *Assessing the consumer decision process in the digital marketplace*, Omega **31** (2003), no. 5, 349–363.

J. Tiihonen and A. Felfernig, *Towards recommending configurable offerings*, Workshop on Configuration Systems (ECAI '08) (Patras, Greece), 2008, pp. 29–34.

N. Tintarev, *Explanations of recommendations*, Proceedings of the 2007 ACM Conference on Recommender Systems (RecSys '07) (Minneapolis, MN), ACM, 2007, pp. 203–206.

N. Tintarev and J. Masthoff, *Effective explanations of recommendations: user-centered design*, Proceedings of the 2007 ACM Conference on Recommender Systems (RecSys '07) (Minneapolis, MN), ACM, 2007, pp. 153–156.

G. Torkzadeh and G. Dhillon, *Measuring factors that influence the success of internet commerce*, Information Systems Research **13** (2002), no. 2, 187–204.

R. Torres, S. M. McNee, M. Abel, J. A. Konstan, and J. Riedl, *Enhancing digital libraries with Techlens*, International Joint Conference on Digital Libraries (JCDL '04) (Tucson, AZ), 2004, pp. 228–236.

A. Töscher, M. Jahrer, and R. Legenstein, *Improved neighborhood-based algorithms for large-scale recommender systems*, Proceedings of the ACM SIGKDD Workshop on Large Scale Recommenders Systems and the Netflix Prize at KDD '08 (Las Vegas), 2008.

E. Tsang, *Foundations of constraint satisfaction*, Academic Press, London and San Diego, 1993.

K. H. L. Tso-Sutter, L. B. Marinho, and L. Schmidt-Thieme, *Tag-aware recommender systems by fusion of collaborative filtering algorithms*, Proceedings of the 2008 ACM Symposium on Applied Computing (SAC '08) (Fortaleza, Ceara, Brazil), ACM, 2008, pp. 1995–1999.

A. Tveit, *Peer-to-peer based recommendations for mobile commerce*, Proceedings of the 1st International Workshop on Mobile Commerce (WMC '01) (Rome, Italy), ACM, July 2001, pp. 26–29.

A. Tversky and D. Kahneman, *Choices, values, and frames*, American Psychologist **39** (1984), no. 4, 341–350.

―――――, *Rational choice and the framing of decisions*, Journal of Business **59** (1986), no. 4, 251–278.

S. Uhlmann and A. Lugmayr, *Personalization algorithms for portable personality*, Proceedings of the 12th International Conference on Entertainment and Media in the Ubiquitous Era (MindTrek '08) (Tampere, Finland), ACM, 2008, pp. 117–121.

L. Ungar and D. Foster, *Clustering methods for collaborative filtering*, Proceedings of the Workshop on Recommendation Systems, AAAI Technical Report WS-98-08, AAAI Press, Menlo Park, CA, 1998, pp. 114–129.

H. van der Heijden, G. Kotsis, and R. Kronsteiner, *Mobile recommendation systems for decision making "on the go"*, Proceedings of the International Conference on Mobile Business (ICMB '05) (Sydney, Australia), 2005, pp. 137–143.

M. van Setten, *Supporting people in finding information: Hybrid recommender systems and goal-based structuring*, PhD thesis, Telematica Instituut, University of Twente, The Netherlands, 2005.

M. van Setten, S. Pokraev, and J. Koolwaaij, *Context-aware recommendations in the mobile tourist application compass*, Proceedings of the 3rd International Conference of Adaptive Hypermedia and Adaptive Web-Based Systems (AH '04) (Eindhoven, the Netherlands), Springer, 2004, pp. 235–244.

P. Victor, C. Cornelis, and M. D. Cock, *Enhanced recommendations through propagation of trust and distrust*, Proceedings of the 2006 IEEE/WIC/ACM International Conference on Web Intelligence and Intelligent Agent Technology (WI-IAT '06) (Hong Kong), 2006, pp. 263–266.

P. Victor, C. Cornelis, M. D. Cock, and A. M. Teredesai, *Key figure impact in trust-enhanced recommender systems*, AI Communications **21** (2008a), no. 2, 127–143.

P. Victor, C. Cornelis, A. M. Teredesai, and M. D. Cock, *Whom should I trust?: the impact of key figures on cold start recommendations*, Proceedings of the 2008 ACM Symposium on Applied Computing (SAC '08) (Fortaleza, Ceara, Brazil), ACM, 2008b, pp. 2014–2018.

V. Vlahakis, N. Ioannidis, J. Karigiannis, M. Tsotros, M. Gounaris, D. Stricker, T. Gleue, P. Daehne, and L. Almeida, *Archeoguide: An augmented reality guide for archaeological sites*, IEEE Computer Graphics and Applications **22** (2002), no. 5, 52–60.

L. V. Ahn, M. Blum, and J. Langford, *Captcha: Using hard AI problems for security*, in Proceedings of Eurocrypt '2003 (Warsaw, Poland), Springer, 2003, pp. 294–311.

J. Wang, A. P. de Vries, and M. J. T. Reinders, *Unifying user-based and item-based collaborative filtering approaches by similarity fusion*, Proceedings of the 29th Annual International ACM SIGIR Conference on Research and Development in Information Retrieval (SIGIR '06) (Seattle), ACM, 2006, pp. 501–508.

⸻ , *Unified relevance models for rating prediction in collaborative filtering*, ACM Transactions on Information Systems **26** (2008), no. 3, 1–42.

A. M. A. Wasfi, *Collecting user access patterns for building user profiles and collaborative filtering*, Proceedings of the 4th International Conference on Intelligent User Interfaces (IUI '99) (Los Angeles), ACM Press, 1999, pp. 57–64.

S. Wasserman and K. Faust, *Social network analysis: Methods and applications*, Cambridge University Press, 1994.

Y. Z. Wei, L. Moreau, and N. R. Jennings, *A market-based approach to recommender systems*, ACM Transactions on Information Systems **23** (2005), no. 3, 227–266.

B. Weiner, *Attributional thoughts about consumer behavior*, Journal of Consumer Research **27** (2000), no. 3, 382–387.

J. Weng, C. Miao, and A. Goh, *Improving collaborative filtering with trust-based metrics*, Proceedings of the 2006 ACM Symposium on Applied Computing (SAC '06) (Dijon, France), ACM, 2006, pp. 1860–1864.

B. Widrow and S. D. Stearns, *Adaptive signal processing*, Prentice-Hall, 1985.

Wikipedia, *Slope one – Wikipedia, the free encyclopedia*, 2008, (Online; accessed June 4, 2008).

D. Wilson and T. Martinez, *Improved heterogeneous distance functions*, Journal of Artificial Intelligence Research **6** (1997), 1–34.

D. Winterfeldt and W. Edwards, *Decision analysis and behavioral research*, Cambridge University Press, 1986.

Workshop on collaborative filtering, online, March 1996, http://www2.sims.berkeley.edu/resources/collab/conferences/berkeley96/agenda.html.

B. Xiao and I. Benbasat, *E-commerce product recommendation agents: Use, characteristics, and impact*, MIS Quarterly **31** (2007), no. 1, 137–209.

Z. Xu, Y. Fu, J. Mao, and D. Su, *Towards the semantic web: Collaborative tag suggestions*, Proceedings of the Collaborative Web Tagging Workshop at the 15th International World Wide Web Conference (WWW '06) (Edinburgh, Scotland), 2006.

G.-R. Xue, C. Lin, Q. Yang, W. Xi, H.-J. Zeng, Y. Yu, and Z. Chen, *Scalable collaborative filtering using cluster-based smoothing*, Proceedings of the 28th Annual International ACM SIGIR Conference on Research and Development in Information Retrieval (SIGIR '05) (Salvador, Brazil), ACM, 2005, pp. 114–121.

I. Yakut and H. Polat, *Privacy-preserving eigentaste-based collaborative filtering*, Proceedings of the International Workshop on Security (IWSEC '07) (Nara, Japan) (Atsuko Miyaji, Hiroaki Kikuchi, and Kai Rannenberg, eds.), Lecture Notes in Computer Science, vol. 4752, Springer, October 2007, pp. 169–184.

Y. Yang and X. Liu, *A re-examination of text categorization methods*, Proceedings of the 22nd Annual International ACM SIGIR Conference on Research and Development in Information Retrieval (SIGIR '99) (Berkley, CA), ACM, August 1999, pp. 42–49.

Y. Yi, *The effects of contextual priming in print advertisements*, Journal of Consumer Research **17** (1990), no. 2, 215–222.

R. K. Yin, *Case study research – design and methods*, 3rd ed., SAGE Publications, 2002.

S. Yoon and I. Simonson, *Choice set configuration as a determinant of preference attribution and strength*, Journal of Consumer Research **35** (2008), no. 2, 324–336.

K. Yoshii, M. Goto, K. Komatani, T. Ogata, and H. G. Okuno, *Hybrid collaborative and content-based music recommendation using probabilistic model with latent user preferences*, Proceedings of 7th International Conference on Music Information Retrieval (ISMIR '06) (Victoria, BC, Canada), 2006, pp. 296–301.

K. Yu, A. Schwaighofer, V. Tresp, X. Xu, and H.-P. Kriegel, *Probabilistic memory-based collaborative filtering*, IEEE Transactions on Knowledge and Data Engineering **16** (2004), no. 1, 56–69.

K. Yu, X. Xu, M. Ester, and H.-P. Kriegel, *Feature weighting and instance selection for collaborative filtering: An information-theoretic approach*, Knowledge Information Systems **5** (2003), no. 2, 201–224.

V. Zanardi and L. Capra, *Social ranking: Uncovering relevant content using tag-based recommender systems*, Proceedings of the 2008 ACM Conference on Recommender Systems (RecSys '08) (Lausanne, Switzerland), ACM Press, 2008, pp. 51–58.

M. Zanker, *A collaborative constraint-based meta-level recommender*, Proceedings of the 2008 ACM Conference on Recommender Systems (RecSys '08) (Lausanne, Switzerland), ACM Press, 2008, pp. 139–146.

M. Zanker, M. Aschinger, and M. Jessenitschnig, *Development of a collaborative and constraint-based web configuration system for personalized bundling of products and services*, Proceedings of the 8th International Conference on Web Information Systems Engineering (WISE '07) (Nancy, France), Springer, 2007, pp. 273–284.

M. Zanker, M. Bricman, S. Gordea, D. Jannach, and M. Jessenitschnig, *Persuasive online-selling in quality & taste domains*, Proceedings of the 7th International Conference on Electronic Commerce and Web Technologies (EC-Web '06) (Krakow, Poland), Springer, 2006, pp. 51–60.

M. Zanker, M. Fuchs, W. Höpken, M. Tuta, and N. Müller, *Evaluating Recommender Systems in Tourism – A Case Study from Austria*, Proceedings of the International Conference on Information and Communication Technologies in Tourism (ENTER) (Innsbruck, Austria), 2008, pp. 24–34.

M. Zanker, M. Jessenitschnig, and M. Fuchs, *Automated semantic annotations of tourism resources based on geospatial data*, Information Technology & Tourism, **11** (2009), no. 4, 341–354.

M. Zanker, S. Gordea, M. Jessenitschnig, and M. Schnabl, *A hybrid similarity concept for browsing semi-structured product items*, E-Commerce and Web Technologies, 7th International Conference (EC-Web 2006) (Krakow, Poland) (Kurt Bauknecht, Birgit Pröll, and Hannes Werthner, eds.), Lecture Notes in Computer Science, vol. 4082, Springer, 2006, pp. 21–30.

M. Zanker and M. Jessenitschnig, *Case-studies on exploiting explicit customer requirements in recommender systems*, User Modeling and User-Adapted Interaction **19** (2009), no. 1–2, 133–166.

M. Zanker and M. Jessenitschnig, *Collaborative feature-combination recommender exploiting explicit and implicit user feedback*, Proceedings of the 2009 IEEE Conference on Commerce and Enterprise Computing (CEC '09) (Vienna), IEEE Computer Society, 2009, pp. 49–56.

M. Zanker, M. Jessenitschnig, and W. Schmid, *Preference Reasoning with Soft Constraints in Constraint-Based Recommender Systems, Constraints*, Springer, **15** (2010), no. 4, 574–595.

M. Zanker, M. Jessenitschnig, D. Jannach, and S. Gordea, *Comparing recommendation strategies in a commercial context*, IEEE Intelligent Systems **22** (2007), no. 3, 69–73.

P. Zezula, G. Amato, V. Dohnal, and M. Batko, *Similarity search – the metric space approach*, Advances in Database Systems, vol. 32, Springer, 2006.

J. Zhang and P. Pu, *A recursive prediction algorithm for collaborative filtering recommender systems*, Proceedings of the 2007 ACM Conference on Recommender Systems (RecSys '07) (Minneapolis, MN), ACM, 2007, pp. 57–64.

S. Zhang, A. Chakrabarti, J. Ford, and F. Makedon, *Attack detection in time series for recommender systems*, Proceedings of the 12th ACM SIGKDD International Conference on Knowledge Discovery and Data Mining (KDD '06) (Philadelphia), ACM, 2006, pp. 809–814.

S. Zhang, J. Ford, and F. Makedon, *Deriving private information from randomly perturbed ratings*, Proceedings of the 6th SIAM International Conference on Data Mining (SDM '06) (Bethesda, MD) (Joydeep Ghosh, Diane Lambert, David B. Skillicorn, and Jaideep Srivastava, eds.), SIAM, April 2006, pp. 59–69.

————, *A privacy-preserving collaborative filtering scheme with two-way communication*, Proceedings of the 7th ACM Conference on Electronic Commerce (EC '06) (Ann Arbor, MI), ACM, 2006, pp. 316–323.

Y. Zhang, J. Callan, and T. Minka, *Novelty and redundancy detection in adaptive filtering*, Proceedings of the 25th Annual International ACM SIGIR Conference on Research and Development in Information Retrieval (SIGIR '02) (Tampere, Finland), ACM, 2002, pp. 81–88.

Y. Zhen, W.-J. Li, and D.-Y. Yeung, *TAGICOFI: tag informed collaborative filtering*, Proceedings of the 2009 ACM Conference on Recommender Systems (RecSys '09) (New York), 2009, pp. 69–76.

C.-N. Ziegler and G. Lausen, *Spreading activation models for trust propagation*, Proceedings of the 2004 IEEE International Conference on e-Technology, e-Commerce and e-Service (EEE '04) (Taipei), IEEE Computer Society, March 2004, pp. 83–97.

————, *Propagation models for trust and distrust in social networks*, Information Systems Frontiers **7** (2005), no. 4–5, 337–358.

C.-N. Ziegler, G. Lausen, and L. Schmidt-Thieme, *Taxonomy-driven computation of product recommendations*, Proceedings of the 2004 ACM CIKM International Conference on Information and Knowledge Management (CIKM '04) (Washington, DC), 2004, pp. 406–415.

C.-N. Ziegler, S. M. Mcnee, J. A. Konstan, and G. Lausen, *Improving recommendation lists through topic diversification*, Proceedings of the 14th International Conference on World Wide Web (WWW '05) (New York), ACM Press, 2005, pp. 22–32.

A. Zimmermann, M. Specht, and A. Lorenz, *Personalization and context management*, User Modeling and User-Adapted Interaction **15** (2005), no. 3–4, 275–302.

Index

CPSIA information can be obtained
at www.ICGtesting.com
Printed in the USA
LVOW04*1716100816

499851LV00012B/97/P

9 780521 493369